OBJECTIVE MEASUREMENT: THEORY INTO PRACTICE

VOLUME 5

Objective Measurement: Theory Into Practice

Mark Wilson, Series Editor

Volume 1
edited by Mark Wilson, 1992

Volume 2
edited by Mark Wilson, 1994

Volume 3
edited by George Engelhard, Jr., and Mark Wilson, 1996

Volume 4
edited by Mark Wilson, George Engelhard, Jr., and Karen Draney, 1997

Volume 5
edited by Mark Wilson and George Engelhard, Jr., 2000

OBJECTIVE MEASUREMENT: THEORY INTO PRACTICE

Volume 5

edited by
Mark Wilson
University of California, Berkeley

George Engelhard, Jr.
Emory University

Printed in the United States of America

Library of Congress Cataloguing-in-Publication Data

(Revised for vol. 5)
Objective measurement.
"Papers presented at successive International Objective Measurement Workshop (IOMW)"—Pref.
Includes bibliographical references and index.
ISBN 1-56750-432-9 (cloth) — ISBN 1-56750-433-7 (pbk.)
1. Psychometrics—Congresses. 2. Psychometrics—Data processing—Congresses. 3. Educational tests and measurements—Congresses. I. Wilson, Mark. II. International Objective Measurement Workshop.

BF39.024 2000
150'.1'5195—dc21 99–16210
CIP

Ablex Publishing Corporation
100 Prospect Street
P.O. Box 811
Stamford, Connecticut 06904-0811

Contents

Preface

The chapters in this new volume in the *Objective Measurement: Theory into Practice* series describe original research concerned with the practice of measurement, and the theory of measurement. The words in the title, "objective measurement," indicate that the chapters are all related to a particular approach to the philosophy and practice of measurement. By objective measurement, we mean that, in a situation where a certain class of stimuli (for example, items) are being used to measure certain individuals:

> The comparison between two stimuli should be independent of which particular individuals were instrumental for the comparison; and it should also be independent of which other stimuli within the considered class were or might also have been compared. Symmetrically, a comparison between two individuals should be independent of which particular stimuli within the class considered were instrumental for comparison; and it should also be independent of which other individuals were also compared, on the same or on some other occasion. (Rasch, 1961, pp. 330–331)

Thus, it should not matter who else is measured when you are measured, nor should it matter which particular measuring instruments are used to measure you, so long as they all belong to the relevant class (Wright, 1968). Rasch named this quality "specific objectivity" (Rasch, 1966), and he showed that if the probabilistic relationship between an individual's responses to an item and an underlying variable were governed by one of a family of mathematical relationships, then specific objectivity would hold (Rasch, 1960). He demonstrated this for a Poisson model of misreadings and one for speed, and also for a logistic model of dichotomous test responses. This last is commonly known as the "Rasch model" in his honor, and measurement that seeks to adhere to the standard of specific objectivity is also known as "Rasch measurement," or "objective measurement." He also demonstrated that, under certain rather essential conditions, these are the *only* models that can result in specific objectivity, a result that was broadened and clarified by Andersen (1973) and Fischer (1974).

The chapters originated in presentations made at a biennial conference devoted to exploring the interface between theory and practice in measurement, particularly *specifically objective* measurement—the International Objective Measurement Workshop (IOMW). The conference brings together practicing professionals and theoreticians in diverse areas of measurement and related fields. Chapters presented here represent a broad sampling across the range in evidence at the IOMW conference held in Chicago, at the International House at the University of Chicago in April 1996; in addition, there is a particular focus in a subset of the chapters on the issues arising in measurement involving raters and judging. The most recent IOMW conference was held in April 2000 in New Orleans in association with the annual meeting of the American Educational Research Association.

The volume begins with a cluster of chapters that embody the most important aspect of measurement—its application to diverse contexts and purposes. The specific contexts studied range from writing assessments in state assessments, to portfolio assessments in English, to the attitudes of science teachers about their assessment practices, to achievement testing in high school science.

The second cluster of chapters focuses on a topic of increasing importance to modern measurement—the study of measurement applications involving raters and judges. These applications have become more important because of a decreased reliance on item formats such as multiple-choice items that use technical strategies to remove human judgment from the measurement process, and a corresponding increase in the usage of item formats, such as performance assessments, that require the active involvement of raters or judges, or both. The issues studied by these chapters range from the study of rater consistency across time and context, to the exploration of a variety of rater errors.

The volume concludes with chapters that explore the domain of theory in measurement. These chapters are characterized by an innovative approach to model-building, and an insistence that theoretical approaches must be sensitive to the application contexts. Even though they are categorized as "theory" papers, most are based on the complicated and interesting problems that arise in quite specific contexts. As such, they represent some of the most original and exciting developments in the field of measurement today.

The chapters are aimed at a broad audience concerned with measurement and assessment issues in education, psychology, and related social science fields.

REFERENCES

Andersen, E. B. (1973). Conditional inference for multiple-choice questionnaires. *British Journal of Mathematical and Statistical Psychology, 26,* 31–44.

Fischer, G. H. (1974). *Einfuhrungin die Theorie psychologischer Tests.* Vienna: Verlag Hans Huber.

Rasch, G. (1960). *Probabilistic models for some intelligence and attainment tests.* Copenhagen: Denmarks Paedagogiske Institute. (Reprinted by University of Chicago Press, 1980)

Rasch, G. (1961). On general laws and the meaning of measurement in psychology. *Proceedings of the Fourth Berkeley Symposium on Mathematical Statistics and Probability,* (Vol. 4, pp. 321–333). Berkeley, CA: University of California Press.

Rasch, G. (1966). An individualistic approach to item analysis. In P. F. Lazarsfeld & N. W. Henry (Eds.), *Readings in mathematical social science* (pp. 89–108). Chicago: Science Research Associates.

Wright, B. D. (1968). Sample-free test calibration and person measurement. In *Proceedings of the 1967 Invitational Conference on Testing Problems.* Princeton, NJ: Educational Testing Service.

part I
Measurement Practice

1

SETTING AND EVALUATING PERFORMANCE STANDARDS FOR HIGH STAKES WRITING ASSESSMENTS

George Engelhard, Jr.
Emory University

Belita Gordon
The University of Georgia

INTRODUCTION

The purposes of this chapter are to describe procedures for (1) setting a performance standard or passing score on a high stakes writing assessment, and (2) evaluating the quality of the ratings obtained from the standard-setting judges. Standard setting is a judgmental process for determining a cut score (standard) on an educational or psychological instrument that answers the question of "How good is good enough?" (Livingston & Zieky, 1982). The essential components of judgmental standard setting processes (Angoff, 1971; Ebel, 1972; Jaeger, 1982 Nedelsky, 1954) consist of a group of expert judges interacting with a set of test

---

Earlier versions of this manuscript were presented at the annual meeting of the Georgia Educational Research Association, Atlanta (November 1996), the Ninth International Objective Measurement Workshop, Chicago (March 1997), and the 10th European Meeting of the Psychometric Society, Santiago de Compostela, Spain (July 1997).

items through the use of a particular standard-setting process (Plake, Melican, & Mills, 1991).

Frequently in the standard-setting literature, the differences between measurement and evaluation are not stressed. It is important to recognize that measurement represents the development and calibration of a set of items or tasks onto a line that represents a latent variable or construct. Evaluation, on the other hand, deals with judgments of value or worth; the standard-setting process can be viewed as a process used to make these value judgments explicit in order to set a cut score on the line that represents the construct. This distinction between measurement viewed as calibration and standard-setting viewed as evaluation is important because different criteria must be used to examine the quality of these two separate processes. The criteria for examining the psychometric quality of measurements are well known and represented by the *Standards for Educational and Psychological Testing* (American Educational Research Association, American Psychological Association, & National Council on Measurements in Education, 1985) while the criteria for evaluating evaluations or meta-evaluations (Joint Committee on Standards for Educational Evaluation, 1994) do not apply directly to standard-setting processes. In fact, no mention of standard setting is made in *The Program Evaluation Standards* (Joint Committee on Standards for Educational Evaluation, 1994). The procedures described in this chapter provide a new approach for evaluating the quality of judgments provided within the context of standard setting for performance assessments.

This study differs from previous research in several ways. First, it extends the Binomial Trials Model used previously to set cut scores on assessments consisting of multiple-choice items (Engelhard & Cramer, 1997; Engelhard & Stone, 1998) to constructed response items and performance assessments. Second, it provides a multifaceted approach for standard-setting judgments with an explicit facet for rounds (time) included in the model. This differs from previous research that examined judgments from each round with separate models (see, for example, Engelhard & Anderson, 1998).

METHODS

Participants

Twenty-seven judges participated in this study. These judges were members of the standard-setting committee for the Georgia High School Writing Test (GHSWT). The GHSWT is used as one of the requirements for graduation from Georgia high schools. The ratings examined in this study were obtained from the operational standard-setting process; they were not collected as a part of a separate judgment study. The operational standard-setting judges included 24 women and 3 men; 7 African Americans, 1 Hispanic, and 19 whites; 16 teachers, 2 assistant principals, and 9 system-level English language arts supervisors. The teachers include one English as a Second Language (ESL) and one special education teacher.

Instrument

The GHSWT is a criterion-referenced test that requires students to respond to a persuasive writing task within a 90-minute, on-demand context. The compositions written in response to this task can be up to 2 pages long, and they are scored by two raters on four domains of effective writing (Content and Organization; Style; Conventions; Sentence Formation) using rating scales with four categories. A detailed description of this instrument is available from the Georgia Department of Education (1993).

Procedures

The standard-setting process consists of several major steps. First, a set of compositions were selected that represented a range in writing quality. These compositions were written by students throughout the state as a part of the field test of GHSWT. Second, compositions that reflect comparable writing quality were grouped into 19 packets with 6 compositions per packet. The packets were created by the director of the statewide assessment program, and the selection of the compositions was confirmed by staff (three expert raters) who supervise operational scoring sessions. The compositions included in the packets were written by students with varied demographic characteristics, as well as students from the following special education categories: visually impaired, deaf, moderately mentally handicapped, and speech and hearing impaired (Gordon & Dabney, 1996). The compositions were grouped into packets in order to represent a variety of writing samples that reflect comparable writing quality. Since analytical scoring is used for the operational assessment in four domains (content/organization, style, conventions, and sentence formation), compositions can differ in their profiles of domain scores and still produce the same total score. Next, the standard-setting judges were asked to respond to the following guiding question for each packet:

> In the content area of writing, think of the line that separates students who should receive a Georgia high school diploma from those who should not, basing the decision on skills and knowledge learned from the Quality Core Curriculum to grade 11. Focus on the student just above that line and the skills and knowledge that student has achieved in the content area of writing. Is the set of papers in Packet X above or below that line? If the set of papers is above the line, that represents a pass. If the set of papers is below the line, that represents a fail.

The final process included three rounds of ratings with the judges provided with summary data and opportunities to discuss their ratings between rounds. A more detailed description of the standard-setting process is provided in a technical report (Test Scoring and Reporting Services, 1994) and by Gordon and Dabney (1996).

The Rasch (1980) measurement model was used to map the ratings from the judges (0 = fail, 1 = pass) to a judgmental scale of writing competence. A three-facet model (round, packet, and judge) was used to evaluate the quality of the ratings based on the procedures proposed by Engelhard and Anderson (1998). The model used to analyze the judgments can be written as:

$$\log [P_{nij1}/P_{nij0}] = \beta_n - \delta_i - \omega_j,$$

where

P_{nij1} is the probability of judge *n* rating composition packet *i* as a pass at round *j* ($x = 1$),
P_{nij0} is the probability of judge *n* rating composition packet *i* as a fail at round *j* ($x = 0$),
β_n is the view of writing competence required for high school diploma from judge *n*,
δ_i is the judged writing quality of packet *i*, and
ω_j is the performance standard at round *j*.

This model reflects the idea that there are three major facets of the standard-setting process that influence the judgments: view of writing competence by each judge, judged writing quality of each packet, and performance standard at each round.

Several indices of model-data fit are available for this model. The mean square error (*MSE*) statistics provide evidence regarding the validity of the model that can be summarized separately over judges, packets, and rounds. Several researchers (Engelhard, 1994; Lunz, Wright, & Linacre, 1990) have suggested that the region for acceptable fit for these *MSE* statistics ranges from .6 to 1.5; since the exact sampling distributions of these statistics are not known, these values should be viewed as "rules of thumb" that have led to meaningful substantive interpretations in earlier research. Given the nature of standard-setting judgments and the lack of independence inherent in the standard-setting process, low MSE statistics are not flagged as problematic; *MSE* statistics greater than 1.5 can be flagged, and quality control (QC) charts based on standardized residuals constructed to examine unusual judgments.

A reliability of separation index, R, and a chi-square statistic, χ, can also be calculated to examine the statistical significance of differences between the judges, packets, and rounds within each facet. The reliability of separation index is comparable to KR20 (Andrich, 1982). The chi-square statistic is analogous to the homogeneity test statistic Q described by Hedges and Olkin (1985). For the judge facet, the reliability of separation and chi-square statistics provide evidence regarding the following question: Do the judges vary in their views of writing competence required for a high school diploma?

Linacre (1989) provides additional details regarding the computational and statistical aspects of these fit statistics for the general FACETS model. Engelhard and his colleagues describe other applications of the FACETS model to judgments obtained within the context of standard setting (Engelhard & Anderson, 1998; Engelhard & Cramer, 1997; Engelhard & Stone, 1998; Stone & Engelhard, 1997). The FACETS computer program was used to analyze the data (Linacre & Wright, 1994).

RESULTS

For these data, there are significant differences between judges regarding their views of the level of writing competence required for a high school diploma. The descriptive statistics for the 27 judges are presented in Table 1. The locations of the judges on a map of the writing competence variable are presented in Figure 1. Views of writing competence range from 6.14 logits (*SE* = .64) for Judge 7, who has the most lenient view (93% pass rate), to –1.98 logits (*SE* = .45) for Judge 12, who has the most severe view (42% pass rate). The overall differences between the judges' views of writing competence are statistically significant, χ^2 $(26) = 211.96$, $p < .01$ with a high reliability of separation index (R = .89).

TABLE 1.
Calibration of Standard-setting Judges

Judge	Measure	*SE*	*MSE*	Pass Rate
12	–1.98	0.45	.39	.42
2	–1.55	0.48	.96	.46
16	–1.32	0.50	4.60[a]	.47
17	–1.06	0.51	.92	.49
25	–.79	0.53	.20	.51
23	–.79	0.53	.32	.51
21	–.79	0.53	.15	.51
18	–.79	0.53	.51	.51
11	–.79	0.53	.78	.51
4	–.79	0.53	.12	.51
3	–.79	0.53	.80	.51
13	–.51	0.54	.98	.53
27	–.20	0.54	.60	.54
14	–.20	0.54	.38	.54
9	–.20	0.54	.24	.54
6	–.20	0.54	.13	.54
1	–.20	0.54	2.06[a]	.54
26	–.20	0.54	.18	.56

TABLE 1 (continued)

Judge	Measure	*SE*	*MSE*	Pass Rate
22	.06	0.54	.44	.56
20	.06	0.54	1.11	.56
19	.06	0.54	.12	.56
8	.06	0.54	.39	.56
24	.62	0.52	9.00[a]	.60
10	.62	0.52	9.00[a]	.60
15	1.16	0.51	.78	.63
5	4.10	0.55	.19	.82
7	6.14	0.64	.16	.93
M	.00	0.53	1.31	.56
SD	1.65	0.03	2.39	.10

[a] Indicates judge with *MSE* > 1.5.

These analyses indicate that there are significant differences in the views of these judges that merit further attention. It is clear in Figure 1 that Judges 7 and 5 have views of writing competence that differ significantly from the other standard-setting judges. The *MSE* statistics for these judges have a mean of 1.31 logits and standard deviation of 2.39 logits. There are four judges with *MSE*s greater than 1.5: Judges 1, 10, 16, and 24.

In order to illustrate the substantive interpretation of these high *MSE* statistics, a QC chart is presented in Figure 2 that identifies packets within rounds that Judge 10 rated in an idiosyncratic fashion. It is clear in Figure 2 that Judge 10 had unusual judgments for the following packets in Round 1 (z scores greater than 2.00 or less than –2.00): Packets 1, 5, 6, 7, 10, 11, and 12. There was an unusual judgment for Packet 12 in Round 2. By Round 3, the views of Judge 10 have become comparable to the views of the other judges that reflects almost perfect agreement regarding the ordering of the packets on the writing competency scale.

The distribution of the packets on the writing competency scale is presented in Figure 1. The descriptive statistics for the 19 packets are presented in Table 2. The packet difficulties range from –4.73 logits (*SE* = 1.01) for Packet 18 (99% pass rate) which is the easiest to judge as a pass (examples of good writing) to 6.52 logits (*SE* = .93) for Packet 5 (2% pass rate) which is the hardest to pass (examples of weak writing). The overall differences between the packets are significant, χ^2 (18) = 690.35, $p < .01$ with a high reliability of separation index (R = .97). The *MSE* fit statistics for packets had a mean of 1.35 logits and standard deviation of 1.98 logits. Five of the packets had large *MSE* statistics that indicate unusual judgments regarding these packets: Packet 5 (*MSE* = 9.00), Packet 1 (*MSE* = 2.99), Packet 3 (*MSE* = 1.52), Packet 7 (*MSE* = 1.58), and Packet 9 (*MSE* = 1.84). The standardized residuals for Packet 5 are presented in Figure 3. Judges 10 and 24 had unusu-

	Standard by round *Lower*	Packet *Weak writing*	Judge *Lenient*
6.0 +			
		5	
5.0 +			7
		1,4	
4.0 +			
		2	
			5
3.0 +			
		3	
		7	
2.0 +		6	
		9	
1.0 +	2	8	15
			10,24
	3		
0.0 +			8,19,20,22,26
	1		1,6,9,14,27
			13
			3,4,11,18,21,23,25
			17
−1.0 +			16
			2
		12	12
−2.0 +		10, 14	
		11,13	
		16	
−3.0 +			
		17	
		19	
−4.0 +			
		15, 18	
−5.0 +			
	Higher	*Good Writing*	*Severe*

FIGURE 1. Variable map for standards, packets, and judges on writing competence scale.

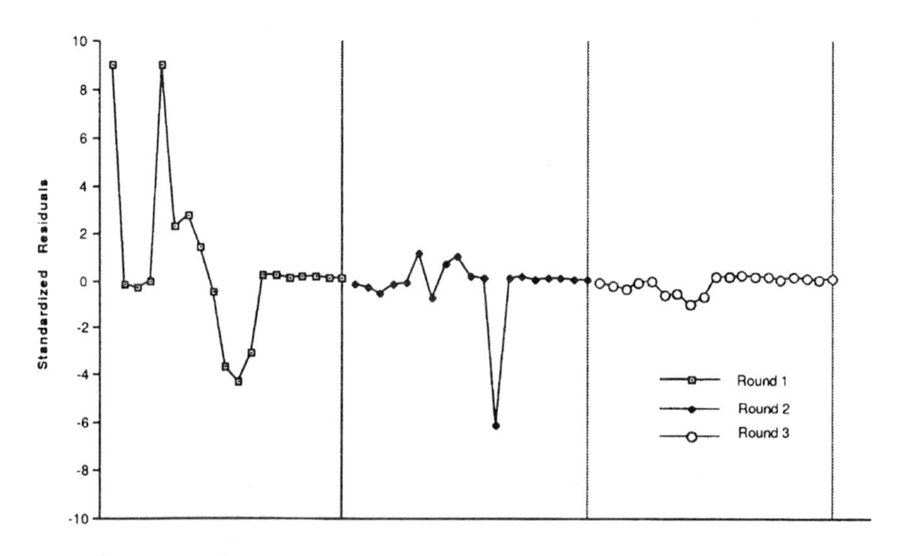

FIGURE 2. Standardized residuals for Judge 10 (*MSE* = 9.00).

TABLE 2.
Calibration of Packets

Packet	Precalibration	Measure	*SE*	*MSE*	Pass Rate
18	–4.33	–4.73	1.01	.83	.99
15	–3.12	–4.73	1.01	.36	.99
19	–4.89	–4.01	0.73	.31	.98
17	–3.88	–3.58	0.60	.59	.96
16	–3.49	–3.02	0.48	.42	.94
13	–2.40	–2.63	0.41	1.13	.91
11	–1.67	–2.63	0.41	.50	.91
14	–2.76	–2.47	0.39	.59	.90
10	–1.30	–2.32	0.37	.71	.89
12	–2.04	–1.96	0.33	1.05	.85
8	–.60	.96	0.27	.97	.35
9	–.95	1.72	0.32	1.84[a]	.23
6	.06	1.93	0.33	1.18	.21
7	–.27	2.30	0.37	1.58[a]	.17
3	1.06	2.98	0.46	1.52[a]	.12
2	1.41	4.15	0.64	.06	.07
4	.72	5.75	0.83	.04	.04
1	1.77	5.75	0.83	2.99[a]	.04
5	.39	6.52	0.93	9.00[a]	.02
M	–1.38	.00	0.56	1.35	.56
SD	2.02	3.80	0.25	1.98	.41

[a] Indicates packet with MSE > 1.5.
Note. Precalibration of packets is based on data obtained from the statewide field test.

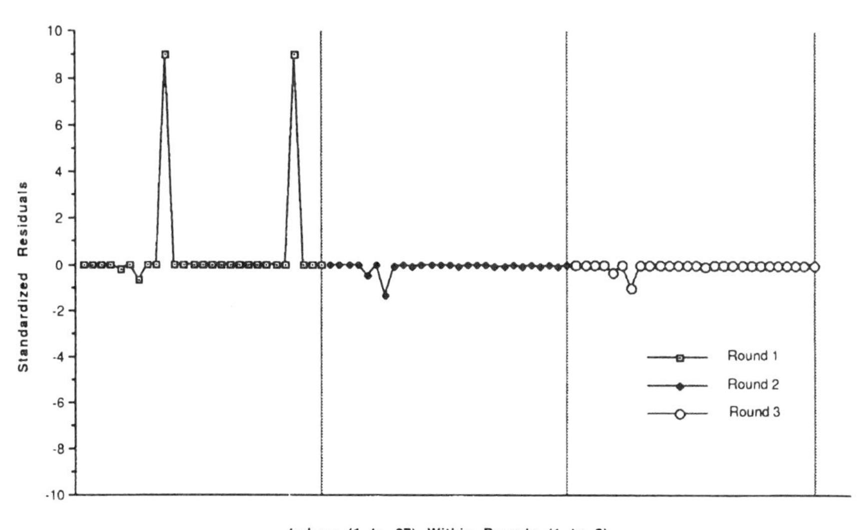

FIGURE 3. Standardized residuals for Packet 5 (*MSE* = 9.00).

al judgments of this packet in Round 1; there were no unusual judgments in Rounds 2 and 3.

The precalibration difficulties of the packets are also shown in Table 2. These difficulties were obtained from the field test of the GHSWT (Georgia Department of Education, 1993). It is interesting to note the high level of correspondence between the judged difficulties of the packets and the precalibrated difficulties (Pearson product moment correlation of .92).

In addition to looking at the quality of the judgments for each judge and packet, the effects of rounds were examined. The judged performance standards by round were as follows: Round 1 = –.34 logits (*SE* = .17) which reflects a pass rate of 51%, Round 2 = 1.03 logits (*SE* = .18) which reflects a pass rate of 60%, and Round 3 = .39 logits (*SE* = .18) which reflects a pass rate of 56%. The overall differences between rounds on the judged writing competency scale are statistically significant, χ^2 (2) = 31.15, $p < .01$ with a high reliability of separation index (R = .90). The *MSE* statistics were 4.0, .4, and .3 over the three rounds respectively. The agreement among the judges regarding the quality of writing represented by the writing packets increased over the rounds. By the Round 3, the judges were almost in perfect agreement regarding the ordering of the packets. The locations of the performance standard for each round on the writing competence scale are presented graphically Figure 1.

In Figure 1, there is a striking discontinuity between packets above Packet 8 and those below Packet 12. Since the compositions included in packets were calibrated with the Rasch (1980) model during the field test, a straightforward approach is available for mapping the judgments examined in this chapter onto the empirical scale as follows: The recommended performance standard is obtained by setting the cut score halfway between Packets 8 and 12 using the empirical precalibrations of the compositions included in these packets that agrees with the performance standard for Round 3. Based on the precalibrations of these packets, as shown in Table 1, the performance standard would be –1.32 [(–2.04 + –.60)/2].

DISCUSSION AND CONCLUSIONS

The approach described in this chapter has several useful features. It allows judges to maintain their differing views of writing competence required for high school graduation. Expert judges are not forced to agree as they sometimes are in other standard-setting processes. Another advantage is that the judges are encouraged to be consistent in their judgments regarding the quality of writing represented within the packets. Evidence of inconsistent and idiosyncratic judgments can be flagged using the mean square error statistics, and appropriate QC charts constructed to examine the standardized residuals in detail. This information can be provided during the standard-setting sessions in order to encourage discussion among the judges. Another advantage is that a variable map, such as Figure 1, can be constructed and presented to the governing board that ultimately sets the cut score on the test. These variable maps can be used to highlight the views of the standard-setting judges, and to identify judges with extreme views that the governing board may want to remove before the final performance standard is set. Finally, since the compositions examined in this chapter were precalibrated with the Rasch (1980) model, the mapping of the cut score to the empirical scale is straightforward. For the data analyzed in this chapter, there was a large discontinuity on the judgmental scale between the packets that the judges viewed as good writing (passes) and weak writing (fails), and the cut score was set between these two groups of packets.

A great deal of work has been done on standard setting for multiple-choice items, but relatively little research has been conducted on how to set passing scores for performance assessments. This chapter focused on the application of recent advances in Rasch measurement theory to problems related to setting passing scores and evaluating the quality of the judgments obtained from standard-setting judges. This study presents a promising method for setting standards on performance assessments based on Rasch measurement theory. This study also contributes to our understanding of how to identify judges with unusual rating patterns that may merit the removal of their ratings.

REFERENCES

American Educational Research Association, American Psychological Association, & National Council on Measurement in Education. (1985). *Standards for educational and psychological testing.* Washington, DC: American Psychological Association.

Andrich, D. (1982). An index of person separation in latent trait theory, the traditional KR.20 index and the Guttman scale response pattern. *Education Research and Perspectives, 9,* 95–104.

Angoff, W. H. (1971). Scales, norms, and equivalent scores. In R. L. Thorndike (Ed.), *Educational measurement* (2nd ed., pp. 508–600). Washington, DC: American Council of Education.

Ebel, R. L. (1972). *Essentials of educational measurement.* Englewood Cliffs, NJ: Prentice-Hall.

Engelhard, G. (1994). Examining rater errors in the assessment of written composition with a many-faceted Rasch model. *Journal of Educational Measurement, 31*(2), 93–112.

Engelhard, G., & Anderson, D. W. (1998). A binomial trials model for examining the ratings of standard-setting judges. *Applied Measurement in Education, 11*(3), 209–230.

Engelhard, G., & Cramer, S. (1997). Using Rasch measurement to evaluate the ratings of standard-setting judges. In M. Wilson, G. Engelhard, & K. Draney. (Eds.), *Objective measurement: Theory into practice* (Vol. 4, pp. 97–112). Norwood, NJ: Ablex.

Engelhard, G., & Stone, G. E. (1998). Evaluating the quality of ratings obtained from standard-setting judges. *Educational and Psychological Measurement, 58*(2), 179–196.

Georgia Department of Education. (1993). *Georgia High School Writing Test: Assessment and Instructional Guide.* Atlanta, GA: Author.

Gordon, B., & Dabney, M. E. (1996, April). *Development of a model for standard setting for a high stakes writing assessment.* Paper presented at the annual meeting of the American Educational Research Association, New York.

Hedges, L. V., & Olkin, I. (1985). *Statistical methods for meta-analysis.* San Diego, CA: Academic Press.

Jaeger, R. M. (1982). An iterative structured judgment process for establishing standards on competency tests: Theory and application. *Educational Evaluation and Policy Analysis, 4,* 461–476.

Joint Committee on Standards for Educational Evaluation. (1994). *The program evaluation standards: How to assess evaluations of educational programs* (2nd ed.) Thousand Oaks, CA: Sage.

Linacre, J. M. (1989). *Many-faceted Rasch measurement.* Chicago: MESA Press.

Linacre, J. M., & Wright, B. D. (1994). *A user's guide to FACETS: Rasch measurement computer program.* Chicago: MESA Press.

Livingston, S. A., & Zieky, M. J. (1982). *Passing scores: A manual for setting standards of performance on educational and occupational tests.* Princeton, NJ: Educational Testing Service.

Lunz, M. E., Wright, B. D., & Linacre, J. M. (1990). Measuring the impact of judge severity on examination scores. *Applied Measurement in Education, 3*(4), 331–345.

Nedelsky, L. (1954). Absolute grading standards for objective tests. *Educational and Psychological Measurement, 14,* 3–19.

Plake, B. S., Melican, G. J., & Mills, C. N. (1991, Summer). Factors influencing intrajudge consistency during standard-setting. *Educational Measurement: Issues and Practice,* 15–22, 25–26.

Rasch, G. (1980). *Probabilistic models for some intelligence and attainment tests.* Chicago: University of Chicago Press.

Stone, G. E., & Engelhard, G. (1997, March). *Informing mastery through binomial trails: A refinement of Objective Standard Setting.* Paper presented at the Ninth International Objective Measurement Workshop, University of Chicago, Chicago, IL.

Test Scoring and Reporting Services (1994). *Technical report: Georgia High School Writing Test.* Athens, GA: Test Scoring and Reporting Services, University of Georgia.

2

FORMATIVE EVALUATION OF A PERFORMANCE ASSESSMENT SCORING SYSTEM

Susan T. Paulukonis
Carol M. Myford
Joan I. Heller
Educational Testing Service

INTRODUCTION

The basis of most performance assessment scoring systems is human judgment: Raters look for evidence of particular aspects of student performance and make judgments about where to place that performance on a rating scale. The development of such a system is qualitatively and quantitatively different from the development of other, more traditional, types of assessment systems, and the cost per student can be significantly higher (Hoover & Bray, 1995). Motivating these labor-

The authors thank the teachers and district administrators who participated in the Pacesetter English portfolio assessment development effort, as well as ETS staff members who contributed to this work, especially Karen Sheingold, Bill Thomas, Barbara Storms, Athena Nuñez, and Jean Wing. An earlier version of this research was presented at the American Educational Research Association Annual Meeting in New York City, April 1996.

intensive efforts is the belief that carefully designed performance assessment has the potential for achieving many of the goals of the educational reform movement (LeMahieu, Eresh, & Wallace, 1992; Resnick & Resnick, 1992).

The development of a scoring system constitutes much of the effort in the design of a standards-based performance assessment system. Extensive work is needed to define the scoring criteria or performance standards against which performances are to be evaluated. Descriptors for each rating category for each dimension along which performances are to be judged must be developed and refined. Performances must be collected (assessments must be administered or student work must be collected to form a portfolio) and scoring sessions held to try out the systems. The data collected from sessions are evaluated and, in an iterative process, the results used to help refine the scoring system. Because of the amount of human effort needed to evaluate and refine such a system, some are constructed and formatively evaluated unsystematically, using simple interrater reliability figures or anecdotal evidence as guides.

In this study, we report the results of one formative evaluation of a standards-based performance assessment scoring system. We were able to reduce the time, and hence the costs, of the development of the system by using a computer program, FACETS (Linacre, 1994b), to conduct a multifaceted Rasch analysis in tandem with analyses of qualitative data collected from raters during a pilot scoring. Others have found this combination useful in evaluating a performance assessment system's effectiveness (Myford & Mislevy, 1995). We found here that neither FACETS nor qualitative data alone provided a full picture of the way the system was functioning. However, the combination of the two was very useful for evaluating the system and for pointing to specific areas that needed refinement.

OBJECTIVES

The context for this work was a portfolio assessment system developed for The College Board's Pacesetter English course. Pacesetter English is a 4th-year high school course that is intended to be rigorous and standards-based. Pacesetter curriculum, instruction, and assessment focus on expanding students' understanding of texts (which are broadly defined to include traditional literature as well as films and speech) from various genres and cultural backgrounds as well as to expand students' skills in creating texts. During the 1994–1995 school year, student achievement in the Pacesetter English course was to be evaluated through four forms of assessment: assessments developed by the teachers for their own classrooms, end-of-unit performance tasks that were common across all classrooms, an end-of-year performance assessment, and assessment of portfolios of student work. In that same year, the Center for Performance Assessment at Educational Testing Service (ETS) participated in an effort to refine the system for standards-based assessment of the portfolios. A collaborative development group was formed, consisting of Pacesetter English teachers from three districts and project staff from the Center. This group worked throughout the school year to refine the standards that an assessment portfolio would address and to develop a system for scoring the portfolios.

The development group based their work on the concept of portfolios as collections of evidence of student achievement in language arts. Each student would decide, in collaboration with his or her teacher, what work best demonstrated the student's achievement with respect to the assessment criteria in the scoring guides developed for portfolio evaluation.[1] There were no required or common pieces and no minimum or maximum number of pieces. Two "course dimensions" had already been established for the Pacesetter English course: Creating and Presenting Texts, and Making Meaning from Texts. The dimensions were intended to provide a framework for the curriculum and assessment and to help both teachers and students focus on what students should be able to demonstrate by the end of the course. Within each of the dimensions, different aspects of student learning and performance had also been identified and defined. Definitions for these dimensions and aspects were refined over the course of the year.

Project staff developed five-category, standards-based scoring guides for the dimensions and aspects (see Appendix A) that were tried out in a pilot portfolio scoring in May 1995. Those who rated the portfolios included many of the participants in the development group as well as teachers and an administrator from another Pacesetter district. The scoring model raters were trained to use involved evaluating student performance across all of the writing samples in the portfolio, rather than evaluating individual pieces of work. Raters were asked to evaluate the assessment portfolios on each of the two dimensions and eight aspects. In addition to the collected rating data, qualitative data collection methods (that is, questionnaires) were employed during the portfolio scoring to gather information about how the scoring system was functioning.

The broad question that focuses this study is:

- How can researchers use qualitative and quantitative data in tandem to effectively and efficiently evaluate a performance assessment scoring system?

Researchers posed two sets of more narrowly focused questions to guide formative evaluation of the scoring system:

- Differentiation among aspects: Do the dimensions and aspects work together to define a meaningful variable that we can recognize (i.e., language arts achievement)? Are there aspects that are overlapping, or redundant?
- Scale usage: Does the rating scale for each aspect function effectively as a five-category scale? Are raters able to differentiate among the categories for each aspect?

METHOD

To help us answer these questions, score data from the pilot portfolio scoring session were analyzed in tandem with qualitative data collected from a questionnaire completed by raters at the same scoring session. We describe these steps in this section.

Portfolio Scoring Session

Description of Raters

A total of 18 raters participated in the 2-day portfolio scoring, including 15 teachers and 2 administrators from the four school districts, and 1 ETS staff member. All of the teachers taught the Pacesetter English course in their high schools, 13 at the 12th grade level and 2 at 11th grade; all but 3 had also been involved in the development effort. No teacher rated portfolios from his or her own class, and each rater rated portfolios from many different classrooms. None of the raters had experience rating portfolios of student work prior to this project.

Training Procedure

Training was conducted during the morning and early afternoon of the first day of the portfolio scoring, and at the beginning of the second day. On both days, training began with a review and discussion of the aspects for each dimension, and moved to a discussion of the scoring guides for the dimensions and the aspects themselves. On the first day, training began with the rating of one benchmark portfolio.[2] Raters worked in pairs, evaluating the portfolio together, then discussing and assigning aspect and dimension ratings. Pairs then shared their ratings with the group, and a discussion of how the scoring guide was used and problems that arose followed. For the second and third rounds of training on the first day, raters worked individually assigning aspect and dimension ratings. They then discussed their ratings in pairs and finally worked as a group to reach consensus on ratings. At the beginning of the second day, the raters and trainers together reviewed the scoring guides and a benchmark portfolio, and a discussion was held about those aspects that project staff believed were problematic to rate on the first day.

Portfolio Rating Procedure

Each rater assigned an overall dimension rating for each of the two dimensions as well as a rating for each of the eight aspects, for a total of up to 10 ratings per portfolio, so that portfolios rated twice received up to 20 ratings. Raters were also given the option of declining to rate a portfolio on a particular dimension or aspect because insufficient evidence was included in the portfolio pertaining to that aspect or dimension. This judgment was referred to as *Not Enough Evidence.*

The intent of the rating design was to have as many portfolios double rated as possible while maintaining a crossed and connected design (Braun, 1988; Linacre, 1994a) to facilitate FACETS analysis. Raters were seated next to each other in pairs to facilitate the movement of portfolios among raters. The two raters in each pair rated the same portfolios but did not discuss them or work together in any way. After approximately half of the total portfolio rating time had elapsed, half of the raters, one from each pair, were moved so that a different set of pairs rated new batches of portfolios in common. Table 1 shows how many portfolios were double rated by each pair. Numbers under columns titled, "First Rater" and "Second Rater" refer to individual raters identified by number. The first pair, for instance, consisted of Rater 1 and Rater 14, who scored six portfolios in common. During the second half of the scoring, Rater 1 was paired with Rater 8, and they scored six portfolios in common. "Other pairs" refers to ad hoc pairings of raters—rater speed varied considerably, and extra portfolios rated by a fast rater in one pair were given to a fast rater in another pair to score. During the final period of the scoring, some portfolios were rated only once as time permitted.

Portfolio Sample Selection

Because time limitations prohibited the rating of every portfolio submitted, a sample (approximately 20% of the portfolios received) was selected to be rated twice in each dimension and aspect. This sample included an approximately equal number of assessment portfolios from each of the four districts included in the development effort. Within a given school district, an equal number of portfolios was drawn from each classroom. Once this sample was rated, additional portfolios were rated as time permitted, resulting in a sample of 137 portfolios representing 26 Pacesetter English classrooms. Of these, 129 of these portfolios were rated twice; 8 portfolios were rated by only one rater.

TABLE 1.
Portfolio Rating Design: Rater Pairs

First half of scoring			Second half of scoring		
First Rater	Second Rater	No. of Portfolios Rated in Common	First Rater	Second Rater	No. of Portfolios Rated in Common
1	14	6	1	8	6
2	3	8	2	7	6
4	5	6	4	3	6
6	7	8	6	5	6
8	9	6	10	9	6
10	11	8	12	11	7
12	13	6	14	13	6
16	17	6	16	15	6
15	18	6	18	17	6
	Other pairs	10		Other pairs	4
	Total	70		Total	59

Data Collection and Analysis

Score Data

The data from all 137 portfolios rated were included in the analyses discussed here. *Not Enough Evidence* judgments were treated as missing data in all quantitative analyses because the absence of evidence does not necessarily reflect the quality of any evidence that may be present, but rather may reflect the student's limited opportunity to demonstrate relevant achievement (LeMahieu et al., 1992). Of the 2,660 possible ratings (129 portfolios each rated by two raters, 8 portfolios each rated by a single rater, each portfolio may have up to 10 ratings from each rater) 2,290 were actual ratings and the remainder (that is, 370 or approximately 14%) were *Not Enough Evidence* judgments. These judgments were not evenly distributed across dimensions and aspects. Making Meaning from Texts had a higher proportion of *Not Enough Evidence* judgments (19%) than Creating and Presenting Texts (9%). The two *Evaluate and reflect* aspects had the highest proportions of *Not Enough Evidence* judgments among the aspects—31% for *Evaluate and reflect* in Making Meaning from Texts and 38% for *Evaluate and reflect* in Creating and Presenting Texts.

Multi-faceted Rasch Model

For the purposes of formative evaluation of the scoring materials and process, we included all the aspect and dimension ratings in our analysis. The rating data were analyzed using FACETS, a rating scale analysis computer program based on an extension of Wright and Master's partial credit model (1982). FACETS has been used to analyze rater behavior and to review patterns of ratings in many and varied performance assessment settings (for example, Engelhard, 1994; Heller, Sheingold, & Myford, 1998; Lumley & McNamara, 1995; Lunz & Stahl, 1990; Myford, Marr, & Linacre, 1996; Myford & Mislevy, 1995). It seemed likely, based on those experiences, that FACETS analyses of rating data would also provide useful information about the various components of the Pacesetter English performance assessment scoring system (that is, the raters, the aspects and dimensions they scored, the scoring guides they used, and the students' portfolios they evaluated).

The FACETS computer program is particularly well suited for analyzing data from performance assessments that are judged by human raters. In performance assessment settings, those in charge of monitoring quality control need information not only about how the students performed but also about the performance of individual raters and the scoring guides. The output from a FACETS analysis can provide useful diagnostic information about the quality of each student's performance, the scoring behavior of each rater, the utility of each aspect and dimension that was scored, and the adequacy of each scoring guide. Having access to such detailed information enables one to pinpoint specific weaknesses or deficiencies in a complex assessment system so that meaningful, informed steps can be initiated to improve the system.

The multifaceted Rasch model extends the partial credit model to incorporate more than the two components typically included in a partial credit analysis (that is, students and items). When applied to analyze data from a performance assessment, the model can be expanded to incorporate additional components of that setting that may be of particular interest. In the Pacesetter English assessment setting, we could include components such as raters, aspects and dimensions, as well as interactions between various components we would specify in the model, such as rater by aspect interactions, rater by student interactions, student by aspect interactions, and so on. Each of these components (that is, raters, aspects, students) is referred to as a "facet." Within each of these facets, each "element"—that is, each individual rater, aspect, student—is represented by a parameter.

FACETS provides a measure for each element of each facet included in an analysis. For each student, the measure is an estimate of that student's proficiency. For each rater, the measure is an estimate of the degree of severity that rater exercised when rating students' portfolios. For each aspect, the measure is an estimate of how easy or hard it was to get high ratings on that aspect. In addition to reporting a measure for each element, FACETS also provides a standard error for the measure (that is, information about the precision of the estimate), and fit statistics (that is, information about how well the data for that element "fit" the expectations of the measurement model that was used). The fit statistics are particularly useful for identifying any students' portfolios, raters, aspects, or dimensions that performed in a manner different from that predicted by the measurement model.

Typically in the scoring of performance assessments, particularly standards-based assessments, students are assigned numerical ratings based on labels or category descriptors found in a scoring guide. The categories defined by the scoring guide are often assumed to comprise an interval scale and are treated as such in statistical analyses (that is, a rating of "2" given by a rater is assumed to be the same distance from a "3" as a "3" is from a "4" on the defined scale). However, such categorical scales often do not function as linear interval scales, but rather as ordinal scales. While the categories in the middle of a scale may exhibit linearity, scales cannot be linear at their ends since each end category represents an infinite range of performance—poor at the low end of the scale, and good at the high end (Linacre, 1994a). The multifaceted Rasch model transforms the nonlinear ordinal categorical scale into a continuous, linear scale, the logit (or log-odds unit) scale. By placing all facets on the same equal-interval scale, we can then make comparisons within and between facets.

In this study, we were particularly interested in determining whether the scoring guides for the various aspects and dimensions were working as five-point rating scales, as intended. We wanted to determine how raters were using individual rating scales, rather than making the assumption that raters used a category on each scale in the same way (that is, that a "1" on the scale of a particular aspect was equivalent in difficulty to a "1" on each of the other aspect's scales). Therefore, in specifying our measurement model, we used an "item-step model" (McNamara,

1996, pp. 286–287) so that we could look separately at the rating scale structure of each of the 10 scoring guides (that is, we did not want to assume that the scoring guides for the eight aspects and two dimensions shared a common scale structure). In addition to facets for student/portfolio, aspect and dimension, and rater, our measurement model included a fourth facet: rating scale step difficulty for the scoring guide for a particular aspect or dimension. Movement up a rating scale to the next category on that scale is called a "step." A five-point scale has four steps—from category "1" to "2," from "2" to "3," from "3" to "4," and from "4" to "5". Rating scale step difficulties depict the relative difficulty associated with moving from one category on a scoring guide to the next higher category on that guide (that is, how much easier or more difficult is it to move up from step "2" to step "3" than to move up from step "1" to step "2").

The multifaceted item-step model we used in this study describes the probability that a specific student (n) rated by a specific rater (j) will receive a rating in a particular category (k) for a specific aspect (or dimension) (i). The mathematical form of this probability (Equation 1) depicts the relationship between these elements in terms of a logistic odds ratio:

$$\log\left(\frac{P_{nijk}}{P_{nijk-1}}\right) = B_n - D_i - C_j - F_{ik} \qquad (1)$$

where:

P_{nijk} is the probability of student n, when rated on aspect (or dimension) i by rater j, being awarded a rating of k,
P_{nijk-1} is the probability of student n, when rated on aspect (or dimension) i by rater j, being awarded a rating of $k - 1$,
B_n is the proficiency of student n,
D_i is the difficulty of aspect (or dimension) i,
C_j is the severity of rater j, and
F_{ik} is the step difficulty (F) of step k for aspect (or dimension) i.

A FACETS analysis produces a number of useful pieces of output. The output most central to this research is shown in Figure 1, a set of "rulers" on which each element of each facet (that is, each student, each rater, each aspect) is placed in relation to the logit (log-odds unit) scale. The figure and each of the rulers included are explained in detail in the accompanying caption. Figure 1 is an overview of the output; the caption includes important general information about each facet. For the purposes of this chapter, our focus will be on a limited number of the rulers (or columns). The column titled "Aspects" shows the location of the eight aspects and two dimensions relative to the logit scale. In this figure, third column from left, the aspects and dimensions are represented in FACETS rulers by codes, CP1 is the first aspect of Creating and Presenting Texts, *Develop and use their own voices,* CP2 is the second aspect of that dimension, *Develop and present texts*, and so on.

CPD and MMD represent the overall dimension scores for the Creating and Presenting Texts and Making Meaning from Texts dimensions, respectively.

The 10 columns on the right side of the figure represent the rating scales for each of the aspects and overall dimensions in relation to the logit scale. They indicate the most probable rating in terms of the original five-category scale for a portfolio at a given level on the logit scale, if rated by a rater of average severity (that is, a rater with a severity measure of 0.0 on the logit scale). The horizontal lines indicate the point at which the likelihood of getting the next higher rating begins to exceed the likelihood of getting the next lower rating.

FACETS output also provides us with measures of infit (the information-weighted mean-square fit statistic) for each of the aspects. Each mean-square infit for an aspect is a summary of the difference between the observed data and the values expected given the parameter estimates. The mean-square infit statistic is based on the mean of the squared standardized residuals of the observed ratings from their expected values. It is a chi-square statistic divided by its degrees of freedom that has an expected value of 1.00 and can range from 0 to ∞ (Linacre, 1996, p. 94).

Qualitative Data

In addition to the ratings collected during the portfolio scoring, qualitative data were collected to help in the formative evaluation. A survey about the scoring process was administered to all raters. The survey was administered immediately after the scoring, before all but one of the raters had left ($N = 17$). Sample questions are provided in Appendix B.

Qualitative data were used to augment quantitative data, in keeping with our goal to make the formative evaluation fast and inexpensive, yet robust and useful. The questionnaire results were summarized and suggestions or problems the raters noted were listed and tabulated for each aspect and dimension. This list was used to generate counts of problems or suggestions noted below.

RESULTS

We found that both the results of the qualitative and the quantitative analyses were useful and informative as we looked at how this performance assessment system was functioning, and that together these two types of data gave a detailed and complex picture of what parts of the system were working well and what parts needed refinement. Each formative research question was answered completely only by using both types of analyses. We structure our discussion of research findings around the specific questions we explored.

In this section, we review the rulers produced by FACETS along with *information-weighted mean-square fit statistics* for each aspect produced by FACETS to determine whether any of the aspects are overlapping or redundant. We examine the *separation reliabilities* for the aspects and for the students to determine to what

Msr	Students	Aspects	Raters	Aspects									
	More proficient	More difficult	More severe	CP 1	CP 2	CP 3	CP 4	CP D	MM 1	MM 2	MM 3	MM 4	MM D
5				5	5	5	5	5	5	5	5	5	5
	*				—								
				—		—		—					
	**						—		—		—		
4												—	—
	*												
										—			
	***												
3	*				4	4		4					
	**								4		4	4	4
				4			4			4			
	*												
2	***												
	**							—		—			—
	**				—	—			—		—		
	*******		10									—	
1	**	CP4 MM4	3 4 12 13				—						
	*****	MM3	5	—									
	*********	MMD	16							3			
	********	MM2											3
0	****	MM1	6 9 11				3		3		3	3	
	*********		14 17		3	3		3					
	******	CP3 CPD	8										
	*******		7 18	3									
–1	*******	CP1	1							—			
	*******	CP2	2 15				—						
	******								—			—	—
	*****			—		—					—		
–2	***				—			—					
	*****												
	*****			2			2			2			
	******								2		2	2	2
–3	**				2	2		2					
	**												
				—									
	**												
–4	***						—		—		—	—	
	***				—	—		—		—			—
	*												
–5													
	*												
	**												
–6	*			1	1	1	1	1	1	1	1	1	1
	Less proficient	Easier	More lenient	CP 1	CP 2	CP 3	CP 4	CP D	MM 1	MM 2	MM 3	MM 4	MM D
Msr	Students	Aspects	Raters	Aspects									

FIGURE 1. All facet rulers from all aspect FACETS analysis.

* = 1

Notes. Column 1 (Measure) is a representation of the logit (log-odds unit) scale against which all facets are measured.

extent the raters were able to use the various scoring guides to spread out the students and the aspects along a latent trait continuum and thus define a meaningful variable. Finally, we review selected *rating scale category statistics* reported as part of FACETS output to compare the structures of the rating scales for each aspect and dimension and to determine whether each rating scale is working properly as a five-point scale.

Differentiation Among Aspects: Do the aspects (and dimensions) work together to define a meaningful variable that we can recognize (that is, language arts achievement)? Are there aspects that are overlapping, or redundant?

In column 3 of Figure 1, aspects next to or overlapping each other could indicate that they may be measuring the same features of language arts achievement, or indicate features of language arts achievement that may be too closely related to differentiate. It is to be expected that aspects within a dimension would be related, and particularly that the overall dimension rating would be related to the other aspect ratings within that dimension. Raters rating a portfolio on related aspects may give similar ratings on those aspects because students who score high on certain aspects may be expected to score high on other aspects within the same dimension.

When we examine the distribution of the aspect and dimension measures along the language arts achievement variable, we see that the measures range from −1.0 logits to 1.0 logits, about a two-logit spread. This implies, then, that it was harder to get high ratings on some of these aspects (for example, MM3 and MM4) than on others (for example, CP1 and CP2). While it would be preferable for the range of dimension and aspect measures to be wider than two logits, the aspects do

---

Column 2 (Students) represents the spread of student portfolio measures against the logit scale. More proficient students are shown toward the top of the figure, less proficient toward the bottom.

Column 3 (Aspects) shows the location of the aspects against the logit scale. Aspects are represented here in order within dimension: CP1 is the first aspect of the Creating and Presenting Texts dimension, *Develop and use their own voices*. CPD represents the overall dimension rating for Creating and Presenting Texts; likewise, MMD represents the overall dimension rating for Making Meaning from Texts. More difficult aspects are shown at the top of the figure (that is, those that are harder to get high ratings on), while easier aspects appear toward the bottom.

Column 4 (Raters) shows the raters' location on the logit scale, with raters represented here by their assigned rater numbers. The raters toward the top of the scale were more severe in their judgments (Rater 10 was the most severe), and those toward the bottom of the scale were more lenient (Raters 2 and 15 were the most lenient).

Columns 5 through 15 show the rating scales for each of the aspects in relation to the logit scale (aspects are abbreviated as in column 3). They indicate the most probable category assignment (in terms of the original five-category scale) for a portfolio at a particular logit measure, if rated by a rater of average (0 on the logit scale) severity. The horizontal lines indicate the point at which a portfolio becomes more likely to be rated at the adjacent category on that rating scale.

appear to define a somewhat limited but discernible line of increasing intensity. The spread of the aspects suggests that the individual aspects may not be measuring identical features of language arts achievement, but rather may be measuring somewhat different features.

In the analysis—within the column titled "Aspects"—the aspects within a dimension tended to cluster with the overall dimension rating for that dimension. With one exception, the aspects within the Creating and Presenting Texts dimension are in close proximity to each other, as is true with those in the Making Meaning from Texts dimension. The aspects within the Making Meaning from Texts dimension are grouped together toward the top of the column, indicating that they were more difficult for students to get high ratings on. Furthermore, aspects of Creating and Presenting Texts are, by and large, separate from Making Meaning from Texts aspects. Aspects of Creating and Presenting Texts represent features of student writing and oral presentation performance, and aspects of Making Meaning from Texts represent features of student analysis, interpretation of, and response to, various kinds of texts. We consulted with language arts content specialists who theorized that because these two broader types of student language arts achievement are related but distinct, aspects should cluster together within dimension, and be less related to aspects in the other dimension. Content specialists also noted that the fact that, as a group, the Making Meaning from Texts aspects were more difficult than the Creating and Presenting Texts aspects was in line with their observations of curriculum and instruction in Pacesetter. They suggested that Pacesetter teachers tended to focus more on the skills and knowledge needed to do well on the aspects within Creating and Presenting Texts than they did on aspects within Making Meaning from Texts. Further, they believed that students had less prior experience with the core ideas in Making Meaning from Texts than in Creating and Presenting Texts.

CP4 and MM4, the two *Evaluate and reflect on work* aspects, are exceptions to these observations. The two aspects share nearly the same measure—0.99 and 1.10 logits, respectively—and are the most difficult to get high ratings on. Project staff suggested that students had little opportunity to demonstrate the knowledge and skills necessary to do well in the *Evaluate and reflect* aspects of each dimension. These two aspects had a significantly higher percentage of *Not Enough Evidence* ratings than other aspects, as mentioned previously. The lack of evidence would also seem to point to a lack of instruction related to these two aspects, which may help explain why it was harder for students to get high ratings on these particular aspects. Staff also suggested that the reason the *Evaluate and reflect* aspect in Creating and Presenting, CP4, had a much higher difficulty measure than the other aspects in that dimension is that students tend to have little experience evaluating and reflecting upon their own work, as compared to the other aspects within that dimension. And even in Pacesetter courses, where evaluation and reflection is considered a core part of the course, the concepts may have been unfamiliar to many of the teachers and therefore not well taught. By contrast, high school students are likely to have experience with the processes evaluated by the other three aspects in the Creating and Presenting dimension.

FACETS reports a separation reliability for the aspects. The separation reliability is a correlation coefficient that indicates the proportion of observed variance not due to measurement error (Wright & Masters, 1982). The separation reliability for the aspects is 0.98, which suggests that 98% of the variance in the aspect measures is true measure variance not due to measurement error. A separation reliability less than 0.5 would imply that the differences between aspect measures are mainly due to measurement error. Similarly, FACETS reports a separation reliability for the students of 0.96, indicating that 96% of the variance in student proficiency measures is true measure variance while only 4% of the variance is attributable to measurement error. Together these findings provide further evidence that raters successfully used the various scoring guides to spread out the students and the aspects along a latent trait continuum, thus defining a meaningful variable (Fisher, 1992).

The aspect mean-square infit statistics provided by FACETS, and shown in Table 2, are also helpful here. In general, aspects with mean-square infit statistics ranging from 0.6 to 1.5 would be considered to show adequate fit to the measurement model (Wright & Linacre, 1994). A mean-square infit statistic less than 0.6 might indicate that the aspect is redundant or is not functioning as expected (that is, raters may have had difficulty distinguishing among aspects that were too closely related, giving the same, or very similar ratings on more than one aspect—possible evidence that a halo effect may have been operating). By contrast, a mean-square infit statistic greater than 1.5 might indicate that the aspect does not "fit" with the other aspects. The aspect may not be working with the other aspects to define a single underlying construct but may, instead, be defining another con-

TABLE 2.
Aspect and Dimension Difficulty Measures, Standard Errors, and Mean-Square Infit Statistics

Aspect	Difficulty Measure	Standard Error	Mean-square Infit
Creating and Presenting Texts			
Develop and use their own voices	–1.05	0.10	1.1
Develop and present texts	–1.30	0.11	0.8
Demonstrate technical command	–0.62	0.11	1.2
Evaluate and reflect on work	0.99	0.13	1.2
Creating and Presenting Texts Dimension	–0.61	0.11	0.9
Making Meaning from Texts			
Respond to texts	0.05	0.11	0.9
Analyze texts	0.19	0.11	1.0
Put texts in context	0.82	0.13	1.1
Evaluate and reflect on work	1.10	0.12	1.1
Making Meaning from Texts Dimension	0.43	0.11	0.8

struct. All the aspect and dimension mean-square infit statistics here were close to the expected 1.0 (range 0.8 to 1.2). From these results we can conclude that the aspects did not function in a redundant fashion but, rather, worked together to define a meaningful variable.

The qualitative data also suggest that raters did not have difficulty differentiating among the aspects. While 7 of the 17 raters who responded noted that there were too many aspects to evaluate at once, only 1 of those 7 raters and 1 additional rater wrote of specific problems differentiating among aspects. Problems cited by these two raters included distinguishing between the two *Evaluate and reflect* aspects and between the Creating and Presenting Texts overall dimension rating and the other Creating and Presenting aspects. Table 3 shows the responses to a related question posed in the questionnaire. Most of the raters responded that each of the aspects and dimensions captured important qualities of student work and that none should be eliminated or combined.

In the case of the first formative research question, then, the quantitative analyses suggest that the aspects were closely related in ways that might be expected, but appeared to be measuring somewhat different features of language arts achievement, especially at the dimension level. Qualitative data suggest that raters found differentiating among the dimensions and aspects relatively straightforward, and that the aspects and dimensions capture distinct qualities of students' work that should be scored separately.

TABLE 3.
Raters' Opinions about the Aspects in Response to a Questionnaire Prompt ($N = 17$)

	Rater's Opinion		
Aspect	Definitely Score This Aspect	Probably Score This Aspect	Eliminate This Aspect
Creating and Presenting Texts			
Develop and use their own voices	15	0	1
Develop and present texts	14	1	1
Demonstrate technical command	15	1	0
Evaluate and reflect on work	13	2	1
Creating and Presenting Texts Dimension	13	1	1
Making Meaning from Texts			
Respond to texts	17	0	0
Analyze texts	17	0	0
Put texts in context	11	4	1
Evaluate and reflect on work	14	2	0
Making Meaning from Texts Dimension	13	2	1

Note. Responses are to the question: "Are these the right dimensions and aspects for evaluating students' Pacesetter English portfolios? Please indicate whether each of the following captures important qualities of students' portfolios and should be scored in the future." Not all raters answered all questions.

Scale Usage: Does the rating scale for each aspect function effectively as a five-category scale? Are raters able to differentiate among the categories for each aspect?

The 10 columns on the right side of the FACETS rulers (see Figure 1) are graphical representations of the rating scales for each of the aspects and dimensions, as described earlier. The individual rating scales appear to share similar scale structures—that is, the area of the logit scale occupied by each score category is roughly equivalent across aspects and dimensions. (A "2" roughly covers the range –4.0 logits to –2.0 logits across scales. A "3" roughly covers the range of –2.0 logits to 1.0 logits across scales, and so on.) However, it appears that there are certain scale categories that function somewhat differently from the other parallel categories when we look across the scales: *Develop and use their own voices* (CP1)–scale categories 2 and 3, and *Analyze texts* (MM2)–scale categories 4 and 5, for example. When we compare the proportion of ratings falling into each of the scale categories (see Table 4), we note that raters are using the inner scale categories more frequently than the outer scale categories as we would expect with a standards-based scoring guide (that is, we would not expect ratings to be equally distributed across each of the five scale categories defining each scale). This means, for instance, that although raters infrequently use the top and bottom categories for some aspects, they appear to be necessary categories. Some work is very poor, and some work far exceeds expectations, and categories are needed to describe such work, even if it seldom appears. If, on the other hand, we saw an infrequently used central category, we might infer that raters found it difficult to differentiate between that category and those surrounding it. We find no such problem here.

As recommended by Linacre (1994b, 1995) and Andrich (1996), we report two additional pieces of information regarding the structure of each rating scale. The

TABLE 4.
Proportion of Ratings in Each Scale Category

	Scale Category				
Aspect	1	2	3	4	5
Creating and Presenting Texts					
Develop and use their own voices	.07	.12	.34	.41	.07
Develop and present texts	.03	.12	.46	.32	.07
Demonstrate technical command	.06	.21	.44	.25	.05
Evaluate and reflect on work	.16	.38	.28	.15	.02
Creating and Presenting Texts Dimension	.06	.16	.52	.22	.05
Making Meaning from Texts					
Respond to texts	.10	.24	.45	.18	.04
Analyze texts	.08	.35	.39	.13	.06
Put texts in context	.11	.26	.47	.13	.03
Evaluate and reflect on work	.16	.31	.39	.13	.02
Making Meaning from Texts Dimension	.10	.28	.47	.11	.03

first is information about "Average Measure Differences" (AMD). The information here is taken from the Category Statistics table generated by FACETS for each of the rating scales. Linacre (1994b) defines "average measure difference" as "the average of the [student proficiency] measures that are modeled to generate the observations in this category" (p. 69). If a scale is working properly, then the average measure differences should ascend in value, from scale category 1 ("Beginning") to category 5, ("Exemplary"). Ascending values signal that higher ratings do correspond to "more" of the aspect (or dimension) being rated. Average measure differences are shown in Tables 5 and 6. If AMDs do not increase—if we see identical values for adjacent categories or one or more descending values—then that suggests that some of the categories are not functioning as intended. The problematic categories may need to be collapsed, reversed in order, or treated as missing if the category in question is found to not be measuring the desired aspect

TABLE 5.
Average Measure Differences (AMDs) and "Most Probable from" Thresholds (MPTs) for Individual Aspects within the Creating and Presenting Texts Dimension

Aspect	Scale Category	Number of Examinees	Average Measure Difference (AMD)	"Most Probable from" Threshold (MPT)
Develop and use their own voices	1	17	–2.96	low
	2	31	–1.81	–2.98
	3	88	–0.19	–2.10
	4	106	1.56	0.51
	5	17	4.18	4.56
Develop and present texts	1	8	–3.82	low
	2	30	–1.84	–4.00
	3	119	0.06	–2.25
	4	83	2.27	1.53
	5	18	4.59	4.72
Demonstrate technical command	1	15	–3.43	low
	2	54	–1.67	–4.17
	3	114	–0.21	–1.76
	4	65	1.78	1.48
	5	13	4.17	4.45
Evaluate and reflect on work	1	26	–4.15	low
	2	63	–2.10	–4.00
	3	47	–0.46	–1.04
	4	25	0.97	0.91
	5	4	2.65	4.14
Creating and Presenting Texts Dimension	1	15	–3.85	low
	2	41	–2.17	–3.95
	3	134	–0.09	–2.31
	4	57	1.91	1.77
	5	13	4.46	4.49

TABLE 6.
Average Measure Differences (AMDs) and "Most Probable from" Thresholds (MPTs) for Individual Aspects within the Making Meaning from Texts Dimension

Aspect	Scale Category	Number of examinees	Average Measure Difference (AMD)	"Most Probable from" Threshold (MPT)
Respond to texts	1	24	–3.47	low
	2	60	–2.23	–3.85
	3	112	–0.24	–1.76
	4	44	1.56	1.62
	5	10	4.12	3.99
Analyze texts	1	18	–3.77	low
	2	78	–1.75	–4.20
	3	87	0.12	–0.93
	4	28	1.24	1.97
	5	13	3.73	3.16
Put texts in context	1	20	–3.76	low
	2	48	–2.09	–3.87
	3	87	–0.47	–1.87
	4	24	1.52	1.77
	5	5	3.01	3.97
Evaluate and reflect on work	1	29	–4.18	low
	2	56	–2.04	–3.89
	3	71	–0.71	–1.61
	4	23	1.29	1.43
	5	4	3.09	3.98
Making Meaning from Texts Dimension	1	25	–4.21	low
	2	67	–2.22	–4.09
	3	112	–0.20	–1.69
	4	27	1.71	2.10
	5	8	3.62	3.68

or raters are using the category in an idiosyncratic fashion (Linacre, 1995). All scales in these analyses show ascending values.

We also report probabilities of categories at particular measures (again, from the Category Statistics table in the FACETS output). By "most probable" we mean, "the lowest [student proficiency] measure at which this category is the one most probable to be observed" (Linacre, 1994b). Similar to AMDs, these "Most Probable" Thresholds (MPTs) should also be ordered, increasing as we move from categories at the lower end of the scale to categories at the higher end of the scale. The word "low" (for each of the first rating categories for the scales) reminds us that this is the lowest rating category, so there is no measure at which this category begins. If a particular scale category never becomes most probable, then the word "no" appears for that category in the output rather than a measure. This would imply that the score category was either so underutilized as not to become most probable, or that raters were using the category so inconsistently that no clear pattern was established. If MPTs do not increase in value, that may signal a need to reduce the number of categories by combining some of them. As Andrich (1996) notes, a scale may show ascending AMDs but not ascending MPTs. There were no instances of this in the output for the 10 rating scales. All AMDs and MPTs were ascending for each scale. The FACETS analysis, then, appears to indicate that the scales are functioning effectively and that the scale categories within each one are distinct.

FACETS provides us with information about the structure of each rating scale and how each scale performed, information helpful in understanding whether these scales were functioning as intended. For an additional source of information about whether raters experienced problems distinguishing among the various categories on each scale, we again considered raters' responses on the questionnaire.

The qualitative data indicated that raters experienced some problems differentiating among scale categories for one or more of the aspects. Of the 17 raters, 7 made comments indicating problems with deciding which category was appropriate. Their comments were quite specific in suggesting the nature of the problem. Two noted problems with categories for particular aspects. One rater wrote:

> If the portfolio was sketchy, I had a difficult time deciding if analysis [*Analyze texts* from Making Meaning from Texts] was at the "Beginning" or "Developing" stage, especially when techniques were not mentioned. On Creating and Presenting, the *Develop and present texts* aspect was difficult because the few pieces that were there may have been focused but seemed too limited in scope or content. (Rater M, Response to Question 2)

Other raters noted more general difficulties distinguishing among rating categories: "To me there is a fine line between the distinctions of 'Promising' and 'Accomplished' when trying to distinguish the aspects of the dimensions" (Rater I, Response to Question 2). "At times performance is a + or – [between categories] in very different ways" (Rater R, Response to Question 2). Three raters specifically commented on problems with distinguishing between "Beginning" and "Developing." A typical response: "[Consider eliminating] 'Beginning.' 'Developing' seems to cover that level to me" (Rater F, Response to Question 6). One rater noted problems with the differences between "Promising" and "Developing": "The wording of some of the levels at times did not seem to be fair to the assessment of the Portfolio. The problem showed up mostly between 'Promising' and 'Developing'" (Rater L, Response to Question 6).

The answer to the second formative research question appears to be that although analyses of the score data suggest that the rating scales do function effectively, some raters found the process of using the scales and differentiating among the rating categories to be problematic.

DISCUSSION AND CONCLUSIONS

We found that both the FACETS analyses (using the item-step model) and qualitative data separately provided valuable information necessary to refine this portfolio assessment system and materials. However, by viewing them together, we saw a more complete picture of how the system was working. Neither type of data collection and analysis was perceived by researchers or project staff to be time-consuming (and therefore costly) or burdensome, yet both yielded helpful results.

Our first formative research question concerned the investigation of aspect differentiation. Both the qualitative data and the aspect mean-square infit statistics from the FACETS output suggest that raters experienced no serious problems differentiating among aspects. However, when we examined the aspect measures in FACETS, we found evidence to suggest that raters may have experienced problems distinguishing among certain aspects. In particular, the fact that the logit measures for the two *Evaluate and reflect* aspects were nearly identical suggested that raters may not have been able to separate these two aspects, giving students very similar ratings on both.

Our second formative question asked whether the rating scales were functioning as intended. From our review of selected FACETS rating scale category statistics output, we concluded that there were no major problems with the rating scales. Each of the scales appeared to function appropriately as a five-category scale. The qualitative data provide a somewhat different view, in that some of the raters reported problems with differentiating among some of the scale categories.

There are plausible reasons for the different stories the two types of data tell. The quantitative and qualitative data represent two different lenses on the system and materials. The FACETS analysis views patterns of ratings across all raters, whereas the qualitative data can tell us only what individual raters were experiencing, and only at the very end of the scoring. The difference is useful. When raters responded to our questionnaire, they reported no difficulties distinguishing among aspects; but when we reviewed the FACETS results, we found that the two *Evaluate and reflect* aspects shared nearly the same measure. It is possible that individually, raters developed what they perceived to be consistent methods for separating the aspects while scoring, but the FACETS results would seem to suggest that in practice the raters may not have been able to distinguish between them. Perhaps the two aspects were measuring the same thing in practice. Project staff, however, expressed the belief that these two aspects could measure two different and useful features of language arts achievement, and so decided to rewrite the two to differentiate and clarify them. These differences will be emphasized in rater training for future scorings. So, although the findings of the analysis of the qualitative data suggested there were no perceived changes needed, findings from the FACETS analysis suggested that there were useful changes to be made.

When we investigated the second formative research question, we found that although some raters reported finding parts of the scoring process difficult, they may have been coping with their difficulties well and thus no problems within the scales were evidenced in the FACETS analysis. The perception of difficulties, even if the problems were dealt with effectively, is important for project staff to be aware of. Any change that makes the scoring process easier for raters is desirable.

It seems evident, in light of the differences in the results of the two types of analyses, that the system and materials were in general working well, and only fairly minor changes are needed. It was the researchers' recommendation, based on these findings, that the two *Evaluate and reflect* aspects be rewritten so that the differences between them could be made clearer, and that the *Evaluate and reflect* aspect in Creating and Presenting Texts be reworked so that it would relate more to other aspects within the same dimension. We further recommended that project staff review and possibly rework the definitions of the aspects within each dimension with an eye to sharpening the distinctions among them. We also pointed out that some raters reported problems with differentiating among categories within aspects, and that the descriptors should be clarified and the differences between problematic categories addressed in training and in materials. Project staff also took the results of these and other quantitative analyses to those charged with curriculum development, and encouraged them to direct their efforts toward more and better instruction in the aspects of Making Meaning from Texts and in *Evaluation and reflection* in both dimensions.

Some of the results presented here suggest further investigation. If we had asked more specific questions of raters about the differentiation among aspects, we might have had results that more closely reflected the results from the FACETS analysis. If we had asked questions earlier in the 2-day scoring, results might have been different. It may be that the raters who perceived problems differentiating among categories were actually having problems with the materials (that is, category descriptors), or their perceived problems may have been related to the collection of portfolios they were assigned. Changes in the qualitative data collection process are suggested by the first two possibilities, and an additional FACETS analysis may suggest the answer to the last, both for the 1995 data and for future work. One possible type of follow-up FACETS analysis is suggested in McNamara and Adams's work (1991); that of modeling the way each individual rater interprets the scoring categories for each scale separately (that is, what McNamara [1996, p. 287] refers to as using a "judge-item-step" model). Using this partial credit model for raters and aspects would allow us to closely investigate each rater's use of each of the 10 rating scales. We would see whether the problems an individual rater described in questionnaires (for example, having difficulty distinguishing between categories on a scale) were in fact borne out in that rater's use of the scale (for example, whether certain categories were never most probable for that rater, whether the rater tended to overuse certain categories while seldom using other categories, and so on).

APPENDIX A
STANDARDS-BASED SCORING GUIDES FOR PACESETTER ENGLISH ASSESSMENT

Voices of Modern Cultures: Pacesetter English Dimension—Making Meaning from Texts

	Exemplary	Accomplished	Promising	Developing	Beginning
Respond to texts	Responds thoughtfully in terms of own background and experiences. Develops interpretations that show understanding of characters and inquiry into issues, at times in original ways. Incorporates thoughtful prediction, inference, opinions, and judgments to develop responses.	Responds to texts, making connections…[a]	Responds in straightforward manner with some connection to own background and experiences. Presents interpretations with some explanation. Makes some reference to own opinions.	Responds in quite broad or narrow ways…[a]	Responds briefly by mentioning connections and referring to isolated facts or events from text. Opinions are brief statements of likes or dislikes, often unrelated for overall response.
Analyze texts	Incorporates analysis of varied voices and points of view in discussion of text. Uses appropriate textual analysis of literary or other media techniques to support discussions. Analyzes relevance of facts, examples and other resources.	Analyzes the effect of voices and varied points of view…[a]	Identifies voices or points of view with some discussion of the effect. Identifies several techniques with some references to the effect. Identifies facts, examples, and other resources that are usually relevant to a topic.	Identifies major voices or points of view…[a]	Mentions concept of voice or point of view. Mentions techniques, facts, examples, or other resources with no discussion.
Put texts in context	Incorporates appropriate context information in developing overall meaning. Connects texts to other texts, characters, situations, or cultures to enhance meaning. Connects texts and recurring themes to show meaning of both texts and themes.	Uses a range of context information to develop meaning…[a]	Identifies context information, with straightforward statements of their effects. Makes some logical connections to other texts, characters, situations, or cultures. Connects texts to recurring themes with brief explanations.	Briefly mentions some context information, at times with inaccuracies…[a]	Makes superficial, inaccurate, or inappropriate comments about context or connections to other texts, characters, situations, or cultures. Connections to recurring themes are inappropriate or nonexistent.
Evaluate and reflect on work	Evaluates efforts to develop an understanding of texts. Reflects thoughtfully on engagement with texts, difficulties encountered, strategies for dealing with those difficulties, and successful ways to create meaning.	Evaluates efforts and reflects on the processes used in making meaning from texts…[a]	Evaluations and reflections tend to focus on surface features or on general impressions of meaning making processes. May include some comments on ways to improve.	Evaluations and reflections, if they do appear, tend to be general…[a]	Any reflections and evaluations are limited to likes and dislikes about assignments, activities, or texts.

[a] These scoring guides have been altered for this publication. Score point descriptors for the some of the score categories have been shortened to save space.

Voices of Modern Cultures: Pacesetter English Dimension—Making Meaning from Texts

	Exemplary	Accomplished	Promising	Developing	Beginning
Making meaning from texts	Responds to texts thoughtfully in terms of own background, experience, opinion, and judgments. Interprets and analyzes texts, showing understanding of characters and issues, with details about voices, perspectives, and techniques. Uses details about author, cultural and historical period, and connections to develop overall meaning. Analyzes relevance to a topic of facts, examples, and other resources. Evaluates efforts to make meaning from texts. Reflects thoughtfully on engagement with texts.	Responds to texts in terms of own background, experience, and opinions; interprets and analyzes texts with the use of detail about voice…[a]	Responds to texts with some reference to own background, experience, and opinions; interprets and analyzes texts with some use of supporting details. Includes some details about author, cultural or historical period, and connections with some reference to the meaning of the texts. Selects facts, examples, and other resources that are usually relevant to a topic. Contains general reflections on or evaluations of how meaning is made.	Responds to texts in quite broad or vary narrow ways with limited references to own background, experiences, and opinions…[a]	Responds to texts briefly, usually in factual ways. Makes few connections to own background. Opinions, analyses, and interpretations are brief and unsupported or long and at times incoherent. Makes inaccurate or inappropriate comments about author, cultural or historical background, or connections. Selects facts, examples, and resources that appear random or illogical. Comments on work in terms of like and dislikes.

Voices of Modern Cultures: Pacesetter English Dimension—Creating and Presenting Texts

	Exemplary	Accomplished	Promising	Developing	Beginning
Develop and use their own voices	Communicates in a personal style that incorporates a variety of voices that represent the student's cultures and points of view, and that is aesthetically pleasing and/or convincing.	Expresses a variety of voices employing a personal style…[a]	Displays a limited variety of voices which at times are not distinct or personal.	Almost always communicates in the same voice…[a]	Uses the same voice no matter the audience or purpose of the text.
Develop and present texts	Communicates in a variety of well-crafted texts in various genre and media. Texts are focused and coherent, and, when appropriate, effectively revised to clarify meaning and enhance voice. Uses diction, including figurative language, effectively. Oral presentations, when included, are polished in both content and technique.	Communicates in various genre and media…[a]	Shows limited range in genres and media. Texts have basic focus with some lapses. Texts show evidence of some revision of ideas or presentation, although revisions tend to focus on surface features. Presents some oral work that uses own voices or mentions the perspectives of others.	Texts consist of almost entirely one genre and one medium…[a]	Texts are very limited both in scope and content. Texts are disorganized and confusing.

Demonstrate technical command	Edits and presents skillfully crafted products. Clear command of grammar, syntax, structure, punctuation, and spelling are evident throughout edited texts.	Edits and presents texts to meet the expectations of the audience…[a]	Shows some concern for the quality of work and how it is presented, although some errors in language command and conventions are noticeable.	Shows only limited attention to editing…[a]	Shows no evidence of editing beyond recopying. First drafts are final drafts. Serious errors that interfere with the meaning of text are common.
Evaluate and reflect on work	Evaluates and reflects on processes for creating texts and final presentation of texts. Discusses difficulties encountered and strategies for dealing with those difficulties. Includes reflections on the student as learner and ways to improve. Uses feedback and criteria to refine texts.	Evaluates and reflects on the processes used to create and present texts…[a]	Evaluations and reflections tend to focus on surface features or on general impressions of the overall quality. May include some comments on ways to improve. Little or no evidence that feedback or criteria are considered when refining texts.	Reflections, if they do appear, tend to be general…[a]	Any reflections and evaluations are limited to likes and dislikes or other surface features.
Creating and presenting texts	Communicates in a personal style that incorporates a variety of voices, and that is aesthetically pleasing and/or convincing. Texts are well crafted, and when appropriate, effectively revised. Texts show clear command of diction, grammar, syntax, structures, punctuation, and spelling. Evaluations and reflections thoughtfully address processes used to create and present texts.	Effectively expresses a variety of voices and perspectives using a personal style…[a]	Uses a few voices well to express own ideas and some ideas of others. Has command of a limited number of genre and media. Texts show basic focus with some lapses in logic or presentation. Uses revision more to correct surface features that to enhance voice or style. Finished texts show some editing. Evaluations and reflections tend to focus on surface features. Other attempts at reflection are very general.	Usually communicates in one voice and in one genre and media. Texts lack focus and may be confusing at times…[a]	Uses the same voice no matter what the audience or purpose of the text. Texts are unorganized and very limited in scope and content. There is no evidence of editing or revision beyond recopying. Evaluations and reflections if they appear are exclusively about likes and dislikes.

[a] These scoring guides have been altered for this publication. Score point descriptors for the some of the score categories have been shortened to save space.

APPENDIX B
PACESETTER ENGLISH PORTFOLIO ASSESSMENT DEVELOPMENT

Scoring Feedback: Sample Questions

Question 2. Please describe as specifically as possible what was most difficult or problematic about scoring the aspects and dimensions of Pacesetter portfolios.

Question 3. Are these the right dimensions and aspects for evaluating students' Pacesetter English portfolios? Please indicate whether each of the following captures important qualities of students' portfolios and should be scored in the future. [Aspects and dimensions followed—raters were asked to circle one of the following responses for each aspect and dimension: "Yes, definitely score," "Probably continue to score," or "No, consider eliminating."]

Question 5. Did you have problems with the number of aspects and dimensions you were asked to score?

❑ I did not have problems with the number of aspects and dimensions I was asked to score.
❑ I was asked to score too many aspects and dimensions, fewer would have worked better.
Please explain:

Question 6. How well did having five scoring levels (Exemplary, Accomplished, Promising, Developing, and Beginning) work for capturing differences among students' portfolios with respect to the dimensions and aspects you were scoring?

❑ The number of scoring levels worked well
❑ There were too many scoring levels, fewer would have worked better. Consider eliminating or collapsing the following score levels:
❑ There were too few scoring levels, more would have worked better (Check all that apply.)
  ❑ I wanted to have another level below the lowest score level.
  ❑ I wanted to have another level above the highest score level.
  ❑ I wanted to have additional levels because many portfolios fell between levels.
  ❑ I wanted to split one or more of the levels into two because the level didn't make important distinctions.
Please explain:

NOTES

1. A scoring guide, sometimes called a rubric, is a detailed and specific set of criteria for rating student work at particular scoring levels. Generally, for each possible score category there will be a description of what should be present (or what is lacking) in a student's work in order for that score to be assigned. Appendix A contains abbreviated versions of the scoring guides used for this research.

2. A benchmark portfolio is a sample portfolio that is used in training to help raters understand what qualities are present in work at a particular score level. Trainers chose the portfolios in advance and agreed upon the scores that should be assigned them, then photocopied them and used them during rater training to help all raters come to a common understanding of the scoring guides.

REFERENCES

Andrich, D. (1996). Category ordering and their utility. *Rasch Measurement Transactions, 9*(4), 464–465.

Braun, H. I. (1988). Understanding scoring reliability: Experiments in calibrating essay readers. *Journal of Educational Statistics, 13*(1), 1–18.

Engelhard, G., Jr. (1994). Examining rater errors in the assessment of written composition with a many-faceted Rasch model. *Journal of Educational Measurement, 31*(2), 93–112.

Fisher, W. P., Jr. (1992). Reliability statistics. *Rasch Measurement Transactions, 6*(3), 238.

Heller, J., Sheingold, K., & Myford, C. (1998). Reasoning about evidence in portfolios: Cognitive foundations for valid and reliable assessment. *Educational Assessment, 5*(1), 5–40.

Hoover, H. D., & Bray, G. (1995, April). *The research and development phase: Can a performance assessment be cost effective?* Paper presented at the annual meeting of the American Educational Research Association, San Francisco.

LeMahieu, P. G., Eresh, J. T., & Wallace, R. C. (1992). Using student portfolios for a public accounting. *The School Administrator, 49*(11), 8–14.

Linacre, J. M. (1994a). *Many-facet Rasch measurement.* Chicago: MESA Press.

Linacre, J. M. (1994b). *A user's guide to FACETS Rasch measurement computer program.* Chicago: MESA Press.

Linacre, J. M. (1995). Categorical misfit statistics. *Rasch Measurement Transactions, 9*(3), 450–451.

Linacre, J. M. (1996). Generalizability theory and many-facet Rasch measurement. In G. Engelhard, Jr. & M. Wilson (Eds.), *Objective measurement: Theory into practice* (Vol. 3, pp. 85–98). Norwood, NJ: Ablex.

Lumley, T., & McNamara, T. F. (1995). Rater characteristics and rater bias: Implications for training. *Language Testing, 12*(1), 54–71.

Lunz, M. E., & Stahl, J. A. (1990). Judge consistency and severity across grading periods. *Evaluation and the Health Professions, 13*(14), 425–444.

McNamara, T. J. (1996). *Measuring second language performance.* New York: Addison Wesley Longman.

McNamara, T. J., & Adams, R. J. (1991, March). *Exploring rater behaviour with Rasch techniques.* Paper presented at the 13th Language Testing Research Colloquium, Educational Testing Service, Princeton, NJ.

Myford, C. M., Marr, D. B., & Linacre, J. M. (1996). *Reader calibration and its potential role in equating for the Test of Written English* (TOEFL Research Report No. 52). Princeton, NJ: Educational Testing Service.

Myford, C. M., & Mislevy, R. J. (1995). *Monitoring and improving a portfolio assessment system* (Center for Performance Assessment Report No. MS 94-05). Princeton, NJ: Educational Testing Service.

Resnick, L. B., & Resnick, D. P. (1992). Assessing the thinking curriculum: New tools for educational reform. In B. R. Gifford & M. C. O'Connor (Eds.), *Future assessments: Changing views of aptitude, achievement and instruction* (pp. 37–70). Boston: Kluwer.

Wright, B. D., & Linacre, J. M. (1994). Reasonable mean-square fit values. *Rasch Measurement Transactions*, *8*, 370.

Wright, B. D., & Masters, G. N. (1982). *Rating scale analysis: Rasch measurement.* Chicago: MESA Press.

3

USING CRITERION-REFERENCED MAPS TO PRODUCE MEANINGFUL EVALUATION MEASURES: EVALUATING CHANGES IN MIDDLE SCHOOL SCIENCE TEACHERS' ASSESSMENT PERCEPTIONS AND PRACTICE

Lily Roberts
University of California, Berkeley

INTRODUCTION

In this chapter, I address a common methodological challenge in program evaluation; that is, to make evaluation meaningful to its various stakeholders. To this end, I provide an example of using criterion-referenced maps to exemplify middle school science teacher change on measures of perceptions of the usefulness of

---

This research was supported in part by a grant from the American Educational Research Association, which receives funds for its AERA Grants Program from the National Science Foundation and the National Center for Education Statistics (U.S. Department of Education) under NSF grant RED-9255347 Further, this project has been supported by NSF grant MDR9252906. Opinions reflect those of the author and do not necessarily reflect those of the granting agencies.

alternative assessment strategies and their actual assessment practices. In terms of teacher enhancement programs, to be useful and meaningful, evaluation information needs to be available in an easily interpretable fashion for teachers and administrators. I argue that maps generated using a partial credit model, an item response theory (IRT) approach to the analysis of ordered response categories (Masters, 1988), offer a solution to this methodological challenge. To illustrate my argument, I draw on an evaluation of the teacher enhancement aspects of a curriculum and assessment development project from the Science Education for Public Understanding Program (SEPUP). I present the use of criterion-referenced maps to illustrate evaluation results from a pre–post survey of teachers' assessment, collegial, and instructional practices. As demonstrated by Monsaas and Engelhard (1996), Rasch measurement can be used to evaluate programs because it provides a "psychometrically defensible" framework for examining change (p. 128).

Maps were generated using the *Quest* program (Adams & Khoo, 1993), then were analyzed by examining the clustering of the item thresholds to identify *criterion zones* that provide meaningful interpretations of teacher location on the various constructs measured. Further, the maps allow one to look across the groups, comparing their means and visually inspecting growth of participants in the aggregate as well as individually. Consequently, these maps can be used for both formative (for example, teacher self-assessment) and summative (for example, program impact) evaluation purposes.

DESCRIPTION OF SEPUP

SEPUP received funds from the National Science Foundation (NSF) to develop and field test a year-long, middle school science course entitled *Issues, Evidence and You* (*IEY*). In addition, funds were designated for the creation of an assessment system. The SEPUP Assessment Project was designed to apply new theories and methodologies in the field of assessment to the practice of teacher-managed, classroom-based assessment of student performance (see, for example, assessment nets in Wilson & Adams, 1996).

Teacher enhancement is central to the principles embodied in the SEPUP assessment system (Sloane, Wilson, & Samson, 1996; Wilson & Sloane, in press). The assessment system provides a set of tools for teachers to use to (1) assess student performance on central concepts and skills in the curriculum, (2) set standards of student performance, (3) track student progress over the year on the central concepts, and (4) provide feedback (to themselves, students, administrators, parents, or other audiences) on student progress as well as on the effectiveness of the curriculum materials and classroom instruction. Initially, managing a new classroom-based system of embedded assessment demands much of the teacher. Empowering teachers (through the provision of the tools, procedures, and support) to collect, interpret, and present their own evidence regarding student performance, is an important step in the continuing professionalization of teachers in the field of assessment (Shepard, 1995).

The SEPUP curriculum is designed to engage students in an "issue-oriented, hands-on" approach to thinking about scientific issues that are relevant to their daily lives (for example, water, waste, energy, and environment) and their understanding is assessed on an ongoing basis. The students' role is changed as they conduct labs and other activities that are designed to help them understand that science is really a way of asking and answering questions and not just a collection of established facts that they are asked to memorize. The teachers' role is that of facilitator in their development. The aim is that assessment information becomes a scaffolding mechanism for instructional change that further facilitates student learning. The long-term goal is for teachers to become autonomous assessors of their students' understanding of science. Facility with these types of maps may be essential to teachers attainment of autonomy with respect to applying the SEPUP assessment system. Teacher professional development is therefore a necessary outcome of SEPUP implementation that must be evaluated.

MEANINGFUL EVALUATION OR MAKING EVALUATION MEANINGFUL

For over two decades, evaluators have been exhorted to conduct useful studies (Patton, 1997) and to be sensitive to the political nature of their activities (Weiss, 1991, 1998). One of the thrusts of this research has been to make evaluation findings more meaningful by working on linking evaluation more closely to assessment. The aim is to present the results in a manner that is substantively interpretable and that can be easily used within the educational context in which it is to be applied. In fact, the approach is to embed evaluation within the design of the program to enhance its usefulness. Rasch measurement provides a medium for addressing this methodological challenge of program evaluation, which will be illustrated in this chapter.

Meaningful Evaluation

Meaningful evaluation serves both formative and summative purposes, but its usefulness is not limited to system-level decision makers as the principal audience. Conducting evaluations strictly for decision makers is too narrow and may well decrease the probability for use. Further, the issues of the politicization of evaluation and misuse are inextricably linked to evaluation for decision makers (Comfort, 1982; House 1993). I assert that if evaluation is to be truly useful in the context of educational change, it needs to be meaningful and equally accessible to practitioners, the public, and policy makers.

Meaningful evaluation, as a conceptual framework, seems most appropriate when viewing the stakeholders of an evaluation from a constructivist perspective. If evaluation is to serve as an educational device with respect to a theory of use and evaluators function as educators (Shadish, Cook, & Leviton, 1995), then viewing stakeholders as learners is consistent with an overarching theoretical framework for program evaluation. Stakeholders, broadly construed as both those who deliver as well as those who consume the services of a program, are those individuals who need to construct an understanding or make meaning of evaluation results.

Use of evaluation results can be promoted via the medium through which the information is communicated (Shadish et al., 1995; Stallworth & Roberts-Gray, 1987). In this chapter, the medium of communication being proposed is empirical evidence presented on criterion-referenced maps generated using item response theory (IRT). These maps are meaningful to the academic audience familiar with IRT, but more importantly, the maps are simple and clear enough to be meaningful to a broad range of stakeholders. The maps were developed using empirical evidence from the 1994–1995 field test of SEPUP's *IEY*, and the various scales were validated using standard applications of IRT (Roberts, 1996).

The SEPUP evaluation has several audiences or stakeholders: the SEPUP staff, including both curriculum developers and the assessment team; the funding agency (that is, National Science Foundation); and ultimately, the teachers who use the SEPUP course *IEY*. The first two audiences will be interested in teacher change as it relates to the teacher enhancement component. They will want the answer to the age-old evaluation question of whether the group that received *X* did better than the group that did not receive *X*. The latter audience of teachers will benefit from the development of criterion-referenced maps that can help teachers self-assess; this is also of paramount interest to the SEPUP assessment team.

Evaluation Research Questions

The two questions addressed in this chapter focus on the effect that program participation has had on teachers.

- Were the changes for the SEPUP teachers greater than those for the non-SEPUP teachers in terms of their perceptions and practices in science assessment?
- Were the changes for the SEPUP Assessment Development Center (ADC) teachers greater than those for the SEPUP Professional Development Center (PDC) teachers in terms of their perceptions and practices in science assessment?

METHODOLOGY

For the overall program evaluation, an integrated mixed methods research design was used (Roberts, 1996, 1997). There were two levels of treatment for teachers;

teachers in ADCs received the "full" treatment while those in PDCs received a "partial" treatment (see later description). Both quantitative and qualitative information was collected for the following three reasons: (1) to be able to triangulate quantitative and qualitative findings given the developmental nature of the SEPUP program; (2) to enable understanding of outliers or discrepant cases; and (3) to identify features of the ADCs that had a direct effect on the use of the assessment system, and consequently, on teacher professional development with respect to their perceptions and practices in science assessment.

Description of the Sample

The sample consists of all teachers involved in the 1994–1995 field test of SEPUP's *IEY* course with embedded assessment system. A portion of these teachers had also participated during the pilot year, but about a third of the SEPUP teachers were new to the program in the field test year.

Source and Number

Thirteen of the 15 original PDCs continued with the field test of SEPUP after the pilot year (1993–1994). Six of these 13 were selected to serve as ADCs and 7 continued to function as PDCs. In each center, the number of SEPUP teachers ranged from two to six (the minimum was due to attrition during the field test year). All centers, except two PDCs, had at least one comparison teacher.

The ADCs were located in Alaska, California (Bay Area), Colorado, Kentucky (Louisville), Louisiana, and Oklahoma. There were 26 SEPUP teachers and 7 comparison teachers in the ADCs. The PDCs were located in California (San Diego), eastern Kentucky, New York City, North Carolina, Pennsylvania, Washington, D.C., and western New York. There were 25 SEPUP teachers and 5 comparison teachers in the PDCs. The participating teachers were from various schools and districts within the centers.

SEPUP Teachers' Roles and the Selection Criteria for ADCs

ADC teachers had different demands placed on them in terms of requirements for participation during the field test year. The ADC teachers were required to use the SEPUP course along with the elements of the SEPUP assessment system. In particular, ADC teachers were to implement the assessment activities embedded in the SEPUP course, to score student papers using the SEPUP scoring guides (specific to the variables in the course), and then to meet regularly as a group to moderate one or two assessment activities (up to at least 10 tasks for the whole year) using a sample of papers from each teacher's classroom. Local assessment moderation is a crucial part of the SEPUP assessment system because it provides a mechanism for ongoing support of teachers engaged in changing their assessment practices (Roberts, Sloane, & Wilson, 1996).

The PDC teachers were only required to use the modified version of the SEPUP course and submit feedback to the curriculum developers at the Lawrence Hall of Science. They were not required to use the assessment system, however, the assessment tasks were integrated as course activities and PDC teachers could choose to use them as such or not. The non-SEPUP or comparison teachers were to teach their regular science course and not use any of the SEPUP activities.

Description of the Instrument

A pre–post teacher survey, SEPUP Inventory of Teachers' Assessment, Collegial, and Instructional Practices (SITACIP), was developed to meet the particular needs of this evaluation research. The items were paneled by the SEPUP assessment team and several science educators to address face validity. The initial SITACIP was pilot tested in 1993–1994, then a modified version was used in 1994–1995.

The SITACIP survey is a 77-item, self-report of instructional and assessment practices and attitudes as well as measures of collegiality and collaboration. The subsections of the SITACIP fall into three basic question types: (1) frequency of use of various instructional and assessment strategies, (2) Likert-type scales for both collegiality and reasons for assessment strategy choices, and (3) attitudes about the usefulness of 13 different assessment strategies for assessing learning, guiding instruction, and grading.

SITACIP Assessment Scales

The two SITACIP scales that measure teachers' perceptions (Assessing Learning, AL) and practices (Assessment Strategy Use) related to science assessment are the focus of this chapter. The Assessing Learning scale was administered as part of the post-SITACIP, with the preadministration having been done during the pilot year. The pilot survey was modified significantly and only some scales were used again during the SEPUP field test. The Assessment Strategy Use scale was administered (pre/post) during the field test year only.

Assessing Learning is an affective scale that measures teachers' perceptions of the usefulness of various assessment strategies for developing an understanding of what students know. A Likert-type scale was used, as follows: useless (0), somewhat useful (1), quite useful (2), and very useful (3). Table 1 presents the nine items that fit the model and also reports the pre and post means along with the t test of mean change for both ADC and PDC teachers.

Assessment Strategy Use is a behavioral scale that identifies the frequency with which teachers use different assessment strategies. Frequency of use was measured on the following scale: never use (0); use less than once per month (1); use once per month to once per week (2); use more than once per week, but not every day (3); and use every day (4). Table 2 summarizes teacher change at the item level for the six items that fit the model.

TABLE 1.
Analysis of Mean Differences: Teachers' Perceptions of the Utility of Assessment Strategies for Assessing Learning

	Assessment Development Centers								Professional Development Centers							
	Pre			Post			*t* Test		Pre			Post			*t* Test	
Assessment Strategies	*n*	*M*	*SD*	*n*	*M*	*SD*	*T*	*df*	*n*	*M*	*SD*	*n*	*M*	*SD*	*T*	*df*
Lab reports	13	3.00	.82	13	3.39	.77	2.13*	12	12	3.33	.49	12	3.00	.74	−1.77	11
Observations of groups of students	13	3.31	.86	12	3.58	.79	.76	11	12	3.17	.58	12	3.50	.74	2.35*	11
Observations of individual students	13	3.46	.66	13	3.77	.60	1.30	12	12	3.17	.72	12	3.42	.67	1.00	11
Performance assessments	12	3.67	.49	12	3.42	.67	−.80	10	11	3.55	.69	12	3.42	.79	.00	10
Personal notes on student performance	13	2.92	.86	13	2.85	.80	−.37	12	11	2.64	.67	10	3.40	.84	2.75*	9
Portfolio or folder of student work	13	3.08	.86	13	3.08	.76	−.00	12	12	3.17	.84	12	3.42	.67	1.39	11
Presentations by students	13	3.08	.64	13	2.92	.49	−.69	12	11	3.09	.83	12	3.42	.67	1.31	10
Questions and activities that promote student reflection	13	3.54	.66	13	3.31	.75	−1.15	12	12	3.58	.52	12	3.42	.67	−1.00	11
Writing assignments	13	3.46	.66	13	3.08	.76	−1.44	12	12	3.00	.74	12	3.42	.67	1.82	11

* $p < .05$.

TABLE 2.
Analysis of Mean Differences: Frequency of Use of Assessment Strategies

	Assessment Development Centers								Professional Development Centers							
	Pre			Post			*t* Test		Pre			Post			*t* Test	
Assessment Strategies	*n*	*M*	*SD*	*n*	*M*	*SD*	*T*	*df*	*n*	*M*	*SD*	*n*	*M*	*SD*	*T*	*df*
Lab reports	22	4.32	.89	22	4.18	.80	–.83	21	10	4.10	.74	11	4.09	.70	.00	9
Performance assessments	22	4.09	.81	22	4.09	.75	–.00	21	11	4.00	1.10	11	3.91	.83	–.36	10
Projects	22	2.82	.96	22	2.82	1.01	.00	21	10	2.70	1.06	11	2.27	.91	–1.08	9
Conferences	22	2.64	.85	22	3.05	1.13	1.62	21	9	3.22	.97	9	3.33	.78	.43	8
Open-ended questions	22	4.27	.63	22	4.41	.80	.83	21	11	4.09	.83	11	3.36	1.03	–2.03	10
Closed-ended questions	22	3.14	1.13	20	2.55	1.19	–2.13*	19	11	3.46	.93	11	3.64	1.03	.38	10

* $p < .05$.

Construct Validation Process

Because the SITACIP instrument was developed for the SEPUP evaluation, measurement characteristics, such as validity or reliability, were not known. Consequently, construct validation was integrated as a preliminary step in this study's analyses. Using IRT techniques, the scales in the SITACIP were examined for fit to the partial credit model (Masters, 1982). Item features such as parameter invariance were also analyzed. Once fit was established for the scales of interest, the anchored case estimates were used in subsequent statistical analyses. The construct validation process is described in detail elsewhere (Roberts, 1996).

The item fit analyses for the Assessing Learning and Assessment Strategy Use scales are summarized next, and the variable structure is described later using threshold maps. Threshold maps provide a graphical representation of both person and item placement on variables simultaneously. At a glance, the pattern of person responses can be compared to item difficulties.

Each of the SITACIP scales underwent separate item analyses using *Quest* (Adams & Khoo, 1993). The *Quest* program was used to obtain traditional internal consistency measures (for example, Cronbach's α) as well as item fit statistics (infit mean square and infit t) for all items of substantive interest in each scale.

The infit mean square and infit t statistics gauge the extent to which response patterns indicate that the data fit the partial credit model (Masters, 1982). By fitted models, I mean the final fit statistics for the fitted model for which the evaluation of fit indicates that the item calibration is valid. To be deemed valid, the infit mean square for the whole scale should have a mean of about 1 and a small standard deviation and the infit t should have a mean of about 0 and a standard deviation near 1 (Wright & Masters, 1982, p. 114). Further, this assessment of item fit identifies whether or not a unidimensional scale or construct is present and being measured. Items with an infit t of greater than +2 or less than –2 are considered to have a lack of fit to the partial credit model. A positive misfit indicates that the responses to an item are not consistent with other responses. An item with negative misfit is overly consistent or too Guttman-like. Items that misfit somewhat, however, can be maintained in a scale if there are strong substantive reasons for doing so.[1] Table 3 presents the item fit statistics for the fitted models from the Assessing Learning and Assessment Strategy Use scales.

TABLE 3.
Summary of the Item Fit Statistics for the SITACIP Assessment Scales

			Infit Mean Square		Infit t	
Scale	*N*	Number Fitted Items	*M*	*SD*	*M*	*SD*
Assessing Learning	104	9	1.00	.16	–.04	1.15
Assessment Strategy Use	104	6	1.00	.21	.04	1.50

N represents the total number of teachers analyzed for each subscale.

For most SITACIP scales, the fit of the substantively relevant items was satisfactory without further analyses. All nine items were maintained in the Assessing Learning scale. Item 4, Reports, was removed from the Assessment Strategy Use scale. On the initial analysis, the Reports item had an infit t of –2.6 and item 7, Closed-ended Questions, had an infit t of 3.6. Since Item 7 was substantively important, Item 4 was removed first, and the analysis was redone. Item 7 still misfit somewhat, but its infit t had been reduced to 2.6 after removing Item 4.

Table 4 presents the measures of internal consistency for the SITACIP assessment scales. Two measures of reliability are presented in this table, including Cronbach's α and the reliability of the case (that is, teacher) estimates. Both of these measures range from 0 to 1, with 0 being totally unreliable and 1 being completely consistent or perfectly reliable. Cronbach's coefficient α indicates how much the items in each scale are measuring the same thing and is based on the raw score. The reliability of case estimates (also called "test reliability of person separation") also indicates how much the items in each scale are measuring the same thing, but this statistic is based on the proportion of observed variance of the scaled scores (that is, not raw scores) which is not due to measurement error (Wright & Masters, 1982, p. 106). The measure of reliability based on the raw scores is very similar to that of the reliability of case estimates generated from the scaled scores in the IRT analysis.

In psychological testing, desirable reliability coefficients are usually in the .80s or .90s (Anastasi, 1988). However, not all social science measurement is or needs to be this precise. In the current example, the ramification of lower reliability coefficients does not carry the same consequences as a psychological test. There are at least two plausible explanations for the reliability coefficients being in this middle range. First, the number of items in each scale is small, and it can be shown that by increasing the number of items (that is, comparable or parallel items) the reliability of the test will increase (Traub, 1994, pp. 102–103). Of course, it is preferable to increase reliability without increasing the length of a test. A second factor is the homogeneity of the sample of teachers. If the population of all middle school science teachers was sampled from, rather than this purposive sample of very similar individuals, then the variance of the true scores would likely be increased, and thus the score reliability would also be larger.

TABLE 4.
Summary of the Internal Consistency Measures for the SITACIP Assessment Scales

Scale	*N*[a]	L[b]	Cronbach's α	Reliability of Case Estimates[c]
Assessing Learning	104	9	.69	.68
Assessment Strategy Use	104	6	.49	.54

[a] *N* represents the total number of teachers analyzed for each subscale.
[b] L represents the number of items in the scale.
[c] Test reliability of person separation.

A brief explanation about how to read the criterion-referenced maps is offered here followed by the presentation of maps for the Assessing Learning and Assessment Strategy Use scales. The left half of the map presents the location of teachers on the variable (each X represents one teacher) and the right half presents the item threshold locations. Items are noted as *n.m*; *n* represents the item number and *m* is the associated threshold (1, 2, 3, or 4, depending on the variable and its possible range). The scale at the far left is based on logits (units of log-odds).[2]

In order to interpret teacher location on the variable maps, refer to Table 5, which includes the probability indicated by selected logit differences. Probabilities in this table are relative to the logit difference between a teacher's location and an item threshold, so that positive logit differences correspond to the teacher being above the item threshold, the reverse being that negative logit differences correspond to the teacher being below the item threshold.

Thurstone first described item thresholds as the upper and lower boundaries for ordered response categories, and identified the difference between two thresholds as the estimated width of a category on the psychological continuum being measured (Masters, 1988). Threshold interpretation is based on the probability of achieving the various response categories. On a four-point scale (0 = strongly disagree, 1 = disagree, 2 = agree, 3 = strongly agree), for example, the thresholds would be interpreted as follows. At the first threshold, the probability of a 0, or strongly disagree, response is as likely as a response of disagree or higher. The second threshold is the point where a response of strongly disagree (0) or disagree (1) is equally likely as agree (2) or strongly agree (3). The third threshold is the point where a response of agree (2) or less is as likely as a response of strongly agree (3).

The groups of asterisks down the center of the map in Figure 1 represent "usefulness clusters" that can assist a stakeholder in understanding teachers' perceptions of the usefulness of alternative assessment strategies for assessing learning. These usefulness clusters or criterion zones were determined by observing the clusters of item thresholds. In Figure 1, note that the first, second, and third thresholds for items cluster into bands. These bands have been identified using asterisks

TABLE 5.
Some Representative Values for Interpreting a Logit Scale

Logit Difference	Probability
3.0	.95
2.0	.88
1.0	.73
0.0	.50
−1.0	.27
−2.0	.12
−3.0	.05

down the middle of the map. Shading of bands could also be used to highlight the different usefulness clusters. Teachers located near the top half of the map in Usefulness Cluster 3 perceive alternative assessments as more useful than those at the bottom in Usefulness Cluster 1. The average teacher estimate is .83 logit, which lies at the bottom of Usefulness Cluster 3. This cluster begins at the point where teachers are more likely to indicate that alternative assessment strategies are very useful for assessing what students have learned.

The purpose of the variable maps in Figures 1 and 2 is to establish the structure of the SITACIP assessment scales, so the maps are based on all responses (pre and post from all teachers) thus maximizing the information used to calibrate the items. Figure 1 presents a graphical representation of the Assessing Learning scale. Only three (Items 3, 4, and 8) of the nine items had a first threshold, indicating that the other six strategies were easier for teachers to find at least somewhat useful for assessing student learning. There were three perfect scores on the Assessing Learning scale, so Figure 1 presents only 101 of 104 cases on the map.[3] On this map, no teachers fall into the lowest criterion zone, or Usefulness Cluster 1, meaning that all these teachers find at least some utility in alternative assessment strategies for assessing what students know.

For example, in Figure 1, note the teachers located just under 1.0 logits; these five teachers are directly across from 5.3 (that is, Item 5's third threshold). The third threshold is the point at which these teachers are equally likely to score very useful as to score quite useful, somewhat useful, or useless on Item 5 (Observations of Individual Students).

Moving up to almost 2.0 logits on Figure 1, these same five teachers have a smaller probability (about .27) of getting to a response of very useful on Item 8 (Personal Notes on Student Performance) because they are located a logit below 8.3 (Item 8's third threshold).

Within Usefulness Clusters 2 and 3, Item 7 (Performance Assessments) is at the bottom of each cluster, meaning that teachers are more likely to find this strategy quite useful before other strategies, such as personal notes on student performance (that is, Item 8). An analogy might facilitate understanding the placement of these item locations. If this map represented a test, then Item 7 would be an easy item to get correct while Item 8 would be much harder.

The next figure presents the frequency of teachers' use of alternative assessment strategies, such as Lab Reports or Performance Assessments. Item 7, Closed-ended Questions, was recoded and included in this scale as well. The Assessment Strategy Use scale has four criterion zones or behavior clusters based on a 5-point scale of never use (0) to use more than once per week (4). In Figure 2, the average teacher estimate is .40 logit, and it is located just above Behavior Cluster 2. This mean is not surprising, however, given that many of the assessment strategies included in the scale require more time to complete, such as lab reports or projects.

Items 1, 2, and 6 are consistently clustered with items at lower thresholds; therefore, these criterion zones or behavior clusters are represented by bands of

```
4.0  Case Estimates in Logits     |    Item Thresholds      Criterion Zones

-- TEACHERS PERCEIVE ALTERNATIVE  |
 ASSESSMENTS AS VERY USEFUL --    |                         -- VERY USEFUL --
                                  |
                                  |
                            XXX   |
                                  |
                                  |
  3.0                             |
                                  |
                                  |
                            XXX   |
                                  |
                                  |
                                  |
                         XXXXXX   |
  2.0                         X   |**      10.3
                             XX   |**       8.3
                    XXXXXXXXXXX   |**
                              X   |**       3.3     9.3
                          XXXXX   |**      13.3               USEFULNESS CLUSTER 3
                              X   |**       4.3
                                  |**
                     XXXXXXXXXX   |**
  1.0                         X   |**      11.3
                          XXXXX   |**       5.3
MEAN = .83                        |**       7.3
                 XXXXXXXXXXXXXX   |
                                  |
                     XXXXXXXXXX   |
                                  |
   .0                   XXXXXXX   |**       8.2
                                  |**
                          XXXXX   |**
                        XXXXXXX   |**       3.2
                             XX   |**       9.2               USEFULNESS CLUSTER 2
                           XXXX   |**       4.2
                                  |**       5.2    10.2
                                  |**
 -1.0                        XX   |**      13.2
                                  |**
                              X   |**      11.2
                                  |**
                                  |**
                                  |**       7.2
                                  |
                                  |
 -2.0                             |**       3.1
                                  |**       8.1
                                  |**
                                  |**                         USEFULNESS CLUSTER 1
                                  |**
-- TEACHERS PERCEIVE ALTERNATIVE  |**
   ASSESSMENT AS USELESS --       |**
                                  |**       4.1
 -3.0                             |                           -- USELESS --
-------------------------------------------------------------------------------------
```

FIGURE 1. Variable map of Assessing Learning scale (N = 104; 101 cases mapped).

```
 3.0 Case Estimates in Logits   |    Item Thresholds      Criterion Zones
                                |
-- TEACHERS WHO USE ALTERNATIVE |     -- FREQUENT USE OF ALTERNATIVE ASSESSMENTS --
  ASSESSMENTS FREQUENTLY --     |
                                |
                                |
                                |
                                |
                                |
                                |
                                |
 2.0                        XX  |**   3.4    5.4        BEHAVIOR CLUSTER 4
                                |**
                                |**   7.4
                                |
                           XXX  |
                             X  |**   3.3
                                |**
                        XXXXXX  |**
                                |**                      BEHAVIOR CLUSTER 3
                        XXXXXX  |**
                                |**   2.4
 1.0                            |**   1.4    6.4    7.3
                    XXXXXXXXXX  |**
                                |**   5.3
                XXXXXXXXXXXXXX  |
                                |
                                |
             XXXXXXXXXXXXXXXXX  |
MEAN = .40                      |
                    XXXXXXXXXX  |
                                |**   3.2    7.2
                                |**
  .0                XXXXXXXXXXXXX  |**
                                |**                     BEHAVIOR CLUSTER 2
                         XXXXX  |**   5.2
                             X  |**   1.3
                         XXXXX  |**   2.3
                                |**
                                |**
                          XXXX  |**   6.3
                                |
                            XX  |
                                |
-1.0                       XXX  |
                                |**   7.1
                                |**   1.2
                                |**   5.1
                                |**   1.1               BEHAVIOR CLUSTER 1
                                |**
                            XX  |**
                                |**   6.2
                                |**
 -- TEACHERS WHO SELDOM USE     |**
    ALTERNATIVE ASSESSMENTS--   |**   2.2
-2.0                            |**   3.1          -- LIMITED OR INFREQUENT USE --
-------------------------------------------------------------------------------------
```

FIGURE 2. Variable map of Assessment Strategy Use scale (N = 104; all cases mapped).

consistently related item thresholds. Item 1, Lab Reports, was a frequently used assessment activity in the SEPUP course. Item 2, Performance Assessments, was also perceived as a more useful strategy for assessing learning. Item 6, Open-ended Questions, was also a common feature of the SEPUP embedded assessment tasks.

Analyses

Within-group *t* tests were used to compare differences between each of the three groups (ADC, PDC, and comparison teachers) on all the SITACIP scales. Between-group differences were analyzed using analysis of covariance (ANCOVA). Only teachers from the four ADCs that complied with the requirements of the field test of the SEPUP assessment system were included in these analyses. Teachers in the other two ADCs did not participate regularly in local assessment moderation and the ADC directors from these two sites left the project early in the year. Further, some of the teachers in these two ADCs either dropped out or never fully participated.

The *t* tests assess pre- to postmean score differences for each group, which was useful for interpreting the direction of change in the groups. ANCOVA was used because the teachers were not randomly selected nor randomly assigned to the centers. In all of the ANCOVAs, the prescale score was used as the covariate. The ANCOVA results examine the differences between SEPUP and comparison teachers as well as between ADC and PDC teachers.

RESULTS: USING MAPS TO COMMUNICATE MEANINGFUL EVALUATION

In this section, a summary of the within- and between-center differences is provided (for full details of the within- and between-group analyses, see Roberts, 1996). These summaries are then followed by an illustration of how criterion-referenced maps can be used to make evaluation meaningful to all stakeholders of an evaluation.

Within-Center Differences: A Summary

The pre- to postmean differences begin to reveal an emerging pattern with respect to the differences between ADC and PDC teachers. The ADC teachers had a decrease on the Assessing Learning scale whereas the PDC teachers had a significant increase ($p < .05$) on this scale. In terms of the Assessment Strategy Use scale, the ADC teachers reported an increase in their use of alternative assessment strategies whereas both PDC and comparison teachers reported a decrease in their use of such strategies.

Between-center Differences: A Summary

With respect to between-center differences, the expectation was that SEPUP (that is, both ADC and PDC combined) teachers' change would be greater than comparison teachers' change overall. Further, it was expected that ADC teachers' change would be greater than PDC teachers on the assessment scales of the SITACIP.

On the Assessing Learning scale, the difference between ADC and PDC teachers was almost significant ($p = .08$) with a decrease for the ADC teachers from pre to post, and a significant increase ($p < .05$) from pre to post for the PDC teachers. No comparison was possible between SEPUP and non-SEPUP teachers on this scale.

On the Assessment Strategy Use scale, ADC teachers changed significantly more than PDC teachers. Note that the ADC teachers had a positive change pre to post on the Assessment Strategy Use scale whereas the PDC teachers had a decrease by the end of the school year. Since the PDC teachers were more similar to the comparison teachers on their Assessment Strategy Use change, the expected differences between SEPUP and comparison teachers was not evident.

Making Evaluation Meaningful: Mapping Teacher Change

In this section, I combine what has been learned about changes within and between groups and the person locations on the variable maps to create a graphical representation of individual teacher change on Assessing Learning and Assessment Strategy Use. With these figures, I wish to illustrate the value of such maps for making evaluation meaningful to various audiences as well as participants. These particular scales were chosen because they exemplify the emerging theme of the rhetoric versus reality of assessment reform in middle school science teachers' perceptions and practice in assessment. The results indicate that PDC teachers continued to perceive of alternative assessments as very useful for assessing learning, but in reality, they did not use them. In fact, PDC teachers increased their use of closed-ended questions over the year while ADC teachers significantly decreased ($p < .05$) their use of such questions (see Table 2). Meanwhile, ADC teachers grappling with the SEPUP assessment system reduced their overall perceptions of the usefulness of alternative assessment strategies for assessing learning while at the same time increasing their use of such strategies.

On the following map, the teachers' locations are indicated by two digits. The first digit represents the center and the second digit is a sequential number to identify the different teachers. This identification is different than the teacher codes used by the project; therefore, these results maintain anonymity for both teachers and centers. For the ADCs, the centers are numbered 1 to 4, and the PDCs are numbered beginning with 0. The ADCs were ordered based on a qualitative analysis of teachers success with local assessment moderation (Roberts, 1996; Roberts, Sloane, & Wilson, 1996). No such ordering is possible for the PDCs.

Figure 3 presents both the ADC and PDC teachers' pre- and postmatched estimates on the Assessing Learning scale. This type of map display is beneficial for both formative and summative evaluation purposes because (1) it presents the pre- and postlocation of individual teachers (centers also indicated), (2) it presents the pre- and postmeans for both treatment groups, and (3) both individual and group change can be interpreted substantively with respect to the usefulness clusters. In Figure 3, note that the ADC premean is higher than its postmean, indicating that

```
                     TEACHERS PERCIEVE OF ALTERNATIVE ASSESSMENTS AS VERY USEFUL

Perfect (∞)                 21 |                  ||               01 |                  | **CLUSTERS**
  Case Estimates               |                  ||                  |                  |
  in Logits                    |                  ||                  |                  |
                               |                  ||                  |                  |
                            43 |                  ||                  |               01 |
                               |                  ||                  |                  |
 3.0                           |                  ||                  |                  |
                               |                  ||                  |                  |
                               |                  ||                  |               41 |
                               |                  ||                  |                  |
                            41 |               21 ||               12 |           12  21 |
 2.0                           |                  ||                  |               31 | **
                               |                  ||                  |                  | **
                               |           42  43 ||               21 |           11  22 | **
                               |                  ||                  |                  | **
                               |               31 ||                  |   Mean = 1.40    | **  USEFULNESS
                               |                  ||                  |                  | **  CLUSTER 3
               Mean = 1.19     |                  ||                  |                  | **
                               |               22 ||           11  41 |               32 | **
 1.0                           |   Mean = 1.00    ||               42 |                  | **
                            32 |                  ||               22 |                  | **
                               |                  ||     Mean = .93   |                  | **
                    11  31  42 |       12  32  41 ||               31 |               51 |
                               |                  ||                  |                  |
                            22 |               23 ||                  |               42 |
                               |                  ||                  |                  |
  .0                        23 |                  ||           32  51 |                  | **
                               |                  ||                  |                  | **
                               |               11 ||                  |                  | **
                            12 |                  ||                  |           02  33 | **
                               |                  ||               33 |                  | **
                               |                  ||                  |                  | **
                               |                  ||                  |                  | **  USEFULNESS
                               |                  ||                  |                  | **  CLUSTER 2
-1.0                           |                  ||               02 |                  | **
                               |                  ||                  |                  | **
                               |                  ||                  |                  | **
                               |                  ||                  |                  | **
                               |                  ||                  |                  | **
                               |                  ||                  |                  | **
                               |                  ||                  |                  |
                       Pre ADC |     Post ADC     ||      Pre PDC     |     Post PDC     |

                     TEACHERS PERCEIVE OF ALTERNATIVE ASSESSMENTS AS LESS USEFUL
-----------------------------------------------------------------------------------------------
```

FIGURE 3. Pre–post, ADC and PDC matched estimates and group means on Assessing Learning scale map.

overall the ADC teachers revised their perceptions of the usefulness of alternative assessment strategies downward. In particular, three teachers' perceptions significantly changed from pre to post (refer to ADC Teachers 21, 41, 43 in Figure 3). On average, the ADC teachers continue to find alternative assessment strategies very useful for assessing student learning. Figures 3 to 6 have been modified slightly for illustrative purposes.

The PDC teachers reported a significant increase ($p < .05$) in their perceptions of the usefulness of alternative assessment strategies by the year's end. Four PDC teachers started out in Usefulness Cluster 2, but only two remained by the end of the field test (PDC Teachers 02 and 33). Only two PDC teachers had a decrease from pre to

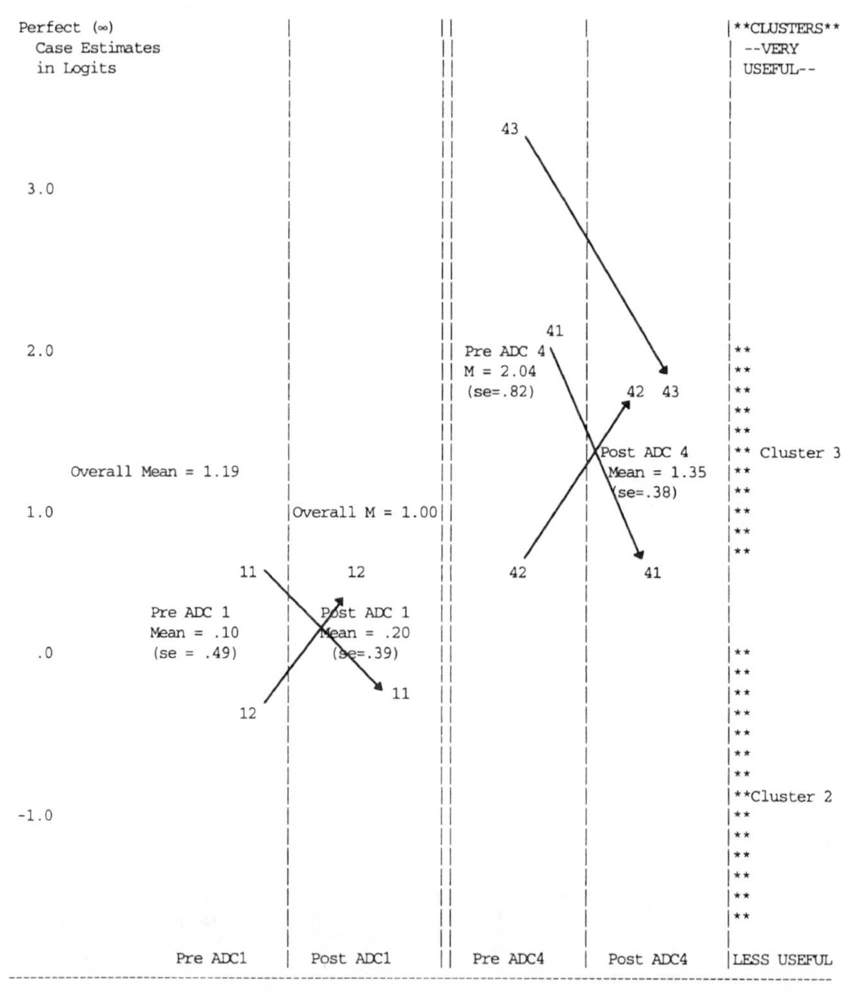

FIGURE 4. Pre–post ADC 1 and ADC 4 matched estimates on Assessing Learning scale map.

post (Teachers 01 and 42). PDC teacher 12 had no change over time. On average, PDC teachers ended up in Usefulness Cluster 3, indicating that they were equally likely to perceive of alternative assessment strategies as being very useful for assessing student learning as they were to find them only quite, somewhat, or never useful.

Figure 4 presents a comparison between the most (ADC 4) and the least (ADC 1) successful ADCs in terms of their implementation of local assessment moderation. Note the positions of individual teachers from pre to post on Figure 4. The two teachers from ADC 1 (11 and 12) have both their pre and post scores below the overall ADC mean. Further, Teacher 12 had an increase of about a logit from pre to post, but Teacher 11 had nearly the same size decrease. Teacher 11 ended up in Usefulness Cluster 2, where teachers may find alternative assessment strategies somewhat useful for assessing student learning. Yet Teacher 12 moved up almost into Usefulness Cluster 3, where teachers may find alternative assessment strategies very useful for assessing student learning.

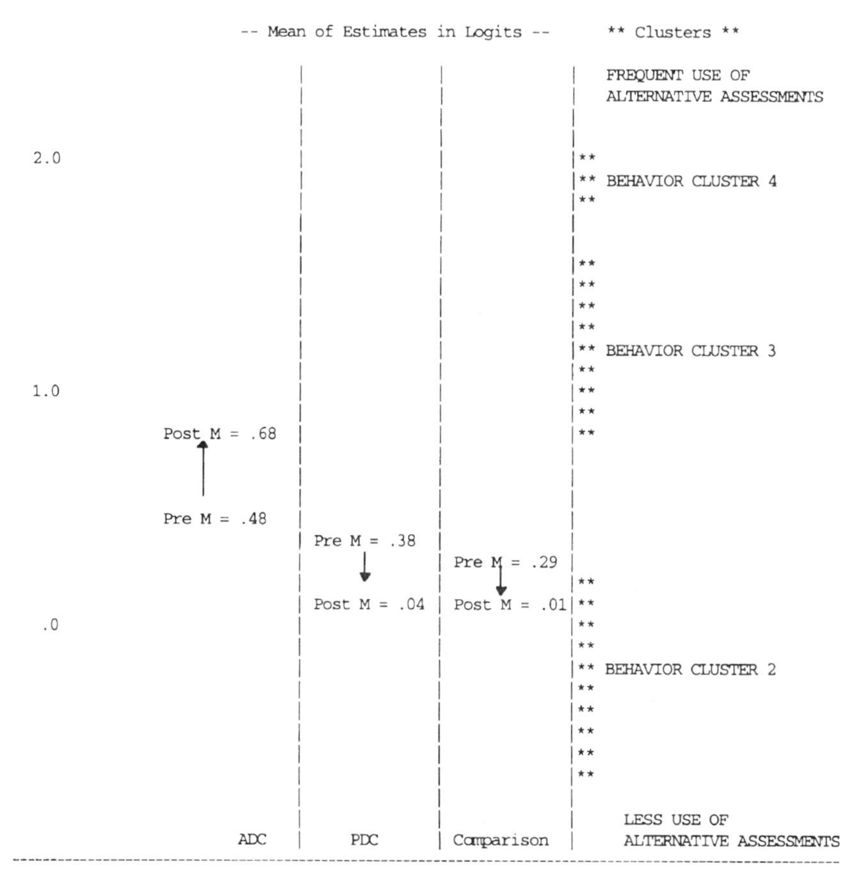

FIGURE 5. Comparing all groups' means on Assessment Strategy Use scale map.

Figure 5 presents the overall ADC, PDC, and comparison teachers' pre- and postmeans on the Assessment Strategy Use scale. This type of map display is beneficial for summative evaluation purposes because (1) it looks at aggregate information by treatment groups, (2) graphically shows the direction and magnitude of change, and (3) presents the change in terms of the behavior clusters. Overall, there was a mean increase of .20 logit for ADC teachers and a decrease of .34 logit for PDC teachers. Neither of these within-center changes were statistically significant; however, given the opposite directions of change, there was a statistically significant difference between ADCs and PDCs. The comparison teachers had a decrease on this scale similar to the PDC teachers, which led to no difference when comparing SEPUP to non-SEPUP teachers.

The postmean for the ADC teachers lies within Behavior Cluster 3, which means that these teachers are most likely to be using alternative assessment on a weekly basis. The postmeans for both the PDC and comparison teachers lie within Behavior Cluster 2, which indicates that these teachers are more likely to use alternative assessments only once a month or less. On average, the ADC teachers used alternative assessment strategies much more frequently than the PDC or comparison teachers.

DISCUSSION AND CONCLUSIONS

One of the prominent themes that emerged from this evaluation was the difference between the rhetoric and the reality of assessment reform as it played out in teachers' minds and in their classrooms. Figure 6 graphically presents this theme; ADC teacher change is noted by the solid arrows and PDC teacher change is noted by the dotted arrows. In this map display, change on both Assessing Learning and Assessment Strategy Use for ADC and PDC teachers is presented laterally to illustrate the rhetoric versus reality theme. The left half of the figure presents the pre and post means for the ADC and PDC teachers on the Assessment Strategy Use scale and the right half presents the same for the Assessing Learning scale.

The SEPUP ADC teachers were intensively engaged in testing the components of the SEPUP assessment system, whereas the SEPUP PDC teachers were only field testing the curriculum. The ADC teachers, while embracing the rhetoric of alternative assessment reform in the beginning, adjusted their perceptions of the usefulness of various alternative assessment strategies over time. Faced with the reality of implementing alternative assessments, ADC teachers revised their perceptions of the usefulness of such strategies for assessing learning. ADC teachers did increase their use of open-ended questions while significantly decreasing their use of closed-ended questions. Thus, ADC teachers' assessment practices did change, but their attitudes seemed to be in a state of some dissonance as they tried to reconcile—through practice—the rhetoric and the reality of assessment reform.

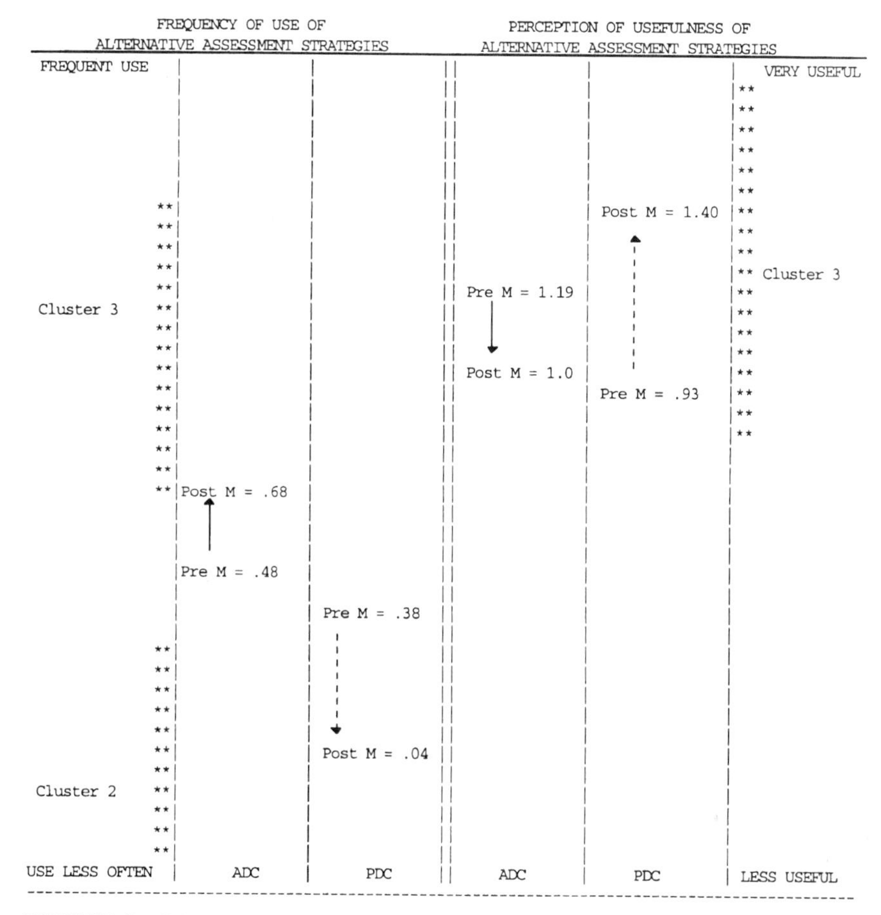

FIGURE 6. Rhetoric versus reality—Comparing group mean change (in logits) on Assessing Learning and Assessment Strategy Use scales.

Meanwhile, the PDC teachers still embraced the rhetoric of reform by the end of the field test, but continued assessment practices in a traditional manner. Even though the SEPUP course has embedded assessment tasks, the results indicate that the PDC teachers resorted to very traditional modes of assessment; in particular, they increased their use of closed-ended questions by the end of the field test year. PDC teachers did have access to the scoring guides for the course variables, but access to scoring guides alone did not result in the use of the embedded assessment tasks.

For the dissemination of the *IEY* course with an embedded assessment system, these results strongly suggest that inservice and ongoing support for teacher change are a necessity. Otherwise, the very heart of the course, which is being valued, may not be used. Evidence at the student level in terms of patterns of student achievement also support this need. Students in ADC teachers' classrooms had significantly greater growth in student achievement compared to either the PDC or comparison students (Wilson & Draney, 1997).

Promoting Use

These maps are clearly a useful mechanism through which teachers may choose to self-assess (formative evaluation) or schools and districts may assess program impact (summative evaluation) on teaching practice. The maps provide meaningful evaluation by serving as a device through which to communicate evaluation results to all stakeholders. Now, the issue to be faced is not the conceptual value of the maps, but rather the methods needed to promote use of the maps. One recommendation is to integrate use of the maps into any SEPUP inservice. By design, the SEPUP assessment system would include maps of student progress (also generated from empirical evidence) and teacher change.

Clearly, the findings suggest that some level of teacher preparation is necessary to promote teacher use of the SEPUP assessment system. The evidence is strong that teachers will not use the embedded assessment tasks, much less the scoring guides and moderation process, without some form of preparation and ongoing support. Adding the maps to this preparation may perhaps be the only way to ensure use of the evaluation results for either formative or summative purposes.

Inservice is only one possibility of facilitating teacher use of the maps. Future research questions include:

- What is the best way to get this type of information into teachers' hands? Possibilities include (1) written materials, such as the SEPUP Assessment Manual; (2) inservice by the publishers of *IEY* as part of the adoption of the curriculum by schools or districts; or (3) local consultation by qualified teachers (for example, SEPUP field test participants).
- What is the best way to get teachers to use the information? Possibilities include (1) promote use of local assessment moderation; (2) provide an integrated inservice approach (that is, the assessment system is embedded in the course, so teachers need to learn about both to have optimal success); or (3) generate administrative support for the evaluation findings (for example, inform administrators that to achieve assessment reform, both in terms of teacher change and, more importantly, student achievement, teacher professional development and ongoing support is necessary).

The evidence from the SEPUP field test strongly suggests that changing teachers' perceptions and practices in assessment is not simple. More questions emerge from this study in terms of identifying how best to facilitate teacher change in light of assessment reform. Nevertheless, one solution is offered herein that may make evaluation meaningful by promoting the use of criterion-referenced maps.

NOTES

1. Positive misfit is associated with an item mean square greater than one; one standard that is used is 4/3 or 1.33. Negative misfit is associated with an item mean square less than one; one standard that is used is 3/4 or .75.

2. "A logit is that distance on the variable that corresponds to odds of success (compared to failure) equal to *e*, the base of the natural logarithms —approximately 2.7:1" (Wilson, Sloane, Roberts, & Henke, 1995).

3. The number of X's presented on the variable maps is not the same as the total number of cases when there are perfect or zero scores. Teachers with zero or perfect scores are not displayed on the variable maps produced by the *Quest* program because they are located at negative and positive infinity.

REFERENCES

Adams, R. J., & Khoo, S. -T. (1993). *Quest: The interactive test analysis system.* Hawthorn, Australia: Australian Council for Educational Research.

Anastasi, A. (1988). *Psychological testing* (6th ed.). New York: Macmillan.

Comfort, L. K. (1982). Evaluation as an instrument for educational change. In *Education policy and evaluation: A context for change.* New York: Pergamon.

House, E. R. (1993). *Professional evaluation: Social impact and political consequences.* Newbury Park, CA: Sage.

Masters, G. N. (1982). A Rasch model for partial credit scoring. *Psychometrika, 47,* 149–174.

Masters, G. N. (1988). The analysis of partial credit scoring. *Applied Measurement in Education, 1*(4), 279–297.

Monsaas, J. A., & Engelhard G., Jr., (1996). Examining changes in the home environment with the Rasch measurement model. In G. Engelhard Jr. & M. Wilson (Eds.), *Objective measurement: Theory into practice* (Vol. 3, pp. 127–140). Norwood, NJ: Ablex.

Patton, M. Q. (1997). *Utilization-focused evaluation: The new century text.* Thousand Oaks, CA: Sage.

Roberts, L. L. C. (1996). *Methods of evaluation for a complex treatment and its effects on teacher professional development: A case study of the science education for public understanding program.* Unpublished doctoral dissertation, University of California, Berkeley.

Roberts, L. (1997, July). *Evaluating teacher professional development: Local assessment moderation and the challenge of multisite evaluation.* Paper presented at the National Evaluation Institute of the Center for Research in Educational Accountability and Teacher Evaluation (CREATE), Indianapolis, IN.

Roberts, L., Sloane, K., & Wilson, M. (1996, April). *Local assessment moderation in SEPUP.* Paper presented at the Annual Meeting of the American Educational Research Association, New York.

Shadish, W. R., Jr., Cook, T. D., & Leviton, L. C. (1995). *Foundations of program evaluation: Theories of practice.* Newbury Park, CA: Sage.

Shepard, L. A. (1995). Using assessment to improve learning. *Educational Leadership, 52*(5), 38–43.

Sloane, K., Wilson, M., & Samson, S. (1996, April). *Designing an embedded assessment system: From principles to practice*. Paper presented at the Annual Meeting of the American Educational Research Association, New York.

Stallworth, Y., & Roberts-Gray, C. (1987). The craft of evaluation: Reporting to the busy decision maker. *Evaluation Practice, 8*(2), pp. 31–35.

Traub, R. E. (1994). *Reliability for the social sciences: Theory and applications* (Vol. 3). Thousand Oaks, CA: Sage.

Weiss, C. H. (1998). *Evaluation: Methods for studying programs and policies*. Upper Saddle River, NJ: Prentice-Hall.

Weiss, C. H. (1991). Evaluation research in the political context: Sixteen years and four administrations later. In M. W. McLaughlin & D. C. Philips (Eds.), *Evaluation and education: At Quarter Century. Ninetieth yearbook of the National Society for the Study of Education*. Chicago: The University of Chicago Press.

Wilson, M., & Adams, R. J. (1996). Evaluating progress with alternative assessments: A model for Chapter 1. In M. B. Kane (Ed.), *Implementing performance assessment: Promise, problems and challenges* (pp. 39–60). Hillsdale, NJ: Erlbaum.

Wilson, M., & Draney, K. (1997). *Developing maps for student progress in the SEPUP assessment system*. Paper presented at the AAAS meeting, Seattle, WA.

Wilson, M., & Sloane, K. (in press). From principles to practice: An embedded assessment system. *Applied Measurement in Education.*

Wilson, M., Sloane, K., Roberts, L., & Henke, R. (1995). *SEPUP Course 1, issues, evidence and you: Achievement evidence from the pilot implementation*. Berkeley, CA: Graduate School of Education, University of California.

Wright, B. D., & Masters, G. N. (1982). *Rating scale analysis: Rasch measurement.* Chicago: MESA Press.

4

USING THE RASCH MODEL TO STUDY LARGE-SCALE PHYSICS EXAMINATIONS IN AUSTRALIA

Andrew Stephanou
Australian Council for Educational Research

INTRODUCTION

This chapter describes findings of work in progress, the primary aim of which is the construction of variables for the assessment of conceptual understanding of physics at senior high school and 1st-year tertiary levels. The work is based on two sets of data collected as follows:

- In a research project designed specifically to investigate a methodology for the assessment of conceptual understanding
- In formal large-scale physics examinations in Australia

A previous version of this manuscript was presented at the Ninth International Objective Measurement Workshop in Chicago, Illinois, March 20–22, 1997 (http://www.rasch.org/iomw.htm). The author wishes to acknowledge Professor Barry McGaw (ACER), for whom the analysis of the 1995 NSW HSC Physics examination data was done in the first instance, for inclusion in the recent review of the Higher School Certificate in NSW (Aquilina, 1997; McGaw, 1997); available online at http://www.boardofstudies.nsw.edu.au/docs_hsc95/mcgaw1_5_4.html; Professor Geoff Masters (ACER) for supervising my work since 1988; and Dr. Richard James (CSHE) and Mr. George Morgan (ACER) for valuable comments and discussions.

The first source of data is the CSHE Physics Project (Masters, 1989), which started in 1988 at the University of Melbourne. The data consist partly of 25 interview transcripts for each of 14 situations that were created to reveal the qualitatively different ways in which students conceptualize physics phenomena. Hierarchical categories of conception were constructed for each situation using phenomenography (Hasselgren, Nordieng, & Österlund, 1997). Currently the Rasch model is being used to describe a variable based on categories of conceptual understanding constructed qualitatively using some interview data and other data obtained recently in the form of written responses. Although a large number of phenomenographic studies have been completed to date (Bruce & Gerber, 1997), none have capitalized on the potential of the Rasch model to extract additional information already present in the data for a statistical validation of the qualitatively constructed categories and for allowing categories of the same task and categories of different tasks to be compared through their positions on a measurement interval scale.

This chapter focuses on the second source of data, originating in formal large-scale examinations. In Australia, while the Rasch model has been used in large-scale examinations at primary and junior secondary levels since 1989 (Masters, 1990), to date none of the six large-scale senior secondary examination programs officially makes use of the Rasch model on a regular basis either for test analysis and reporting or for addressing research issues. This chapter shows how the Rasch model may be used to extract information that cannot be obtained with classical test analysis. The first data file analyzed comes from the 1990 Higher School Certificate (HSC) Physics examination in Victoria and the second from the 1995 HSC Physics examination in New South Wales (NSW). Examples of reporting Rasch analysis findings are shown with the former, and a methodology to investigate what was measured with a test, thus allowing a qualitative interpretation of examination scores, with the latter.

LARGE-SCALE SENIOR PHYSICS EXAMINATIONS IN AUSTRALIA

Every year more than 20,000 students at the end of their secondary education in Australia take one of five external written physics examinations prepared and administered independently by six different state or territory organizations. Figure 1 shows the six states and two territories on a map of Australia, and the organizations responsible for state- and territory-wide assessment. (The Northern Territory Board of Studies uses the examination papers prepared in South Australia, and the assessment of students in the Australian Capital Territory and in Queensland is school based only.)

The analysis and reporting of the results is also done independently by these organizations: No routine cooperation takes place; furthermore, no study has been undertaken recently to produce an Australia-wide picture of the assessment of physics in large scale programs that takes place annually at the end of secondary education. A study published in 1991 (Peterson) summarized physics curricula in

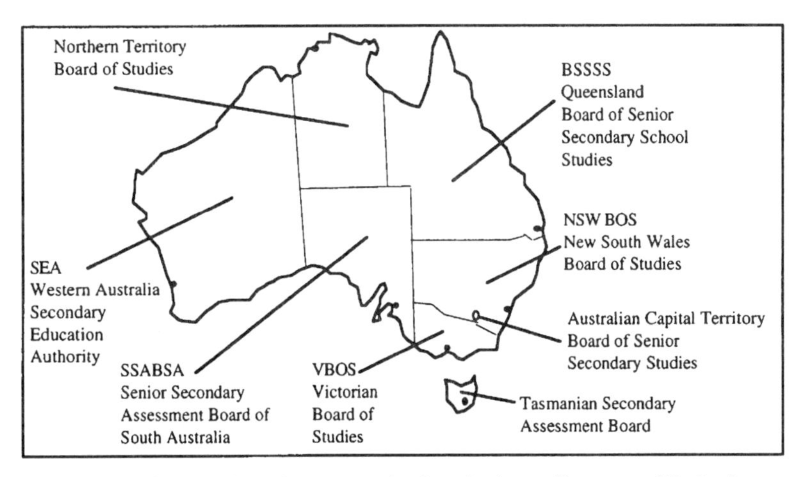

FIGURE 1. State and territory organizations in Australia responsible for large-scale assessment at the end of secondary education.

Australia but did not address the assessment of physics. Studies such as the following, which can be attempted using the Rasch model, have never been undertaken: the construction of a continuum of physics achievement for each paper; the expression of student achievements originating in different examinations on a common continuum so that achievement in one examination may be compared with those in other examinations, taking into account the different difficulties of the papers; time trends of student achievement; and paper content.

At the present time, the data available are only partially analyzed and the examination papers, the data used in the analysis, and the results are not readily accessible. To find information on the assessment in each state or territory at the end of secondary education, it is necessary to contact eight different organizations in the country.

These examination papers contain multiple-choice, calculation, short answer, and extended answer questions marked according to an agreed marking scheme. The routine analysis carried out with the data collected by each organization is adequate for ranking the students on test scores and for informing teachers on the performance of their students on the whole paper, on sections of the paper, or on individual questions. The aim of ranking is satisfied when the average score on the test is neither too high nor too low, and the spread of the test scores shown by the standard deviation is adequate for grading. The primary aim of these examinations is not to provide each student with diagnostic information about his or her preparation and suitability for specific tertiary courses, nor to provide information on the standards achieved in each subject area.

Consequently, a large amount of interesting and useful information remains buried in large data files obtained at high costs. However, this information can realistically be extracted. It could support decision making; it could help to achieve a better knowledge and understanding of physics assessment through similar and parallel experiences, whereby the experience in one system could illuminate developments in another system.

Published examiners' reports usually contain score distributions, item facilities and point biserial correlations, distracter analysis, some measure of internal consistency, comments, and recommendations. Some examiners' reports, papers, and other relevant information may now be accessed on the Internet (for example, NSW BOS at http://www.boardofstudies.nsw.edu.au/ or Tasmanian Secondary Assessment Board at http://www.tassab.tased.edu.au/).

Physics is but one of many subject areas in which examinations are conducted annually at the end of secondary education in Australia and other countries. It is envisaged that the general approaches described in this chapter with data from physics examinations may be applied to other subject areas and to other educational systems.

RASCH ANALYSIS OF THE 1990 HSC PHYSICS EXAMINATION IN VICTORIA: HOW IT MIGHT HAVE BEEN FORMALLY REPORTED

Linacre (1993) has stressed that "the greatest challenge to practitioners of Rasch measurement is not test construction or analysis, but **communication**." Andrich (1988) has highlighted the importance of the use of graphical displays for showing misfits of subgroups of people on a question and the immunity of graphical displays to being too sensitive to the detection of misfits due to large sample size. The use of graphical displays may be stand-alone or accompanied by calculated statistics.

Various levels of technical detail may be displayed in Rasch analysis reports to satisfy the interests and match the background of intended users. Reports may show:

- Person performance (test scores, that is, the number of questions answered correctly or the total number of marks awarded) and question facilities or difficulties (the percentage of persons answering a question correctly or incorrectly) with their distributions or some information about their distributions such as mean, standard deviation, and percentiles.
- Person and question estimated parameters ("abilities" and "difficulties") and their distributions on the continuum; person parameters correspond to test scores and question parameters to difficulties; actual values do not always need to be shown on the scale.
- Person profiles showing the questions along the continuum that contribute to the achieved test score, or question profiles showing which persons along the continuum contribute to the observed facility
- Distributions of scores or parameters, of subgroups of persons or questions
- Item characteristic curves showing model probabilities and observed proportions
- Distribution of questions along the continuum, grouped by topic
- Description of the measured variable on the continuum divided into bands

Figures 2 through 10 comprise four sets of displays from the analysis of the 1990 HSC Physics examination in Victoria, as follows:

- Test profile and characteristic curve (see Figure 2)
- Profile and characteristic curve of a binary question (see Figure 3)
- Profiles and characteristic curves of a multilevel question (see Figures 4 through 9)
- Person profile and characteristic curve (see Figure 10)

These displays are presented at a level intended to be accessible to nonpsychometricians. However, it is assumed that users of the reports have been exposed to appropriate explanations of the transformation of scores into measures on a continuum, and to the separability of student parameters from question parameters, and that they understand that these parameters are expressed in the same unit of measurement and therefore can be displayed on the same continuum. There is no need to resort to complex equations and nonessential technical detail. The displays alone should be adequate to convey pictorially and intuitively the essential findings of the statistical analysis of interest to parents, students, and teachers.

TEST PROFILE AND CHARACTERISTIC CURVE: THE 1990 HSC PHYSICS EXAMINATION

The 1990 HSC Physics examination, taken by 7,132 students, consists of 64 wrong/right questions scored 0,1, and four extended answer questions scored 0,1,2,3,4, according to a marking scheme established to reward partial completion of complex tasks, making the total possible score on the examination equal to 80. After elimination of 0 and perfect scores, 79 groups of students (score groups) were formed according to the test score achieved. The Rasch analysis was performed with Quest (Adams & Khoo, 1993), using the partial credit model (Masters, 1988).

A 0,1 question is represented on the achievement continuum by a position, available with its estimation error not shown in this display. This location is called *threshold* and indicates how difficult it is to answer the question correctly. It is called threshold because all students with an achievement parameter greater than the question parameter have a probability greater than 0.5 for answering the question correctly, while those students with a smaller parameter have a probability less than 0.5. By definition, the difficulty of a question is equal to the parameter of those students who have a 0.5 probability for correctly answering the question.

Four thresholds are required for a 0,1,2,3,4 question: one for the probability of a score of 1 or higher, one for the probability of 2 or higher, one for the probability of 3 or higher, and one for the probability of 4. For example, the lowest of the four thresholds divides students into two groups, those with a probability greater than 0.5 to obtain a score of 1 or higher and those with a probability less than 0.5. Summarizing, the total number of difficulty thresholds is equal to the total possible score for this examination: 80 (64 for the wrong/right questions and 16 for the four extended answer questions).

The four thresholds of each multilevel question are not independent of each other, as required by the Rasch model for wrong/right questions (Andrich, 1988). Any four 0,1 questions are required by the measurement model to be independent of each other. A student who obtains a score greater than 1 for a 0,1,2,3,4 question is assumed to have overcome all of the thresholds below it. For example, having achieved a score of 4, a student must have completed the tasks corresponding to scores 1, 2, and 3. The partial credit model (Masters, 1988) takes this into account by treating adjacent scores as if they were the 0,1 scores of wrong/right questions, and obtains parameters (deltas) from which the difficulty thresholds (gammas) are derived. The delta parameter for a score s is equal to the parameter of students who have equal probability to score s or $s - 1$; the gamma parameter for a score is equal to the parameter of students who have a 0.5 probability for achieving a score s or higher (Masters, 1988).

The test profile in Figure 2 shows the distribution of person parameters (corresponding to test scores) and question thresholds on the same continuum. Such a display cannot be produced with traditional test analysis because test scores and question difficulties, expressed as percentage of students answering a question incorrectly, belong to two different ordinal scales. The test score shows the *number of questions* answered correctly by a student and the question difficulty the *number of students* who answered the question incorrectly. It is meaningless to compare the performance of the student with the difficulty of the question even if the numbers are expressed as percentages.

Person and question parameters are expressed in the same units on the scale of the measured variable, so they can be meaningfully compared directly, but they are separated in the sense that the *same* person parameter (ability to do well on the examination, or achievement) will be estimated if only some of the questions in the test are used, and the *same* question parameters (difficulty thresholds) will be calibrated if only a subset of the students is used, provided the data fit the model. This cannot happen in traditional test analysis where scores, which are counts, rather than model parameters are used. If only the more difficult questions are included in a test, the test score will be less than the test score when easier questions are included. If only the best students are used in the calibration, question difficulties expressed as percentage of students answering incorrectly would be greater than those obtained with poorer achievers (Wright & Stone, 1979).

The test profile also shows the question thresholds by topic and identifies each question with its identification number in the test. The easiest question is Question 43 in Electricity & Magnetism, and the most difficult score to obtain was a score of 4 on Question 62 in the same topic.

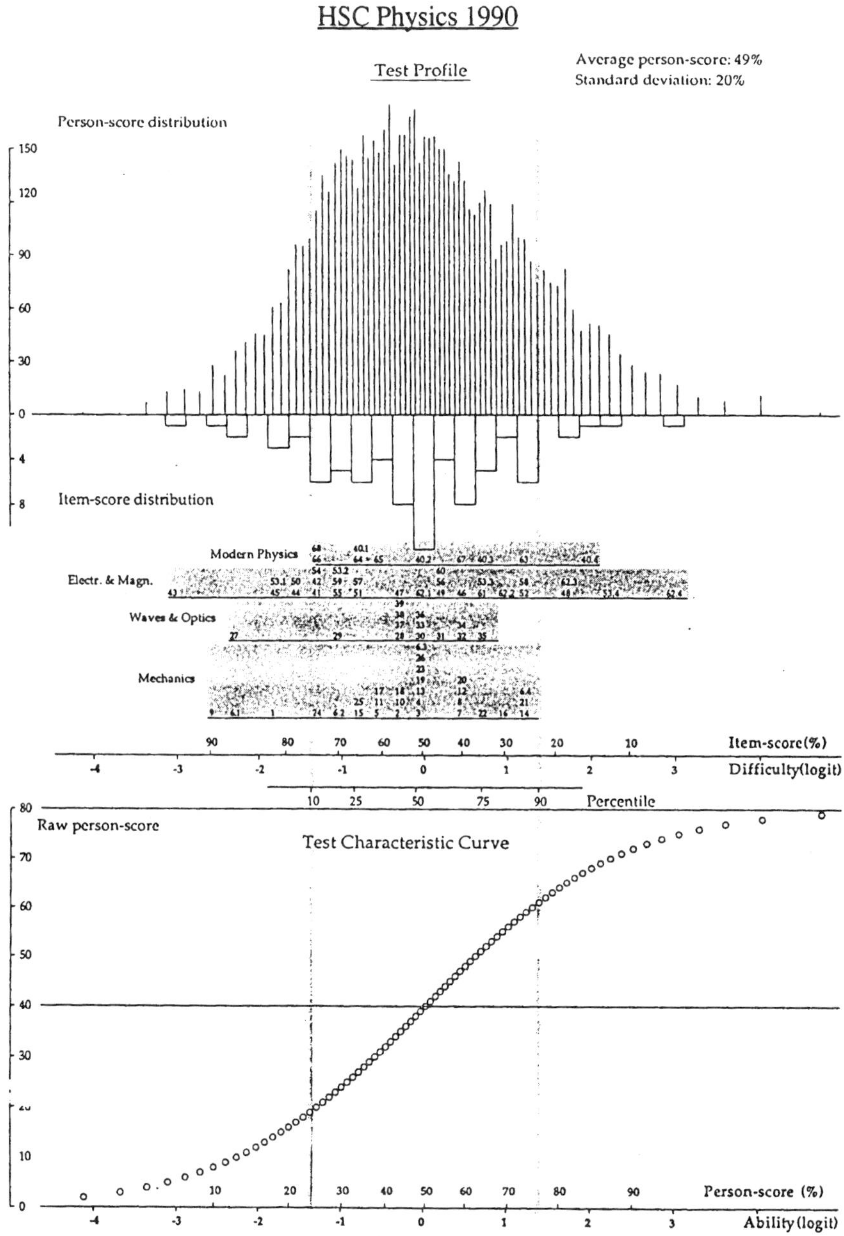

HSC Physics 1990

Test Profile

Average person-score: 49%
Standard deviation: 20%

Person-score distribution

Item-score distribution

Modern Physics

Electr. & Magn.

Waves & Optics

Mechanics

Item-score(%)

Difficulty(logit)

Percentile

Raw person-score

Test Characteristic Curve

Person-score (%)

Ability (logit)

FIGURE 2. Test profile and test characteristic curve.

The test characteristic curve shows how the test scores are mapped onto the continuum. It is important to notice that achievement estimates, or ability to do well on this examination, maintain the rank order of test scores, and that from test scores of about 20% to about 80%, the transformation of scores into measures may be considered to be approximately linear.

PROFILE AND CHARACTERISTIC CURVE OF A BINARY QUESTION: QUESTION 1

The first question in the Victorian 1990 HSC Physics paper is reproduced below.

At time t = 0 a car, travelling along a straight line at a constant speed of 20 m s^{-1}, passes a motorcyclist who is at rest. At the instant the car passes, the motorcyclist takes off in pursuit of the car. The graph below shows the speed of the car and the motorcycle as a function of time.

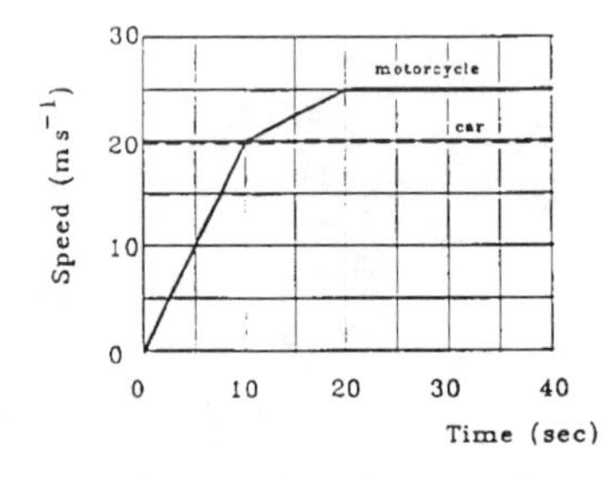

Question 1
Which of the statements below best describes what happens to the acceleration and the velocity of the motorcycle at the instant t = 10 s?

A. The acceleration increases while the speed decreases.
B. The acceleration decreases while the speed increases.
C. The acceleration and the speed both decrease.
D. The acceleration and the speed both increase.

Question 2
At what time will the motorcycle overtake the car?

The length of each bar in the item profile for Question 1 shown in Figure 3, is proportional to the number of students in the corresponding score group. The length of the bar above the horizontal line is proportional to the number of students who answered correctly. The open circles show the expected position of each bar according to the model. The expected number correct was obtained by multiplying the probability for answering correctly and the number of students in the group. The display shows the ability distribution of the students who answered correctly above the horizontal line and of those who did not below the horizontal line. The horizontal axis

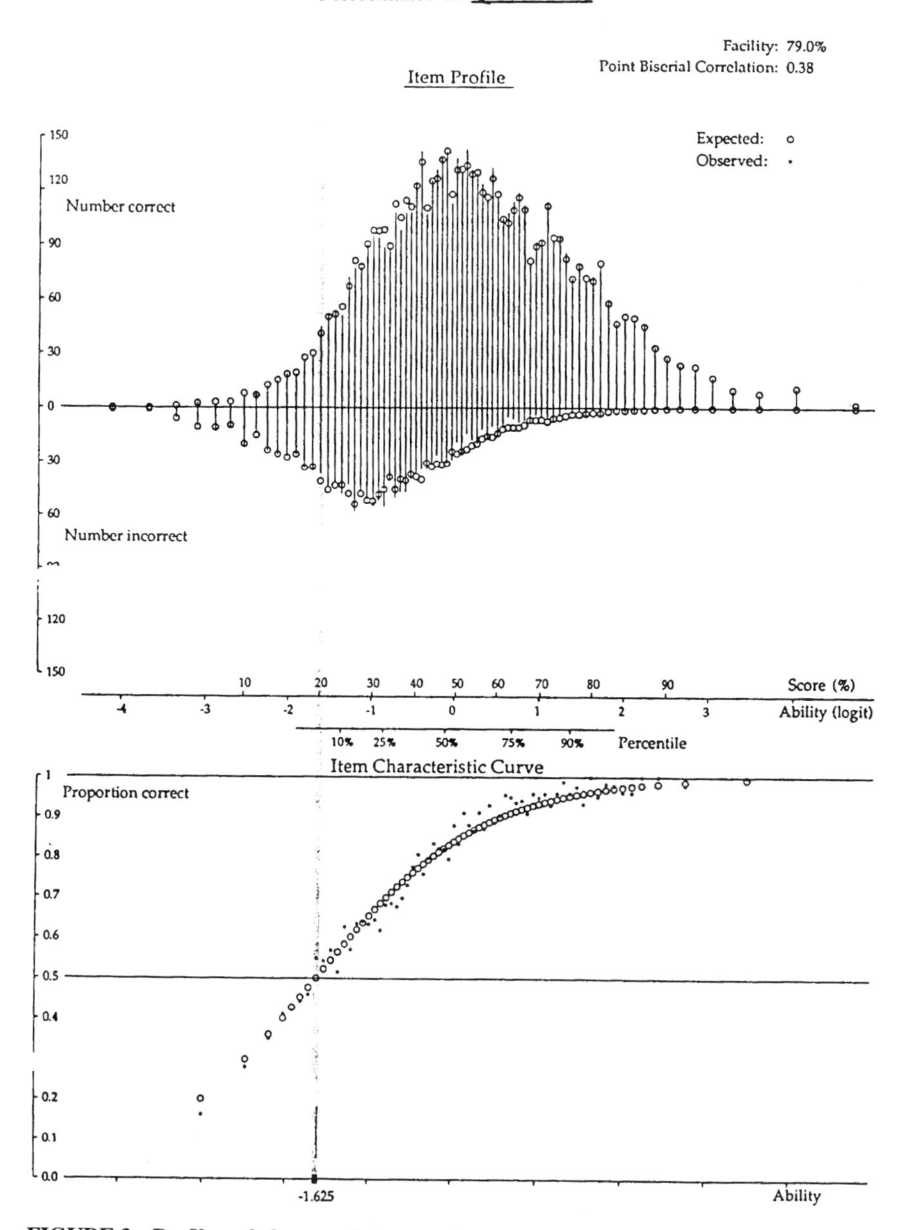

FIGURE 3. Profile and characteristic curve for a binary question.

shows the measured variable in terms of test scores expressed as percentages, clearly squeezed together around 50%, and also in terms of the person parameters ("ability") in logits. A secondary horizontal axis, immediately below the ability measures, locates percentiles for the population tested. The difficulty threshold, or item calibration, which targets students with a 0.5 success probability is shown by the vertical rule. This can be confirmed by eye, because it corresponds to the location on the variable where the succeed and fail portion of each score group bar are of equal length.

The item characteristic curve (ICC) allows one to compare model probabilities for each score group, shown by open circles, with observed proportions of students who answered correctly, shown by dots. The empirical ICC is somewhat fuzzy but it is close to the model line. There are fewer than 79 groups in this display because groups with fewer than 30 cases have been combined with adjacent groups to form new groups with more than 30 cases. The ability parameter of the new score groups was calculated as the average of the abilities of the component groups.

According to the model, the better the performance of a score group on the whole test, the greater the expected proportion of students correctly answering a question. A comparison of the position of the open circles with the position of the dots shows that this multiple-choice question works satisfactorily for all score groups and that there is no evidence of a lower "guessing" asymptote as one may have expected from a four-distracter multiple-choice question. It appears that the preparation of these students is such that there was no need to resort to guessing for answering this question.

Lower asymptotes not approaching 0 have been observed in other questions. The performance on each of the 64 wrong/right questions in the examination has been documented in this way. Some displays show misfits consisting of higher discrimination than that of most of the other questions in the examination, or lower, if any at all, for low achievers.

PROFILES AND CHARACTERISTIC CURVES OF A MULTILEVEL QUESTION: QUESTION 6

The first of the four questions scored 0,1,2,3,4 in the Victorian 1990 HSC Physics examination, reproduced in the following extract, is used to illustrate how Rasch analysis of a multilevel question may be reported.

> **Question 6** (This is an extended answer question.)
> On the moon Rosita drops a package with a mass of 1.0 kg from the spaceship to Mark who is standing 5.0 m vertically below. Mark then slides the package across a horizontal, teflon-coated, friction-free surface to where Raymond stops it. Explain why it would have felt very different for Mark to catch the package on the moon compared with the Earth, whereas for Raymond the action of stopping it would have felt much the same.

Students' responses were scored 0,1,2,3,4 according to a marking scheme. Their performance is shown in the following six figures. Figure 4 contains the model probability

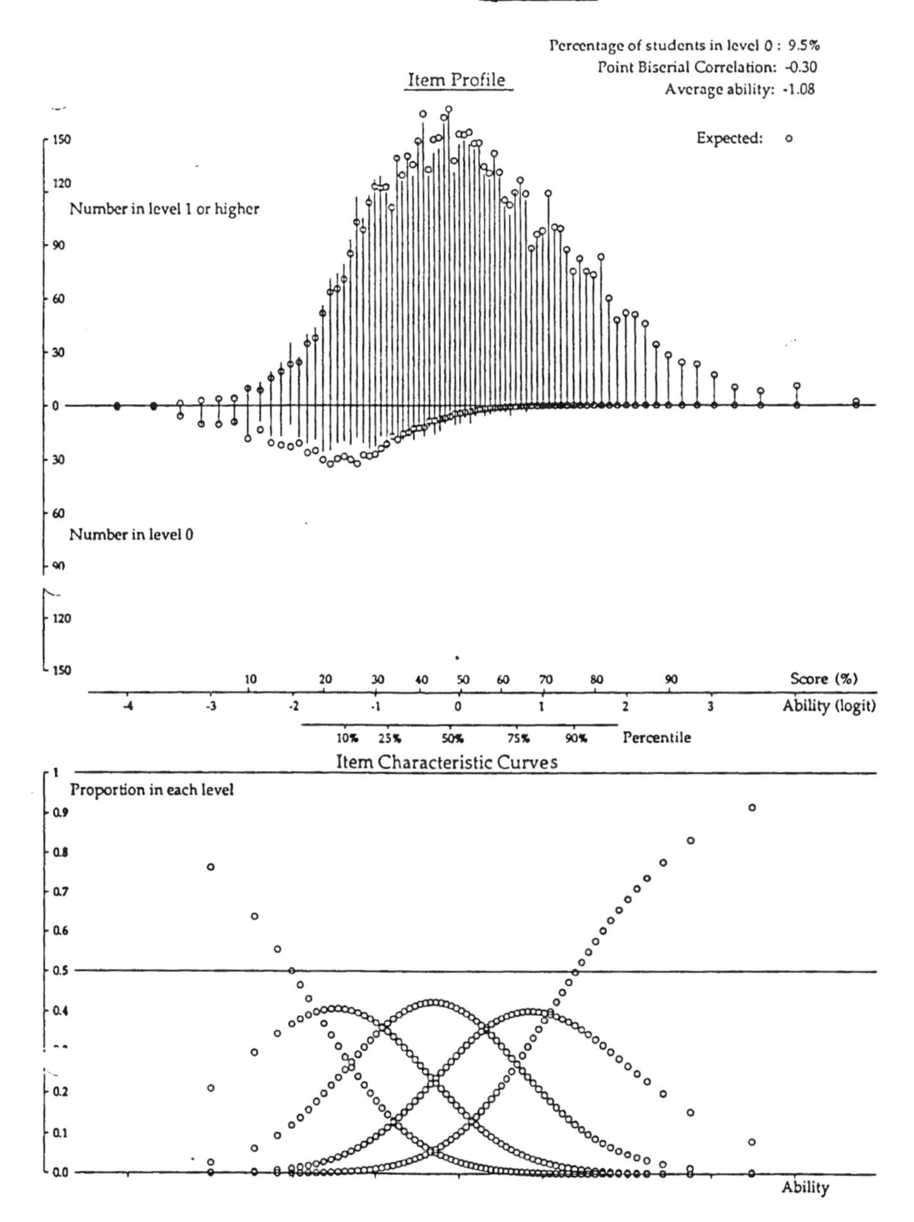

FIGURE 4. Profile and model probability curves for a multilevel question.

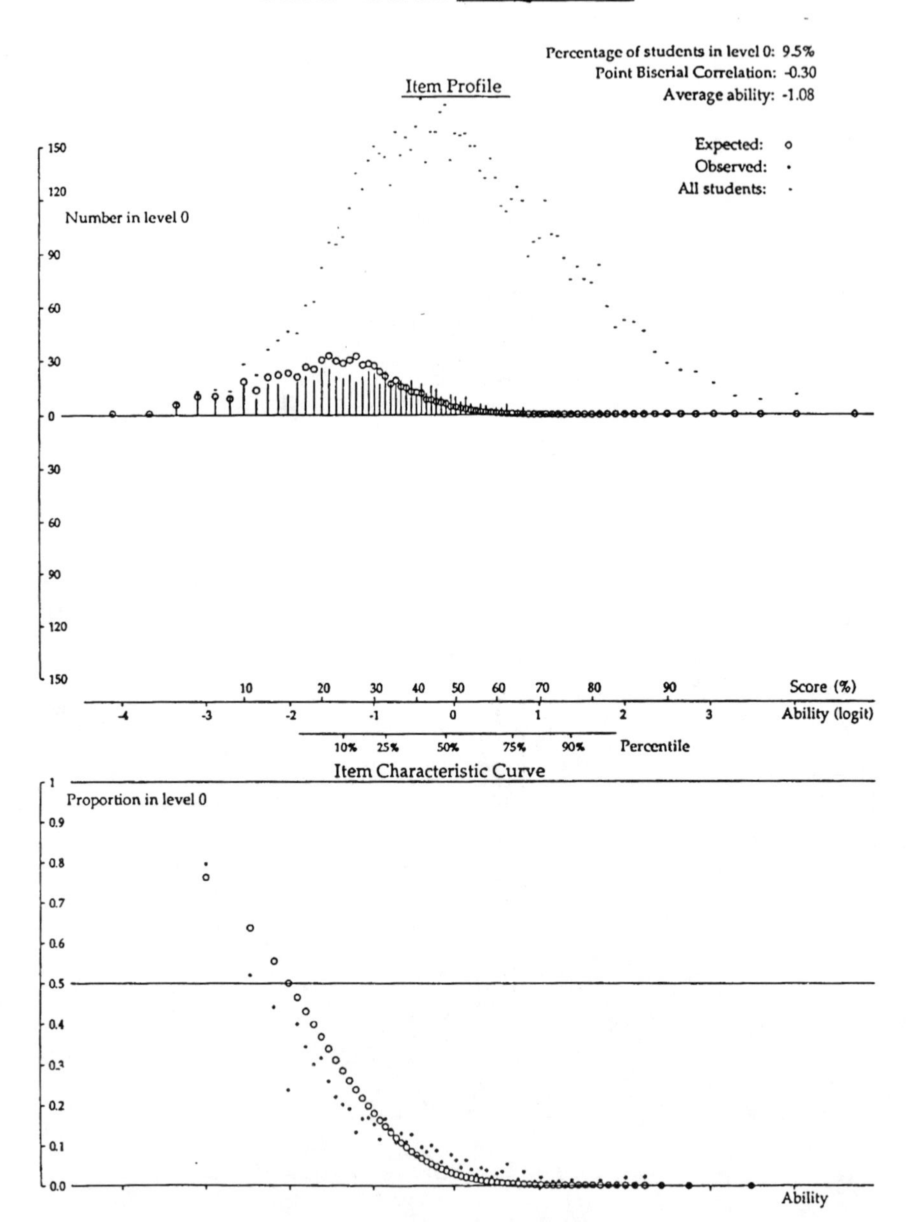

FIGURE 5. Profile and characteristic curve for a multilevel question—score 0.

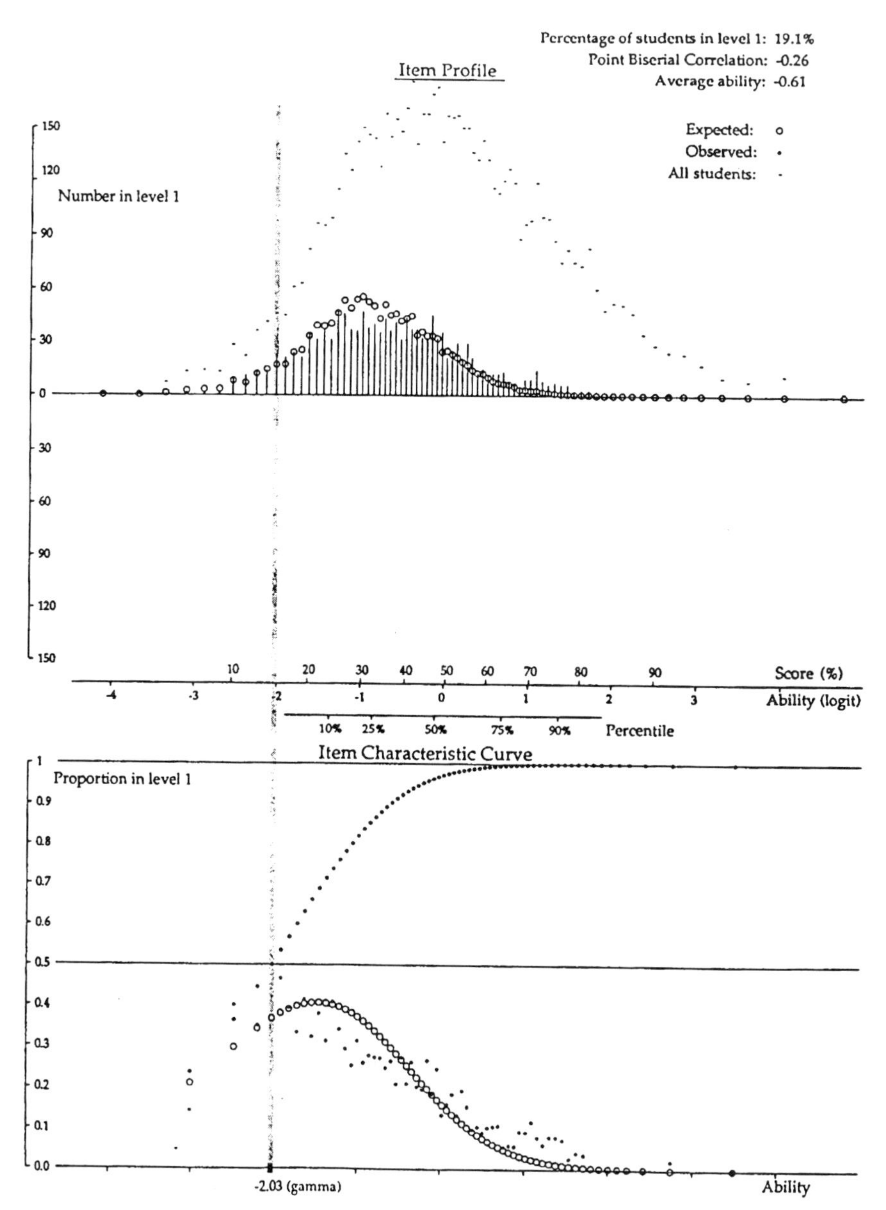

FIGURE 6. Profile and characteristic curve for a multilevel question—score 1.

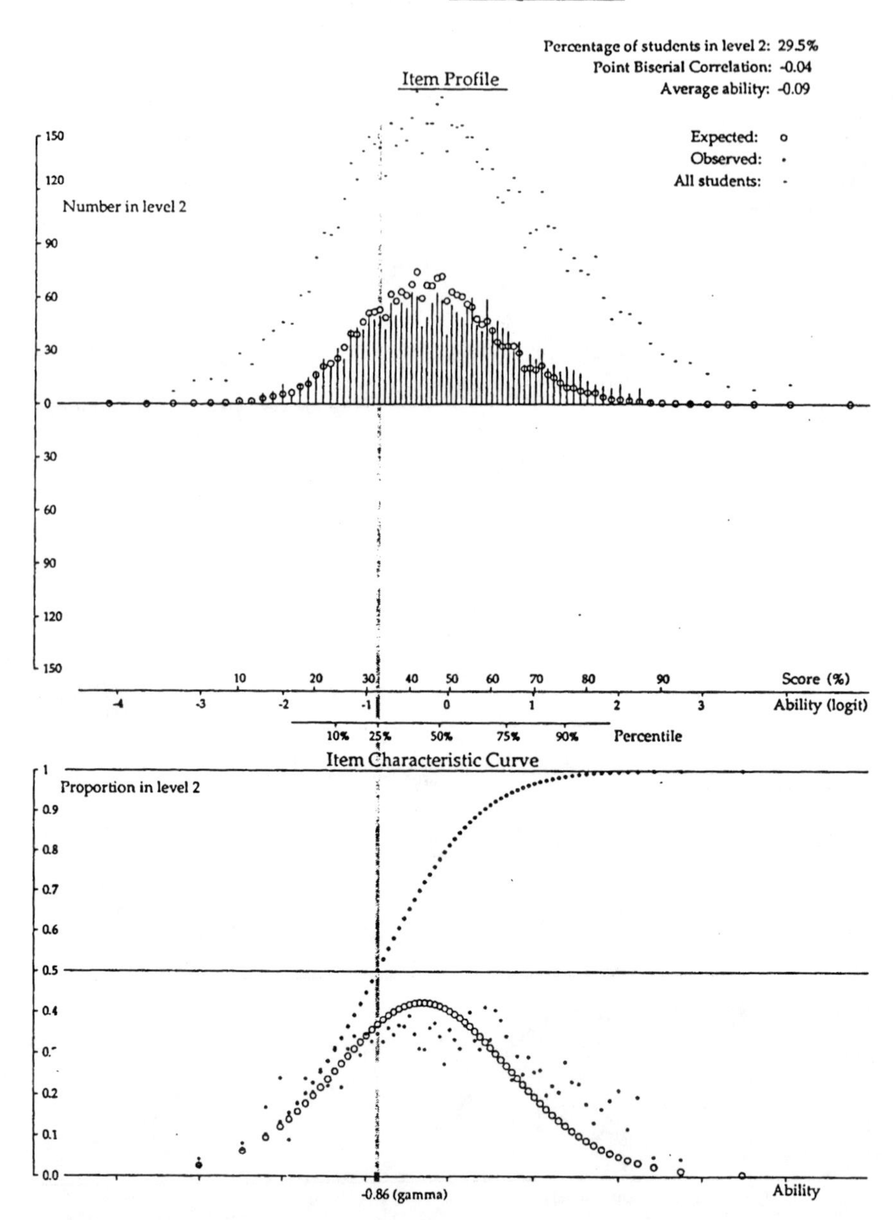

FIGURE 7. Profile and characteristic curve for a multilevel question—score 2.

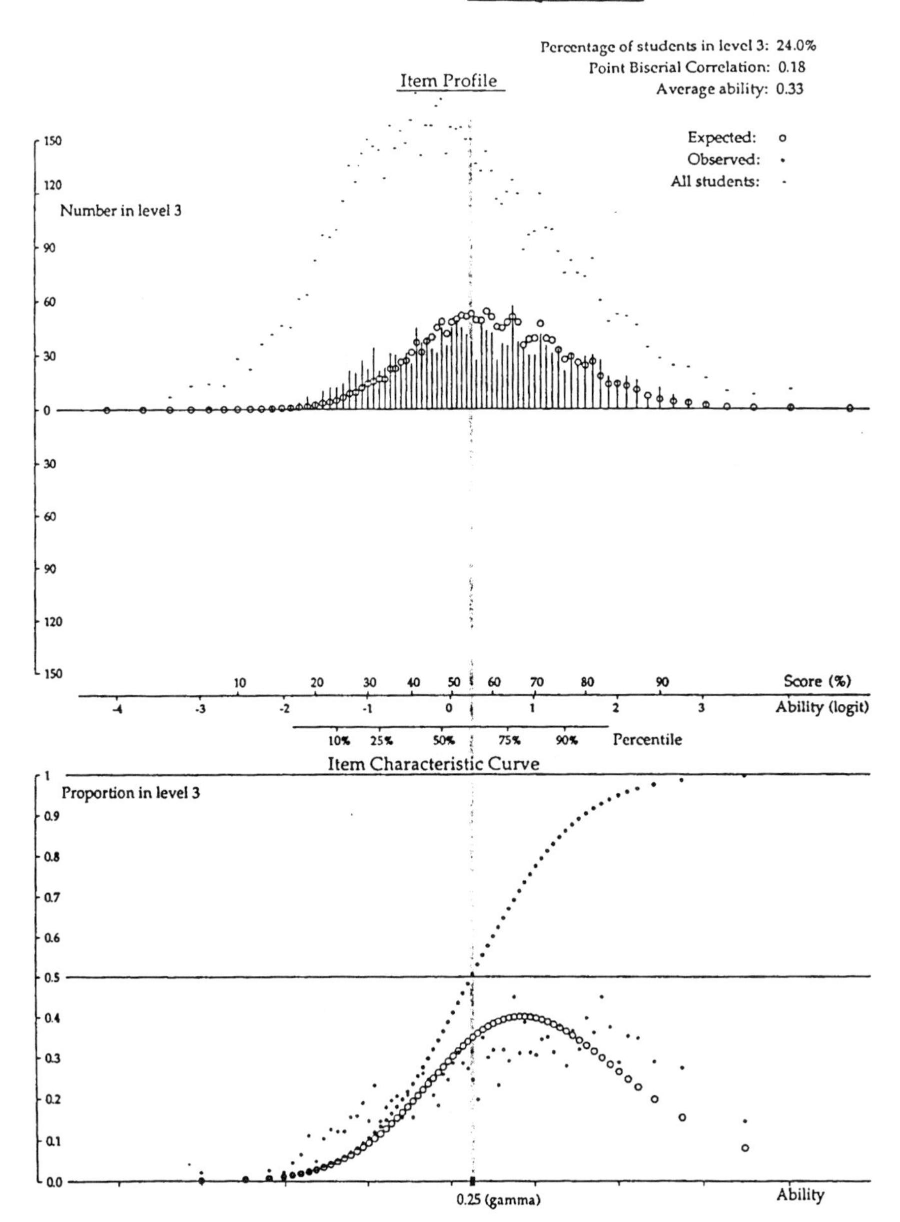

FIGURE 8. Profile and characteristic curve for a multilevel question—score 3.

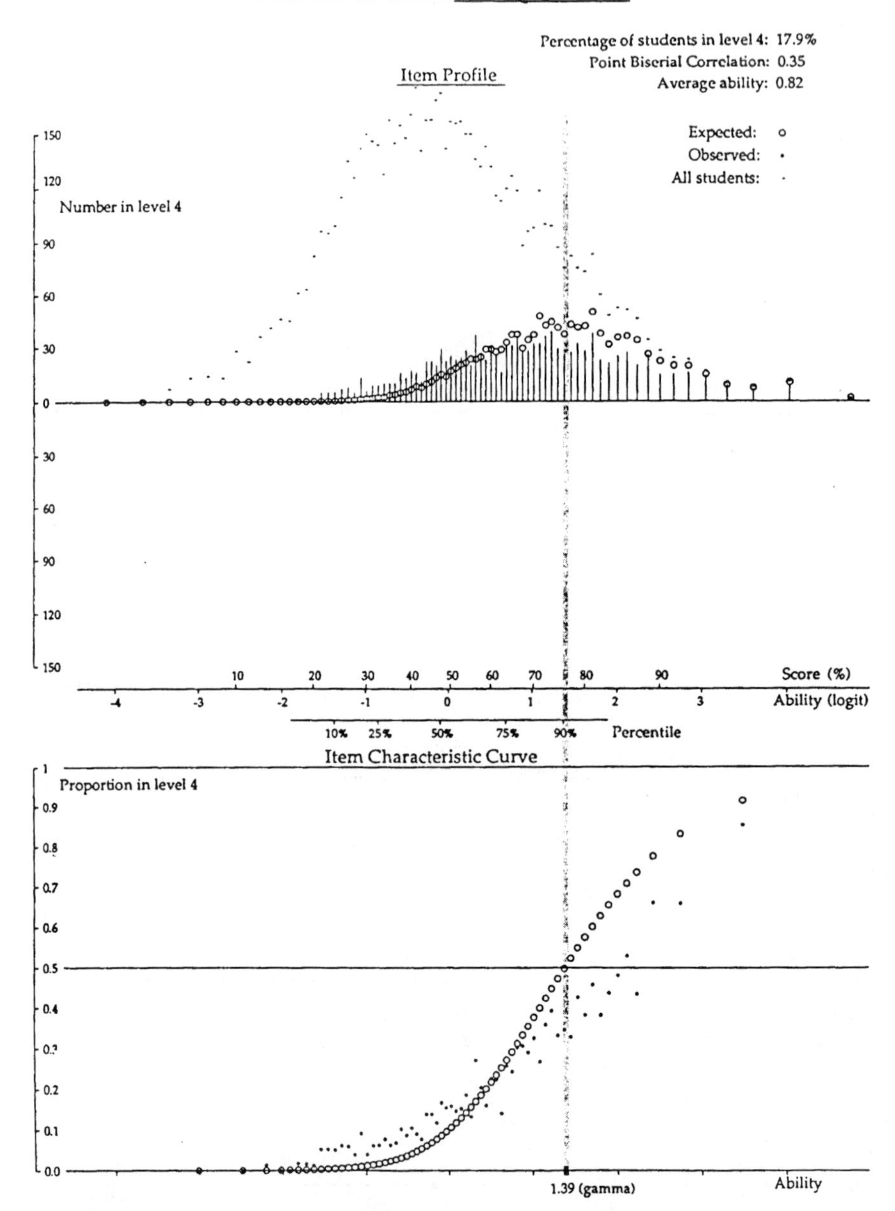

FIGURE 9. Profile and characteristic curve for a multilevel question—score 4.

versus ability curves for each of the possible five scores. Figure 5 shows the profile of the question for a score of 0; that is, the distribution of the test scores of the students who achieved 0 on this question, as a subset of the distribution of the test scores of all students, and the ICC. Figures 6 through 9 show question profiles and ICCs for the other possible scores for this question. The model probability for obtaining a score s or greater for this question is also shown, because of its role in defining the difficulty threshold of each of the possible scores other than 0 for this question: the point on the continuum where the probability for achieving at least that score is 0.5. Similar displays cannot be obtained using scores instead of measures on an interval scale.

PERSON PROFILE AND CHARACTERISTIC CURVE: CASE 25

The model symmetry existing between students and questions may be utilized to display the achievement of a student similarly to the display of the performance of a question, even if there are 7,132 students for each question and only 80 difficulty thresholds for each student.

The distribution of thresholds shown by topic along the continuum is divided into two groups: those achieved by the student and those not achieved. The measure of the achievement of a student is defined as the question threshold for which the probability of success is 0.5. Figure 2 shows that there is one-to-one correspondence between test scores and ability estimates; however, the same test score may be obtained in many different ways, more or less consistently with the model. The way a particular test score is obtained may reveal important diagnostic information on the preparation and general characteristics of a student. A response pattern containing many easy questions answered incorrectly is unusual and may indicate faulty preparation or carelessness.

A vertical line passing through the location of the person on the continuum divides the plane, where thresholds achieved and thresholds not achieved are shown, into four quadrants. Most of the thresholds are expected to be located in the two quadrants of easy thresholds achieved and difficult thresholds not achieved. The *perceived* difficulty of a question depends on the student's location on the continuum, according to the model probability of successful outcome when a person of parameter b attempts a question of threshold d; this probability depends on the difference $b - d$.

The person characteristic curve shows the probability for a correct answer on each selected group of question scores calibrated on the continuum and the corresponding observed proportion. Nine groups have been formed with the 80 thresholds. The location of each group on the continuum is taken as the average difficulty of the question thresholds in that group. These groups may define bands such as those described qualitatively on a continuum in the second part of this chapter, provided there are sufficient thresholds in the band to be able to justify the calculation of observed proportions of question scores achieved by a student. The model expects that the more difficult the questions in the group, the smaller the expected

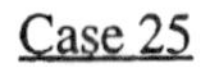

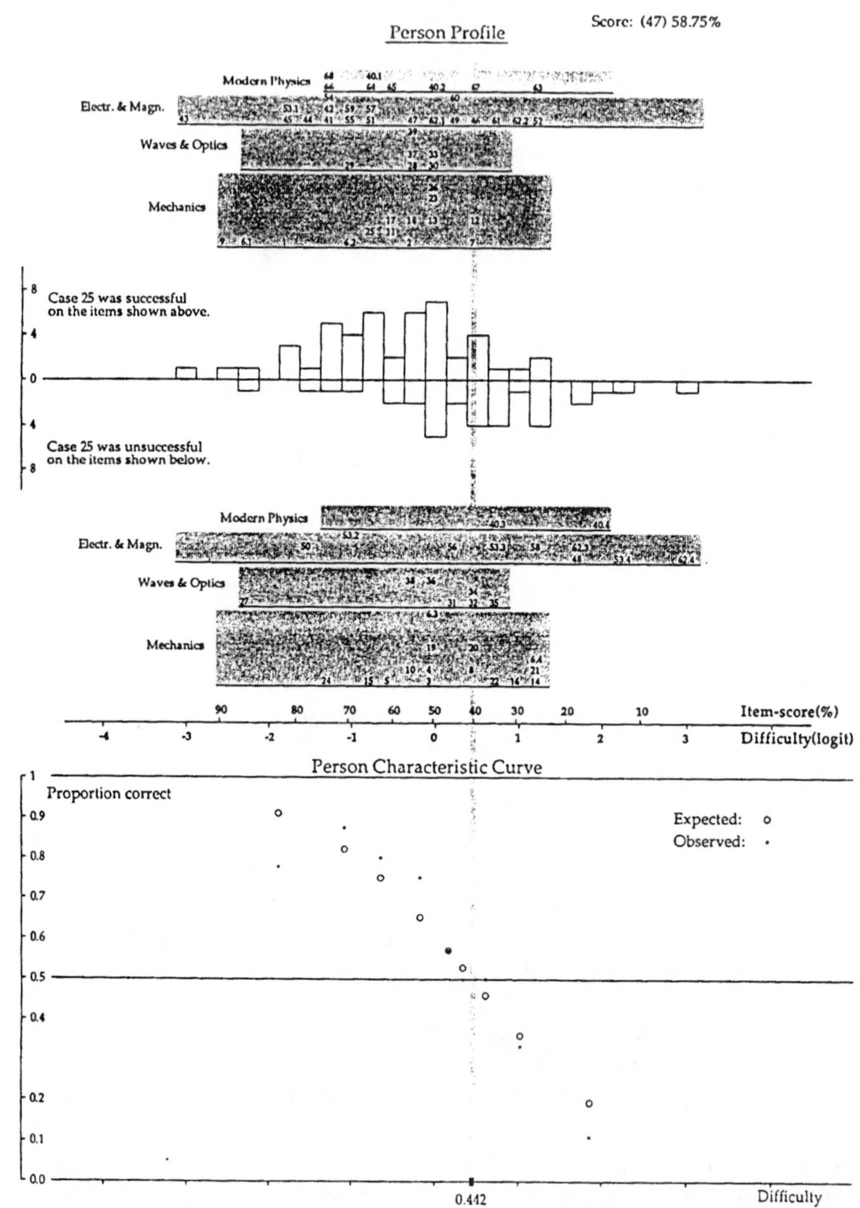

FIGURE 10. Person profile and characteristic curve; an example of above-average ability and good fit.

proportion of correct answers for that group. Expected proportion correct may be interpreted as the mastery level of the student for a group of questions. The mastery level decreases with increasing difficulty of the questions in the group, according to the model probability.

Case 25 in Figure 10 shows an above-average performance on the test with good fit to the model. Although estimation errors are large owing to the small number of question scores in each category, the general trend of decreasing proportions of achieved thresholds with group difficulty can be observed in all cases; group misfits that may be interpreted case by case are often present. These displays cannot be produced with test results expressed as scores.

WHAT WAS MEASURED WITH THE 1995 HSC PHYSICS EXAMINATION IN NEW SOUTH WALES?

The Data

The data used in this section come from the 1995 NSW HSC Physics examination and consist of the test given to the students, the marking scheme used in marking students' written answers, and the results of a statistical analysis of the marks awarded on each question to the 9,593 students who attempted the paper. The test consists of 15 multiple-choice questions scored 0,1, and 16 calculation/explanation-type questions, 10 of which were scored 0,1,2,3 and 6 that were scored 0,1,2,3,4,5. Each student was awarded 31 independent scores, making the greatest possible test score equal to 75 ($15 + 10 \times 3 + 6 \times 5$). The statistical analysis was performed with Quest (Adams & Khoo, 1993), using the partial credit model (Masters, 1988).

Overview of the Methodology

The methodology described in this section illustrates how a standards-referenced scale may be developed by investigating what is measured with a test consisting of binary and multilevel questions. The transformation of test scores and question scores into interval measures is followed by a qualitative analysis of each question to identify the skills required to provide answers that are rewarded according to a marking scheme, and to understand the factors that affected the data whose analysis resulted in the calibration of the 75 difficulty thresholds shown later in Figure 12.

HSC examination score distributions are primarily intended for grading and ranking of students. We thus associate scores to various percentiles, we compare them to the average test score, and we regard as satisfactory those scores that are above 50%, or some other fraction of the maximum possible score. Traditional processing of examination data includes calculation of biserial correlations to check the consistency of each question with the rest of the test, and distracter analysis for multiple choice questions. This information is useful but it does not reveal what is assessed by the test as a whole, the difficulty of the test in relation to the conceptual structure of the subject area, the meaning of the various test scores in terms of mastered skills, and whether the test is able to assess the aims of the course.

One way of addressing these issues is to analyze the data further, both statistically and qualitatively, and use the results of the analysis to create an interpretable scale with which to show what was measured with the test, and how achieved levels of performance may be interpreted. We would do something similar with temperature if we decided to describe regions of its scale that have been conveniently selected: the regions around –10°C (14°F), 0°C (32°F), +15°C (59°F), + 40°C (104°F), for example. Such descriptions are based on common experiences and familiar examples, and can be used without an adequate understanding of the scientific meaning of the variable temperature (Baierlein, 1990). Temperature can be sensed and easily measured with simple apparatus and is used by most people whether they have a grasp of its scientific meaning or not.

Quantitative Analysis

The aim of the additional statistical analysis is to evaluate the fit of the data to a model that satisfies the requirements of scientific measurement and to transform test and question scores into measures. Scores are counts rather than measures and belong to a bound scale with floor and ceiling distortions. Classical test theory deals with scores. The Rasch model used in test analysis can transform scores into parameters on an interval scale, satisfying the properties of measures used in the physical sciences (Andrich, 1988). The location of each test score on the scale is estimated following the calibration of the 75 difficulty thresholds on the same measurement continuum.

An examination of the fit of the data to the Rasch model provides strong evidence that a variable has been measured with this physics test, but complementary qualitative analysis is required to identify and describe *what* has been measured. For convenience, let us refer to the measured variable as "zesty," meaning hotness in Greek. The actual name is not important, provided it is not misleading; what matters is the description of the measured variable along the continuum. As in the case of temperature, it is not essential to understand the scientific meaning of the measured variable to be able to use and make sense of the scale. In this case, the outcome of the test is the measurement of the zesty associated with the performance of each student as expressed by the count of the 31 scores in the test, and with the threshold(s) for each question showing how difficult it is to earn full or partial credit. All students and all difficulty thresholds are located somewhere on the zesty scale; the precision of the location is determined by estimation error. When a student and a difficulty threshold have the same zesty, it means that in an interaction of the two (student attempting a question), the probability for the student to achieve the score associated with the threshold is 0.5. A student with a higher zesty than that of the threshold has a probability of success greater than 0.5 according to the probability law of the model (Andrich, 1988).

Qualitative Analysis

The aim of the additional qualitative analysis is to identify the tasks and what is required to complete them, and to make sense of the position of each threshold on the continuum by asking what makes a question threshold as difficult as shown by its position on the scale. The tasks represented by each threshold are then compared with those of nearby thresholds. Similarities of groups of nearly thresholds can be observed as well as differences below and above each group of thresholds of similar description. In this way, a description of the scale in bands may be obtained (Figure 11). To construct the zesty scale, it is necessary to account for the zesty of each threshold (that is, to find out what is required for the completion of the corresponding task[s]), and to describe the various parts of the scale (bands) in terms of the skills required to deal correctly with tasks calibrated to that part of the scale (Figure 12).

Each wrong/right question has one value for its zesty corresponding to the percentage of incorrect answers (difficulty) and is associated with the skills required to answer it correctly, as revealed in the qualitative analysis.

Multilevel questions consist of a stem followed by a series of related tasks for which students may obtain 0, full marks (4 on a 0,1,2,3,4 question), or partial credit (1,2,3 on a 0,1,2,3,4 question), depending on how their answers have been judged according to the marking scheme. The marking scheme used to award marks to students' answers in HSC-type physics examinations allows the same score on a question to be obtained in more than one way. The manner in which each partial credit score has been obtained has not been recorded in the data, making the interpretation of partial scores in each case rather complicated.

In the analysis of multilevel questions, a threshold corresponds to a full or partial score. The marking scheme specifies the possible ways of obtaining each score other than 0. Question scores correspond to individual tasks or sets of tasks. For example, a score of 2 on a three-mark question consisting of tasks A, B, and C corresponds to a value of zesty but not necessarily to a unique set of completed tasks. A score of 2 may be obtained in a number of different ways: success on tasks A and B, B and C, or A and C, or some other way. The way each student has been awarded partial credit has not been recorded in the data. Therefore, the description of the skills associated with the zesty of partial credit items becomes complicated. The maximum score does not pose any problems because it can only be achieved by completing all tasks. If the score of 1 mark corresponds to a zesty that is much lower than that of the other possible scores for the question, it may be inferred that

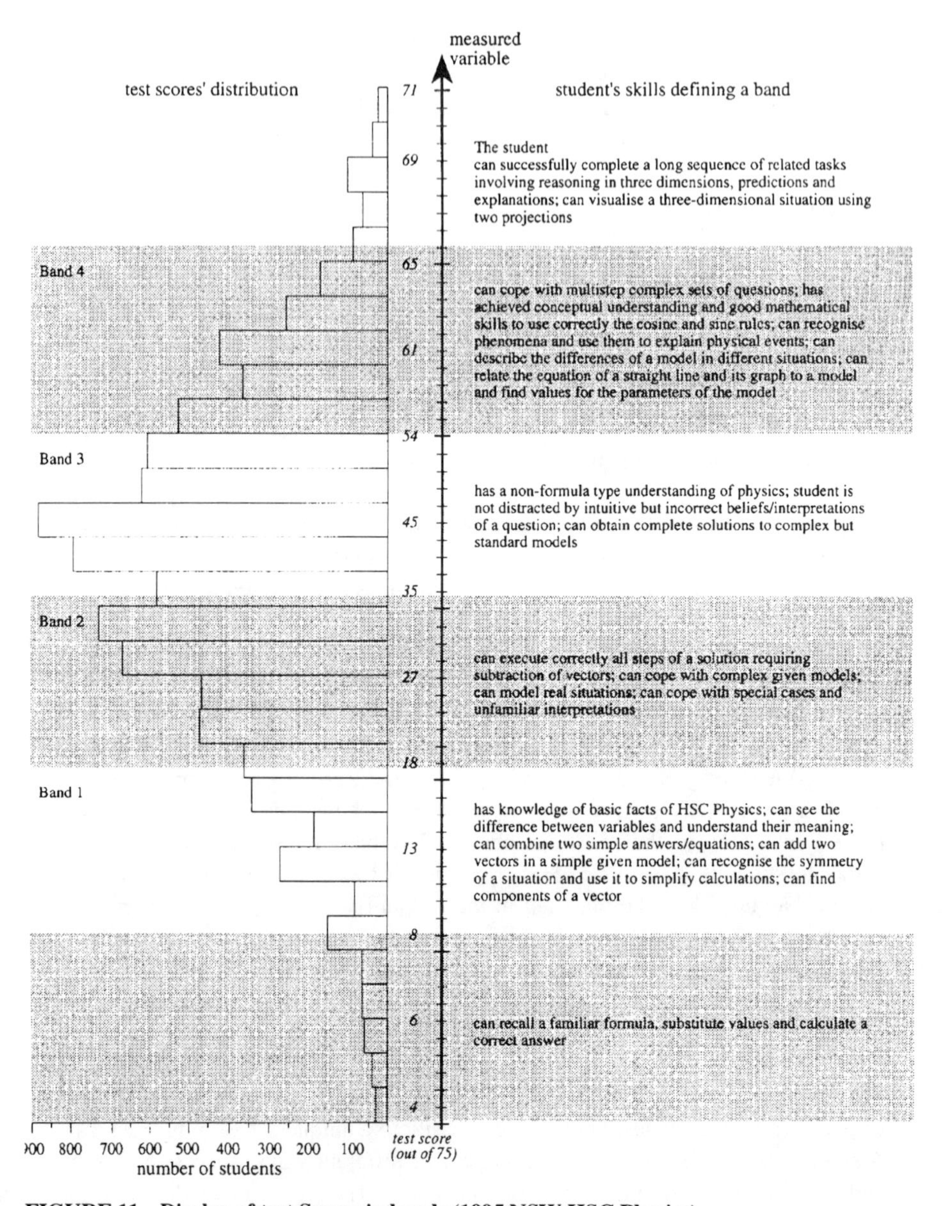

FIGURE 11. Display of test Scores in bands (1995 NSW HSC Physics).

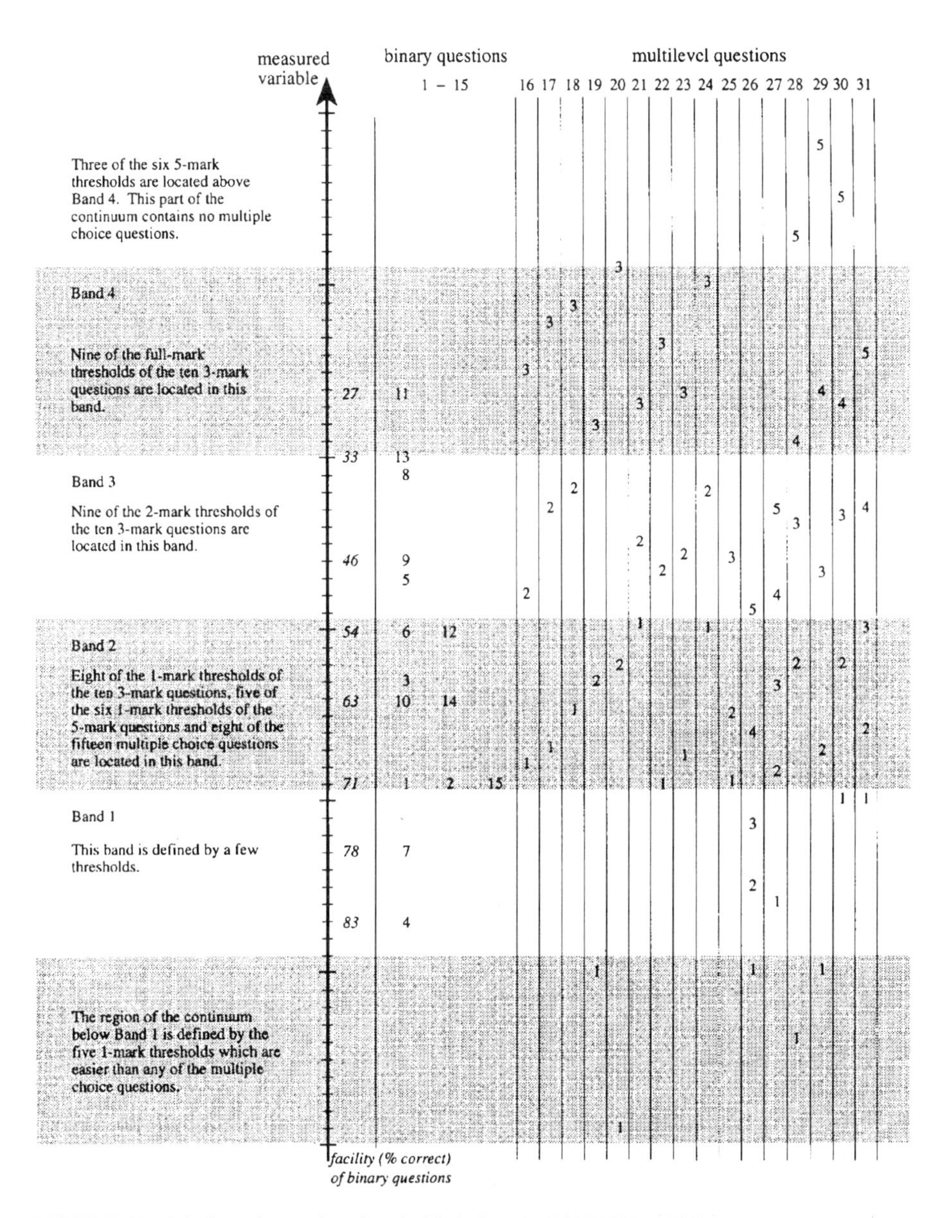

FIGURE 12. Display of question thresholds in bands (1995 NSW HSC Physics).

one of the tasks of that question is much easier than any of its other tasks. If such a task can be identified, then a score of 1 mark can be associated with that task.

The description of the lower part of the scale could be based on the five 1-mark scores whose zesty is lower than that of any of the multiple-choice questions, the description of the central part on the multiple-choice questions, and of the upper part on the ten full mark questions which are highest on the zesty scale. In this way the entire scale could be described. All full credit scores could be used in the description of the scale. After the calibration was completed and the results displayed as shown in Figure 12, the qualitative analysis could start with the multiple-choice questions and continue with the suitable 1-mark scores and finally with the full mark scores of the multilevel questions.

Summarizing, the qualitative analysis consists of:

- Answering each question in writing, going through all steps required to complete the task(s)
- Consulting the marking scheme to understand how students' written answers have been rewarded
- Consulting the distracter analysis to understand students' and question weaknesses
- Understanding what contributes to the zesty of a question in terms of the number and type of steps necessary for the completion of the task, the focus given in the stem of the question, and the relation of this focus to what the student is expected to focus on to complete the task

The focus proposed by the question; the knowledge, concepts, and mathematical techniques the student is expected to focus on for completing the task(s); and a general description that transcends the content area, have been obtained for each of the 31 full credit scores and also for the partial credit scores that could be clearly associated with a specific task.

The construction of the bands shown in Figure 11 has been based on these task descriptions. Careful inspection of the description of the tasks corresponding to each threshold and comparison with those of nearby thresholds resulted in groupings that show a logical progression from a simple to a more sophisticated understanding of and ability to do physics. The lowest band on the continuum has been the easiest to describe: The five thresholds defining this band show the ability to recall a familiar formula, substitute given values, and calculate an answer. This is the .5 mastery level of a student with a test score of 6 out of 75. Students in this band are expected to answer correctly 50% of such questions. There are no strict or absolute rules for the categorization of the thresholds from which the scale is divided into bands.

The interpretation of the zesty values for students and questions is of a probabilistic nature. For a binary question, half of the students having a zesty equal to its threshold are expected to answer it correctly. In other words, a student with a zesty equal to that of a binary question is expected half the time to answer that question correctly. This may be called the 0.5 mastery level. A student with a score in a certain band is expected to be more successful on tasks in lower bands and less successful on tasks in higher bands. The lower (higher) a question threshold is from the position of a student on the scale the more (less) likely it is for that student to achieve that threshold. Cases of students and thresholds have been observed to deviate from this typical behavior. The observation of misfits can often reveal faults in questions and features in the preparation of individual students, thus achieving a diagnostic outcome for the test that often is neglected in large-scale formal examinations; the large-scale physics examinations in Australia are no exception.

The bands of the continuum are defined operationally by the questions in the examination and the marking scheme, and may be described at various levels of detail. The description of the bands given in Figure 11 may be sufficient for an overview of the variable measured with the test. For a fuller understanding of the zesty scale, however, it is necessary to refer directly to the questions and the marking scheme, then to detailed solutions, to the tasks corresponding to the thresholds in each band, to the statistics, and to the skills required for completing the tasks as identified in the qualitative analysis.

Each question threshold used to describe the scale has been illustrated with the following information:

1. The focus proposed by the question; that is, the situation and the task(s)
2. The knowledge, concepts, and mathematical techniques the student is expected to focus on for completing the task(s)
3. A general description of each question that transcends its content area

This type of description is included here for one of the questions, to provide an example.

Question 3

An acrobat is walking along a tightrope as shown in the figure below. The acrobat has weight *W*.

The tension in the wire between the supporting posts is

A. much more than *W*.

B. approximately *W*.

C. approximately $\frac{W}{2}$.

D. much less than $\frac{W}{2}$.

In this question, the question threshold information is as follows:

1. Person standing on a rope; the weight of the person and the tension in the rope.
2. At equilibrium, the vector sum of all forces is 0. ➛ Forces acting on the rope at "the point of contact" with the person: Two forces act along the rope and the force the person exerts on the rope (equal in magnitude to the weight) acts vertically downward ➛ Force diagram to model a three-force equilibrium situation (closed triangle) ➛ Tension is equal to the magnitude of one of the two forces acting along the rope.
3. Modeling a real situation with a vector diagram; use of rules for vector addition.

The position of the 0 on the zesty scale depends on the units used, in the same way as in the case of the physical temperature, which can be expressed in various units such as Celsius and Fahrenheit. On the other hand, the relative positions of the zesty values of test scores and question difficulties on the interval scale are independent of the unit used. For these reasons, no values for a meaningful unit of zesty have been shown on the two displays. Instead, equally spaced divisions on the interval scale, test scores, facilities for the binary questions, and band descriptions have been shown together with the distribution of test scores and the 75 difficulty thresholds. The scale of the measured variable appears in Figure 11 as well as in Figure 12 for convenience and not because there is one scale for test scores (students' performance) and another one for question scores (questions difficulty). Students and the questions in the test have been measured in the same units.

THE POSITION ON THE SCALE OF MULTILEVEL QUESTIONS

Each binary question corresponds to a position on the scale showing its threshold. The position of multilevel questions on the scale is more complicated. The facility of a binary question is defined as the percentage of students who answered the question correctly. A multilevel question has a facility for each possible score except 0. For example, the facilities for a 0,1,2,3 question are defined as follows:

Facility for a score of 1 = percentage of students who scored 1 or higher
Facility for a score of 2 = percentage of students who scored 2 or higher
Facility for a score of 3 = percentage of students who scored 3

The calibration of the difficulty of binary questions is not based on the same equations used for the calibration of the difficulty of the full and partial credit scores of multilevel questions. Therefore, the threshold of the score of a multilevel question is not expected to be equal to that of a binary question of the same facility. In practice the difference between the two thresholds may be small but real data have to be inspected before any claim of negligible difference may be made.

Figure 13 shows the thresholds or gammas (Masters, 1988) of all question-scores versus their facility, obtained in the analysis of all cases and questions with Quest. The scattergram shows that in these data, differences in difficulty of more than one logit may occur for the same facility, and differences in facility of up to 10 percentage points may occur for the same difficulty threshold.

A score of 1 on a multilevel question corresponds to a different location on the continuum from that of a score of 1 on the same task in a binary question. According to the scattergram in Figure 8, the five 1-mark scores that describe the region below Band 1 would be positioned lower on the scale if they were binary questions. This observation is important because the construction of a variable is carried out by describing the tasks corresponding to question scores belonging to various parts of the continuum. If tasks are described in isolation, independently from the other tasks of the multilevel question to which they belong, they would be treated as binary questions and, thus, as not belonging to that part of the continuum. The facility of a score of 1 on a multilevel question using scored data is equivalent to collapsing the levels of the question. This operation may significantly move the task along the continuum.

The description of the tasks corresponding to each question score must include the situation in which the tasks have been attempted by the student and the situation (marking scheme) in which the student's answers have been marked. The data available for the analysis are the marks awarded to students' answers according to a marking scheme. The collapsing of scores is not allowed in the Rasch model (Andrich, 1995b) and results of dubious interpretation may be produced when it is done. The analogy that follows may assist in justifying the phenomenon. In an election where three candidates compete and no aggregation of votes is planned after the election, the number of votes for the third candidate cannot be expected to remain the same if the first two candidates had decided to present themselves as

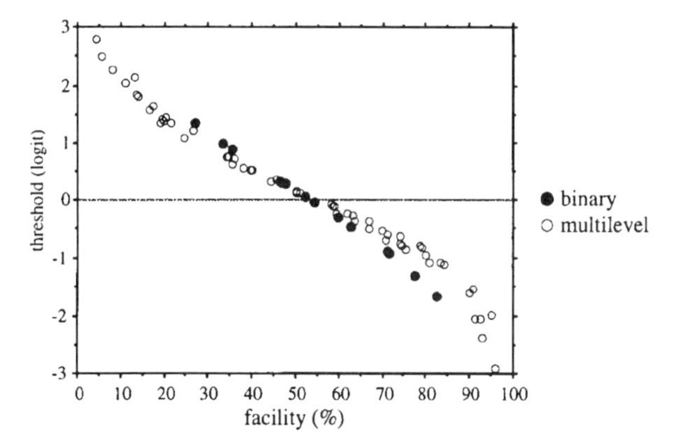

FIGURE 13. Difficulty threshold versus facility (1995 NSW HSC Physics).

a team. Electors who do not like both candidates in the new team would probably vote for the third candidate, if they decide to vote at all.

The gammas obtained from the deltas of the partial credit model are not like the thresholds one would obtain by treating each question-score as a binary question. Such an analysis would violate one of the conditions of the Rasch model; that is, the condition of question independence.

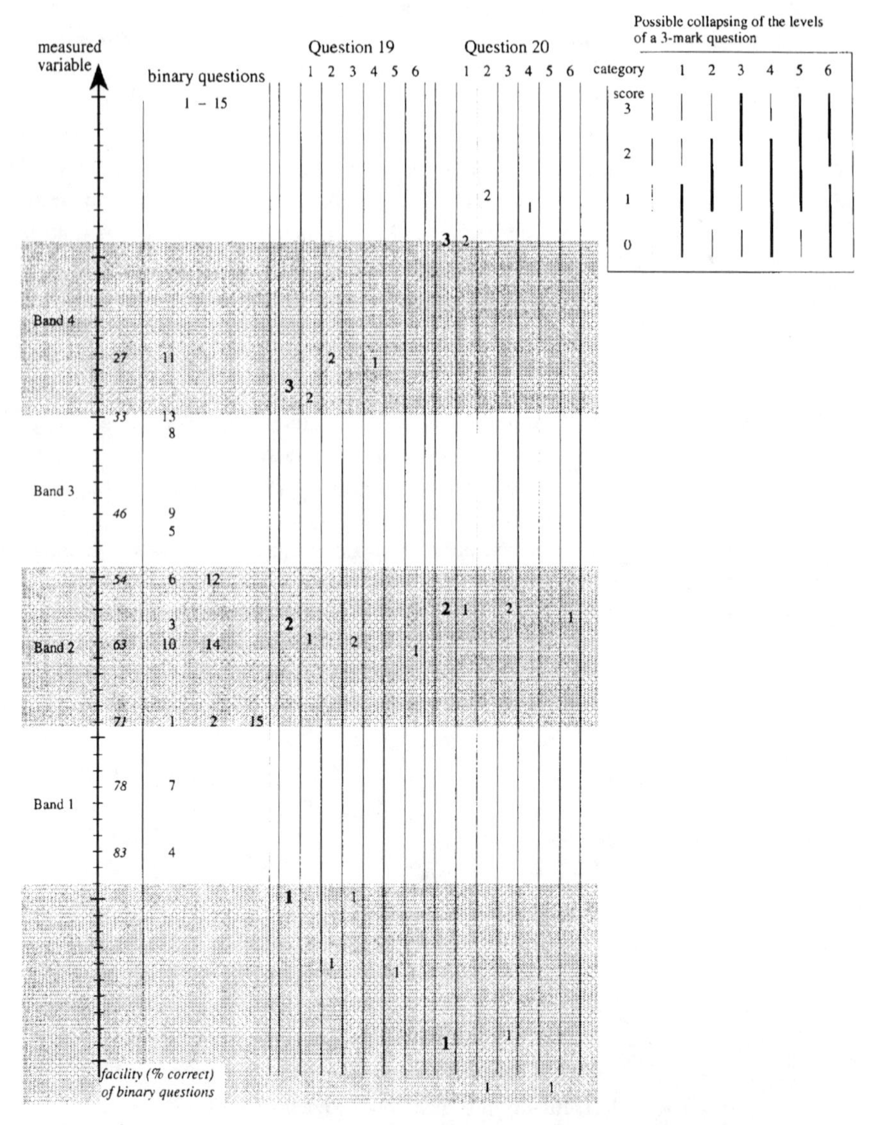

FIGURE 14. Collapsing levels in Questions 19 and 20 (1995 NSW HSC Physics).

Collapsing scores appears likely to produce results that are closer to the results that would be obtained if students' answers had been marked in the first place with the corresponding marking scheme. To check the extent to which thresholds corresponding to the same facilities change, calibrations of the difficulties of two of the 3-mark questions with scores collapsed in all possible ways have been done. Six independent calibrations were required, each time collapsing some of the scores of the question under study and keeping all other questions as in the original marking scheme. The six possible ways to collapse the categories of a 0,1,2,3 question and the results are shown in Figure 14. Changes beyond the boundaries of estimation error occur in most cases.

The issue of collapsing preconstructed categories is discussed further by Andrich and Roskam (Andrich, 1995a; Roskam, 1995).

CONCLUSION

The use of the measurement model does not imply the rejection of familiar results of traditional test analysis, but provides a more accurate expression of those results on an interval scale and allows one to show what has been measured with the description of a variable. The ranking of test scores and question difficulties (percentage of incorrect responses) survive, but now a given difference in test scores corresponds to the same difference in performance anywhere on the measurement scale. Grading is thus more accurate and meaningful because it can now be done on distributions of measures of performance on an interval scale which shows a description of the measured variable, rather than on the distorted score scale with no qualitative information about the content of the questions associated with it.

The preparation of the displays presented in this chapter at one time was prohibitively time consuming, but with today's software and hardware this is no longer the case. However, in the methodology for the construction of a variable, the computer can support the statistical analysis but not the qualitative analysis. This latter must be based on subject area expertise and experience with student learning.

Because the aim of an analysis of this kind is unfamiliar to most people, the methodology presented in this chapter for the construction of a variable is difficult to carry out. Nevertheless, the final result is simple to understand and use, and the component operations of the analysis are familiar. In fact, authors can write detailed and illustrated solutions to test questions; teachers can explain the questions, the underlying physics, and possible solutions; and examiners can discuss the quality of the questions and the adequacy of the marking scheme with the results of statistical analysis. Indeed, all of these operations are required in the methodology but all are focussed toward the description of a continuum through understanding of the factors that contribute to the "zesty" of each question and the presentation of each difficulty threshold in terms that transcend the specific content of the questions.

The description of the scale shown in Figure 11 is the first step toward the formulation of a theory of achievement in the learning of physics. Findings in the CSHE Physics Project, the outcome of which will be a continuum of conceptual understanding of physics, are expected to contribute to the development of one or more scales for assessing understanding of and ability to do physics.

REFERENCES

Adams, R. J., & Khoo, S. -T. (1993). *Quest: The interactive test analysis system.* Hawthorn, Australia: Australian Council for Educational Research. Available online: http://www.acer.edu.au

Andrich, D. (1988). *Rasch models for measurement.* Thousand Oaks, CA: Sage.

Andrich, D. (1995a). Further remarks on nondichotomisation of graded responses. *Psychometrika, 60*(1), 37–46.

Andrich, D. (1995b). Models for measurement, precision, and the nondichotomisation of graded responses. *Psychometrika, 60*(1), 7–26.

Aquilina, J. (1997). *Securing their future—The NSW government's reforms for the Higher School Certificate.* New South Wales, Australia: NSW Ministry for Education and Training.

Baierlein, R. (1990, February). The meaning of temperature. *The Physics Teacher, 28*(2), 94.

Bruce, C., & Gerber, R. (1995). *Phenomenographic research: An annotated bibliography* (3rd ed.) [On-line]. Available: http://www.fit.qut.edu.au/InfoSys/bruce/anabib/title.html

Hasselgren, B., Nordieng, T., & Österlund, A. (1997). The land of phenomenography. *Higher Education Research & Development, 16*(2), 253–256. Available online: http://www.ped.gu.se/biorn/phgraph/home.html

Linacre, M. (1993). Four introductions. *Rasch Measurement Transactions, 7*(2), 290–291. Available online: http://www.rasch.org/rmt/index.html.

Masters, G. M. (1988). Measurement models for ordered response categories. In R. L. Langeheine & J. Rost (Eds.), *Latent trait and latent class models* (pp. 11–29). New York: Plenum Press.

Masters, G. M. (1989). The measurement of understanding [On-line]. Available: http://www.acer.edu.au/Rasch/MCUP/CSHEPP.html

Masters, G. M. (1990). Profiles of learning—The basic skills testing program in New South Wales 1989. Hawthorn: Australian Council for Educational Research.

McGaw, B. (1997). *Shaping their future—Recommendations for reform of the Higher School Certificate.* New South Wales, Australia: Department of Training and Education Co-ordination.

Peterson, K. M. (1991). An overview of secondary physics education in Australia. *Physics Education, 26.*

Roskam, E. (1995). Graded responses and joining categories: A rejoinder to Andrich "Models for measurement, precision, and the nondichotomisation of graded responses." *Psychometrika, 60*(1), 27–35.

Wright, B. D., & Stone, M. H. (1979). *Best test design.* Chicago: MESA Press.

part II

Applications Involving Raters and Judges

5

RATERS AND SINGLE PROMPT-TO-PROMPT EQUATING USING THE FACETS MODEL IN A WRITING PERFORMANCE ASSESSMENT

Yi Du
Edina Public Schools

William L. Brown
Lansing Community College

INTRODUCTION

In assessing writing performance, multiple prompts are usually needed for different genres because students are expected to be able to write in more than one genre. Because of the amount of time required and the cost of the assessment, each student is usually restricted to responding to one or two prompts. It seems evident that test scores derived from different genres will not generally be equivalent, even

A previous version of this paper was presented in the Ninth International Objective Measurement Conference, April 1997, in Chicago.

when efforts are made in the test construction process to make different prompts as nearly equivalent as possible. However, these efforts are often not sufficient to ensure test score equivalence across different prompts. In addition, rater severity is another key source of variation that makes student scores neither equivalent nor comparable. Unless each rater scores every paper, part of each student's score will be dependent on who grades the paper. Therefore, test equating is often used to adjust test scores so that the scores on different prompts, and from different raters, are more nearly equivalent.

A variety of equating models, such as raw score linear equating and equipercentile equating, were considered and tried in this study. However, these equating models were developed primarily for machine-scannable multiple-choice assessment; they can equate prompts very well, but not raters. Because both rater and prompt are primary sources of variation making student scores incomparable, it is not appropriate to apply these models to writing assessment.

The FACETS equating model meets the complex requirement for equating writing performance assessment across both raters and prompts. The FACETS model "can provide a framework for obtaining objective and fair measurements of writing ability that are statistically invariant over raters, writing tasks, and other aspects of the writing assessment process" (Engelhard, 1992, p. 173).

This study is based on an equating of the 1996 writing performance assessment in the Minneapolis Public Schools (MPS). In this assessment, raters and prompts were equated simultaneously using the FACETS model. By presenting the results based on the 1996 assessment, this study presents two conclusions: First, reliable results of equating both rater and prompt can be obtained using the FACETS model scores. Second, single prompt-to-prompt equating is feasible if the appropriate design and equating model are selected.

DATA

About 3,000 grade 5 students and 3,000 grade 7 students participated in this writing assessment. Three prompts, representing narrative, persuasive, and informative writing within a common topic, were assessed at grades 5 and 7. Each student wrote to one of the three prompts. Students were assigned randomly to specific prompts.

The three prompts for different genres were intended to be equivalent. Several strategies were used to make them equivalent. First, the three prompts were under the same topic; for example, grade 5 students were requested to write on topic of "Museum" and Grade 7 students responded to the topic "Space Traveler." Second, more than two prompts in each genre went to field tests. Prompts with little genre difference were selected for the actual tests. Third, the scoring features focus on the common elements of the different genres. In the scoring guide, each scoring feature was explained on a common basis for all three genres. A strong effort was made to ensure that the scoring process did not reflect unique features of any particular genre.

About 30 raters were selected from the population of MPS teachers. The three prompts were scored during three separate sessions in the following order: narrative, informative, and persuasive. Within each session, raters were trained before they scored papers. For each prompt, a representative sample (about 40%) of all papers was scored by two raters. These papers were distributed spirally from rater to rater; that is, each rater was paired with every other rater at least once (most were paired significantly more often than once). After raters were trained, they rescored the double-rated papers. This pattern was consistent for all prompts, ensuring that all raters graded all three prompts of papers and every raters was linked with all others across these prompts. Figure 1 shows the linkage among raters when they scored the double-rated papers.

A uniform scoring rubric was used to score the three groups of papers. The scoring rubric included three domains: Purpose and Voice, Organization and Details, and Conventions of Writing. Under each dimension, multiple features were included in the scoring guide. All the scoring features were rated on a "1-to-4" scale. The framework of the scoring rubric is shown in Table 1.

TABLE 1.
The Framework of the Scoring Rubric

	Domain	Scoring Feature	Scale
1	Purpose and Voice	Purpose	1–4
		Voice	1–4
2	Organization	Main Idea	1–4
		Organization	1–4
		Details	1–4
3	Conventions	Sentence Structure	1–4
		Spelling	1–4
		Punctuation/Capitalization	1–4
		Grammar/Usage	1–4
		Legibility	1–4

Rater 1

Rater 2		A	B	C	D	E	F	G
	A							
	B	X						
	C	X	X					
	D	X	X	X				
	E	X	X	X	X			
	F	X	X	X	X	X		
	G	X	X	X	X	X	X	

FIGURE 1. Linkage of Raters Used in Scoring 40% of Papers

An analytical scoring method was used in this assessment to provide detailed information about each student's writing, compared with the district standards, to report to teachers, students, and parents. The scores in the three domains ("Purpose and Voice," "Organization," and "Conventions") were grouped and averaged, yielding three mean scores on a scale of 1 to 4. A total raw score was then obtained by adding the three scores together. Generally, the overall raw score is derived from these features according to the following formula:

Raw score = average (Purpose + Voice) + average (Main idea + Organization + Details)
+ average (Sentence + Spelling + Punctuation + Grammar + Legibility)

Given that all these writing features were scored on a scale of 1 to 4, based on this formula the raw score ranged from 3 to 12. (Only grade 5 student data is presented in this study, because the results from grades 5 and 7 are very similar.)

EQUATING DESIGN

The random-groups design was used in this assessment, in which different prompts were administered to different but randomly equivalent groups of students. Under the random-groups equating design, student groups who write to different test prompts are regarded as being sampled from the same population. Thus, the population of grade 5 students was divided into three random groups. One of three different prompts (persuasive, narrative, and informative) was administered to each group during the testing period. The common rater group links the three individual student groups. Every rater was paired with all of the other raters at least once. A uniform scoring rubric was used to score all the three prompts. Figure 2 shows the general design of raters, students, scoring features, and prompts.

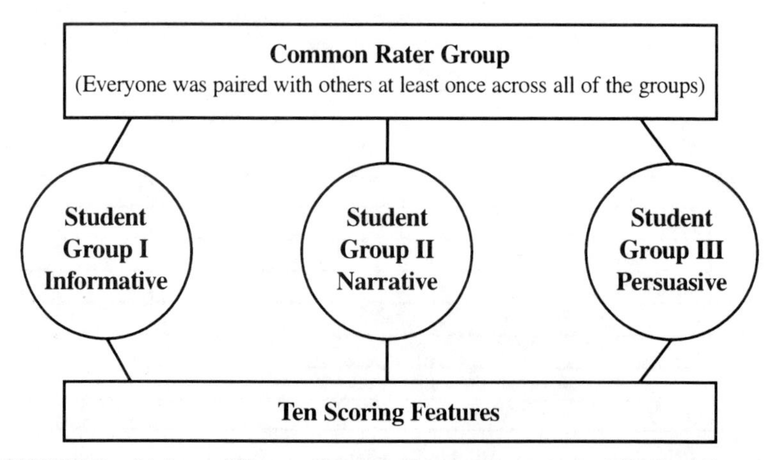

FIGURE 2. Linkage of Raters, Prompts, Scoring Features, and Student Groups

FACETS MODEL

An extension of the Rasch model to include multiple facets (FACETS) was used in equating to determine the transformation rules. For the MPS writing assessment, the primary FACETS model includes four facets: student, scoring feature (scoring component), rater, and prompt:

$$\log \frac{P_{nijmk}}{P_{nijmk-1}} = B_n - D_i - A_m - F_k , \tag{1}$$

where:

P_{nijmk} is the probability of student n being graded in category k by rater j on scoring feature i and topic m,

$P_{nijmk-1}$ is the probability of student n being graded $k - 1$ by rater j on scoring feature i and topic m,

B_n is the writing ability measure of student n,

D_i is the difficulty calibration of scoring feature i, C_j is the severity measure of rater j,

A_m is the difficulty calibration of prompt k, and F_k is the difficulty calibration of grading category $k - 1$ relative to category K.

The rating scale is $k = 0, k$.

Within the FACETS model, the three student groups were anchored to the same group mean. Thus, equating was controlled by the adjustment made for the three student groups based on prompt differences. Because the three equivalent student groups share the same scale with the same group mean and same measurement units, the differences among the prompts can be attributed to the differences of the difficulty level of the prompts and sample errors. Thus, adjustment is made for student measures based on the difficulty of the prompts. Had we not anchored the three groups to the same group mean, students who responded to easier prompts would have appeared to be more able, and students who responded to harder prompts would have appeared to be less able.

In addition, variance analyses related to rater and prompt interaction were carried out in two ways. First, the generalizability analysis was used. Three student papers were randomly selected from each prompt, respectively. Each rater rated these nine papers during one sitting during the rater training session. A generalizability analysis was conducted based on these data. The results show that the interaction between raters and prompts is not a significant consideration (about 0.0001). Second, FACETS analysis of the interaction between prompt and rater was conducted. The analysis revealed that several individual raters tend to favor one prompt over others. These raters received more training. Thus, there is no significant variance between rater and prompt. Only student groups were anchored in this study.

PROMPT DIFFICULTY EQUATING AND ADJUSTMENT

As discussed earlier, student raw scores cannot be assumed to be comparable if they responded to different prompts. Finding that prompts differ substantially in the degree of difficulty can make test developers aware of the prompt differences, and allow them to adjust student scores in accordance with prompt difficulty.

The FACETS model produces a measure of the difficulty level of each prompt. Table 2 shows these prompts rank-ordered from the easiest at the top to the most difficult at the bottom. The narrative prompt was easiest, the informative prompt was hardest and the persuasive prompt was in between. All fit statistics are between 1.0 and 1.1, which indicates that the data from the topics fit the model well enough for measuring student ability. The difficulty differences between the prompts are significant, χ^2 (2) = 4,997.1 and 2,939.5, $p < .001$ with a high separation reliability (R = 1.00). This implies that an equating procedure is necessary to adjust the prompt difficulty for student scores.

Figures 3 and 4 show the differences in difficulties of prompts and how the FACETS equating adjusted these differences. In Figure 3, three ogive curves represent the three student groups who produced informative, narrative, and persuasive writings, respectively. The conversion between raw scores and the Rasch measures indicates that raw score is dependent on the prompts. Students with the same writing ability receive unfair higher raw scores on narrative writing and unfair lower scores on persuasive and informative writing because of the difficulty of the prompts. After equating, the FACETS model adjusted the difficulty of the prompts for student measures. Thus, student measures for different groups are equivalent and comparable. One may notice that there is little difference between students with greater than 6 logits on the Rasch scale. That may imply that the 1 to-4 scale has a ceiling effect so that these high achieving students are able to write very well to any of the three prompts. Exploration of these possibilities is beyond the scope of this study.

To make the Rasch measures more easily understood, these measures were transformed linearly to a scale ranging from 3 to 12. Although the new reporting scale looks like the raw scale, it is quantitatively and qualitatively different from the raw score. The reporting scale retains the sound measurement properties of the Rasch scale—prompt difference adjusted, calibration invariance, and equal interval—so that student scores are "accurate" and comparable.

TABLE 2.
Prompts Calibration and Analysis

Prompt	Rasch Measure	*SE*	Infit Mean Squares	Outfit Mean Squares
Narrative	–0.22	0.01	1.1	1.1
Persuasive	–0.07	0.01	1.0	1.0
Informative	0.29	0.01	1.1	1.1
Overall	0.00	0.01	1.1	1.1

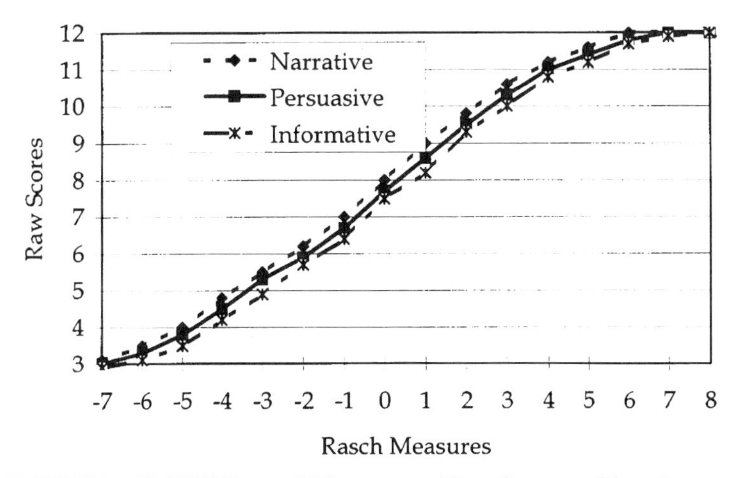

FIGURE 3. FACETS Equated Measures and Raw Scores on Three Prompts

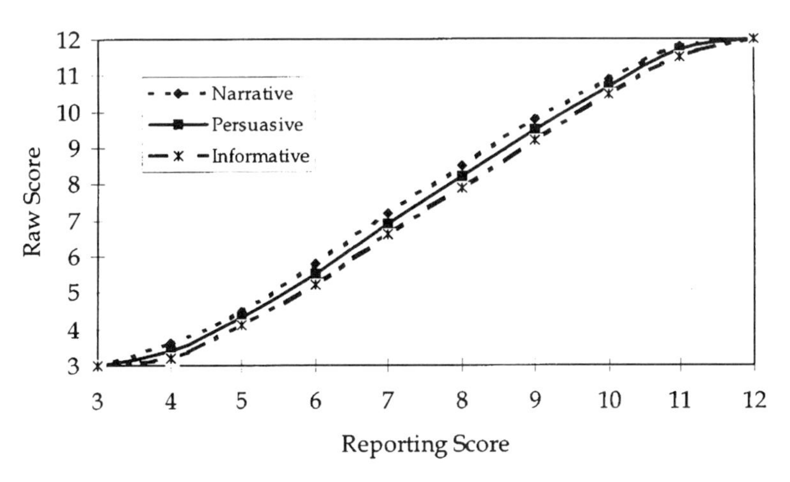

FIGURE 4. FACETS Equated Measures and Reporting Scores on Three Prompts

It will be easier to identify prompt variations in raw scores, and to adjust for them through equating, if the rater variation is controlled. Table 3 exhibits student pairs who wrote to different prompts but were rated by the same raters. This table shows how prompt difficulties affect raw scores and how the FACETS equating removes prompt difficulty differences from student measures.

Students "394540" and "835015" were graded by the same raters, and earned the same raw scores on the narrative and informative prompts. However, their ability measures are –3.47 and –2.96 logits respectively. The substantial differences of

TABLE 3.
Prompt Equated and Adjusted on Rasch Scale (Same Raters)

Student	Prompt	Calibration	Raw Score	Rasch Measure
394540	Narrative	–0.22	4.7	–3.47
835015	Informative	0.29	4.7	–2.96
	Difference	0.51		0.51
075329	Informative	0.29	5.4	–2.22
798274	Narrative	–0.22	5.4	–2.72
	Difference	0.51		0.50
073933	Persuasive	–0.07	6.5	–1.43
591471	Narrative	–0.22	6.5	–1.58
	Difference	0.15		0.15
791185	Persuasive	–0.07	5.5	–2.67
047130	Narrative	0.29	5.5	–2.31
	Difference	0.36		0.36
799301	Informative	0.29	11.8	6.04
012067	Persuasive	–0.07	11.8	5.65
	Difference	0.36		0.39
598687	Persuasive	–0.07	11.6	4.85
791208	Narrative	–0.22	11.6	4.70
	Difference	0.15		0.15
397206	Informative	0.29	8.9	1.06
081213	Narrative	–0.22	8.9	0.57
	Difference	0.51		0.49

Note. Standard errors for all the prompts are 0.01.

0.51 logits occurred because the difficulties of the prompts are different by 0.51 logits. The raw score of the first student (narrative writing) was overestimated because of the easier prompt; the second student (informative writing) was underestimated because of the harder prompt. The student measures, which are corrected for differences in prompt difficulties, provide fair assessment for the two students.

The other pairs of student measures demonstrate similar patterns. These results show that the raw scores were affected by differences in prompt difficulties, and that the FACETS equating process adjusts for student measures based upon these differences.

Table 4 uses the reporting scale score to compare with the raw score, instead of the Rasch measures. This table shows that the reporting scores follow the same pattern as the Rasch measures and that the reporting score removes prompt difficulty differences from student scores.

Table 5 shows the comparison of group distributions before and after equating. The results indicate that for the different student groups, the means, standard devi-

TABLE 4.
Prompt Equated and Adjusted on Rasch Scale (Same Raters)

Student	Prompt	Calibration	Raw Score	Rasch Measure
394540	Narrative	7.7	4.7	5.5
835015	Informative	8.1	4.7	5.9
	Difference	0.4		0.4
075329	Informative	8.1	5.4	6.3
798274	Narrative	7.7	5.4	5.9
	Difference	0.4		0.4
073933	Persuasive	7.8	6.5	6.9
591471	Narrative	7.7	6.5	6.8
	Difference	0.1		0.1
047130	Narrative	8.1	5.5	6.3
791185	Persuasive	7.8	5.5	6.0
	Difference	0.3		0.3
012067	Persuasive	7.8	11.8	11.7
799301	Informative	8.1	11.8	12.0
	Difference	0.3		0.3
598687	Persuasive	7.8	11.6	11.2
791208	Narrative	7.7	11.6	11.1
	Difference	0.1		0.1
081213	Narrative	7.7	8.9	8.3
397206	Informative	8.1	8.9	8.7
	Difference	0.4		0.4

TABLE 5.
Comparison between Raw Scores and Scale Scores

	Raw Score (Before Equating)			Scale Score (After Equating)		
	Informative	Narrative	Persuasive	Informative	Narrative	Persuasive
N count	998	996	999	998	996	999
Mean	7.8	8.1	8.0	7.9	7.9	7.9
SD	2.09	1.09	1.97	1.76	1.65	1.66
Kurtosis	0.08	–0.25	0.30	0.14	0.49	0.52
Skewness	–0.53	–0.13	–0.30	0.28	0.19	0.16

ations, spreads, and shapes of distributions are equivalent and comparable after equating. Without equating, students have very different probabilities of success when they write to different prompts.

RATER EQUATING AND ADJUSTMENT

Student raw scores may be underestimated if their paper happens to be rated by relatively more severe raters. Merely looking at *discrepant* ratings, however, may not be an appropriate or adequate method for resolving this issue. Two severe raters may agree in their ratings of a student, but unless you know that the two raters are significantly more severe than other raters, you would have no basis for questioning these ratings. Finding that raters differ substantially in the degree of severity exercised suggests a need to address such differences in rater training, or to consider the feasibility of adjusting students' scores in accordance with the severity or leniency of the raters, or both.

The FACETS model produces a measure of the degree of severity of each rater. Table 6 (see the column labeled "Severity Measure") rank-orders these raters from

TABLE 6.
Rater Severity Analysis

Rater ID	Severity Measure	*SE*	Infit Mean Squares	Outfit Mean Squares
43	0.46	0.02	1.0	1.0
37	0.41	0.02	1.2	1.1
17	0.34	0.02	1.0	1.0
11	0.28	0.01	1.2	0.7
14	0.28	0.02	0.7	1.0
39	0.26	0.02	1.1	0.9
25	0.20	0.02	0.9	0.9
20	0.18	0.01	0.9	1.1
36	0.14	0.02	1.1	1.1
30	0.10	0.02	1.1	0.9
33	0.05	0.02	0.9	1.2
42	0.04	0.02	1.2	0.7
40	0.01	0.01	0.6	0.9
35	–0.03	0.02	0.9	1.1
21	–0.06	0.02	1.1	1.1
32	–0.13	0.02	1.1	0.9
34	–0.13	0.02	0.9	1.1
22	–0.15	0.02	1.1	1.0
15	–0.16	0.02	1.0	1.1
19	–0.17	0.02	1.0	1.2
27	–0.17	0.01	1.0	1.6
13	–0.18	0.02	1.2	1.1
18	–0.20	0.02	1.6	1.2
38	–0.23	0.02	0.9	0.8
16	–0.28	0.02	1.2	1.0
23	–0.30	0.02	1.3	0.9
12	–0.31	0.02	0.8	1.1
31	–0.47	0.01	1.1	1.3
28	–0.49	0.02	1.3	1.4
26	–0.53	0.02	1.5	1.5
Overall	0	0.01	1.1	1.1

the most severe at the top to the most lenient at the bottom. To the right of each rater severity measure is the standard error of the estimate, indicating the precision with which it has been estimated. Other things being equal, the more observations an estimate is based on, the smaller its standard error. The rater severity ranges from –0.53 to 0.46 at grade 5. The spread is 0.99 logits. This represents a mean score discrepancy of approximately 0.4 on the four-point scale. All of the raters are between –1.00 and +1.00 logit in severity.

Figures 5 through 7 show the raw scores plotted against the Rasch measures within each prompt. These figures illustrate that raw scores unadjusted for rater severity can mask variability in writing competence.

It is easier to see rater severity differences and adjustment if we control for prompt difficulties. Table 7 shows how rater severity affects raw scores, and how rater severity is removed from student measures when prompt difficulties are controlled. The student pairs in Table 7 wrote to the same prompts, but were graded by different raters. These students were selected for comparison of the measures given by different raters.

Student pair "004107" and "780815" earned the same raw scores from different raters, but their ability measures are –0.30 and +0.34 logits respectively. The substantial difference of 0.64 logits occurred because the severity levels of the raters are different (by 0.64 logits). Student "004107" had a more lenient rater, while Student "780815" had a more severe rater. The rater severity difference made the two students' raw scores the same. The Rasch measures removed the effects of rater severity and provided fair and comparable estimates of writing ability. The same can be said for the other pairs of students.

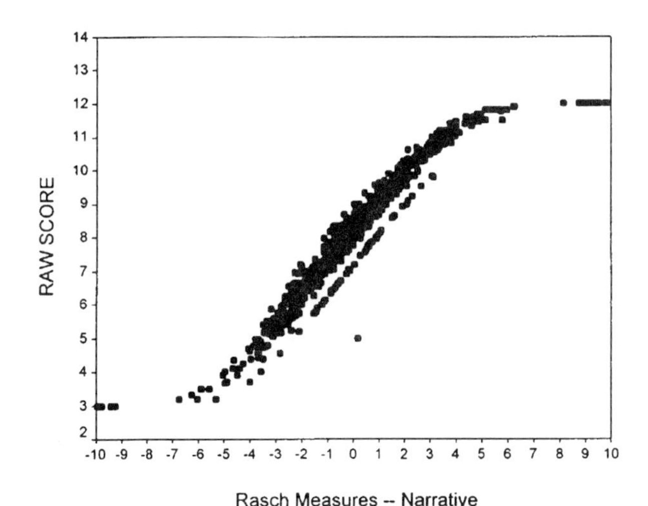

FIGURE 5. Raw scores and the Rasch Measures on Narrative Writing

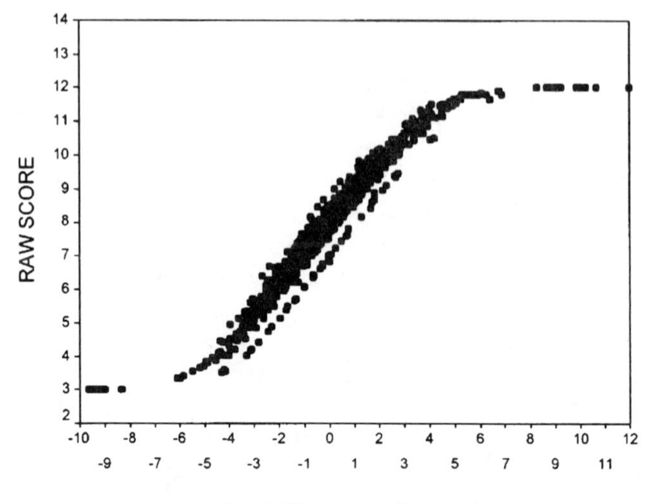

FIGURE 6. Raw scores and the Rasch Measures on Persuasive Writing

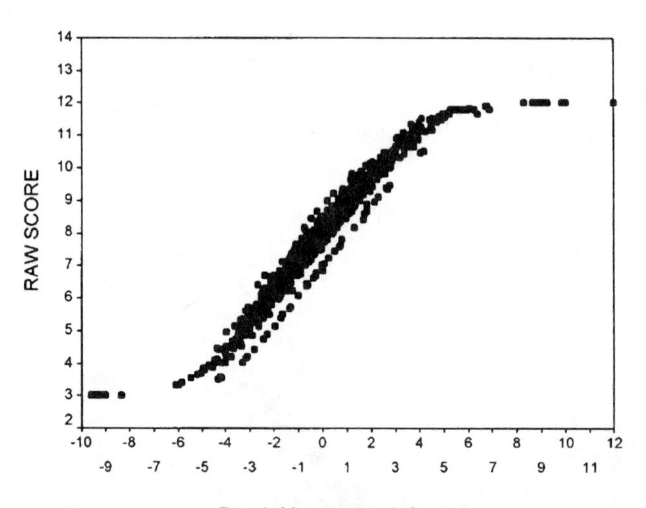

FIGURE 7. Raw scores and the Rasch Measures on Informative Writing

TABLE 7.
Rater Severity Equated and Adjusted

Student	Rater 1 with Severity	Rater 2 with Severity	Average Severity	Raw Score	Rasch Measure	Prompt
004107	23 (–0.30)		–0.30	7	–1.07	Informative
780815	17 (0.34)		0.34	7	–0.43	Informative
Difference			0.64		0.54	
691478	16 (–0.28)		–0.28	4.3	–5.05	Narrative
397613	14 (0.28)		0.28	4.3	–4.49	Narrative
Difference			0.56		0.56	
793336	17 (0.34)		0.34	13	5.97	Persuasive
397613	28 (–0.28)		–0.49	13	5.14	Persuasive
Difference			0.83		0.83	
012690	33 (0.05)		0.05	11.7	4.88	Narrative
598627	31 (–0.47)		–0.47	11.7	4.37	Narrative
Difference			0.52		0.51	
781379	32 (–0.13)		–0.13	7.1	–0.83	Informative
080844	14(0.28)		0.28	7.1	–0.36	Informative
Difference			0.41		0.47	
243309	23 (–0.30)	24 (–0.73)	–0.52	7.3	–1.01	Informative
591402		25 (0.20)	0.20	7.3	–0.28	Informative
Difference			0.72		0.73	
399286	43 (0.46)		0.46	8.5	1.25	Informative
595063	18 (–0.20)		–0.20	8.5	0.60	Informative
Difference			0.66		0.65	
691478	16 (–0.28)		–0.28	3.9	–5.05	Narrative
397613	14 (0.28)		0.28	3.9	–4.49	Narrative
Difference			0.56		0.56	

OVERALL RESULTS

The overall results for students, raters, prompts, scoring dimensions, and scoring features are shown graphically in Figure 8. The FACETS program calibrates all facets so that they are positioned on a common scale. That scale is in logit units which, under the model, constitute an equal-interval scale with respect to appropriately transformed probabilities of responding in particular categories. The figure enables one to view all facets of the analysis simultaneously, summarizing information about each facet.

Figure 8 shows that the student distribution spreads from –7 to +8. All rater severity levels are located between +1 logit and –1 logit, which means none of them are extremely severe or lenient. The informative prompt was the hardest, while the narrative was the easiest. For the MPS students, English conventions (including grammar, sentence structure, organization, spelling, and punctuation) are the most difficult elements for writing, while the Main idea and Details, as well as Purpose and Voice, are

Measr	+ Student	- Rater	- Prompt	- Dimension	- Scoring Feature
8	***				
7	.				
	.				
	.				
6	.				
	.				
	*				
5	*.				
	*.				
	.				
4	*.				
	**.				
	**.				
3	***.				
	***.				
	***.				
2	*****				
	****				
	*****				
1	******.				
	*******.	*			Gram/Usage Organiz Sentence Spell
	********.	*****	Inform	Conventions	Punc/Capit
0	********.	*****.	Persuade	Main idea & details	Details
	********.	*****	Narrate	Purpose & Voice	Purpose
	********.	*			Main idea
–1	*********.				Legible Voice
	********.				
	******.				
–2	******.				
	*****.				
	***.				
–3	***.				
	**.				
	*.				
–4	.				
	*.				
	.				
–5	.				
	.				
	.				
–6	.				
	.				
	.				
–7	*.				
Measr	* = 21	* = 2	- Prompt	- Dimension	- Scoring Feature

FIGURE 8. Map of overall results

relatively easier. The results are consistent with the results from other writing performance assessments used previously in the Minneapolis District. This suggests that classroom instruction needs to focus more on training students in the basic writing skills, such as grammar, sentence structure, organization, spelling, and punctuation.

DISCUSSION AND CONCLUSIONS

The purpose of this chapter is to illustrate how the FACETS model can be used to provide reliable results of equating for both raters and prompts, and to show that prompt-to-prompt equating is feasible if appropriate design and equating models are selected. The advantages of the FACETS mode—sample independence, calibration invariance, equating more than one facet at the same time, and flexibility in the sample size for examinees and items—make equating both raters and prompts feasible and ensures accurate and stable results.

One of the significant features of this research is that it presents a strategy for modeling ratings obtained from a design with equivalent groups for equating prompts—by using a random-groups equating design, with each student responding to only one prompt.

As performance assessment has been widely used in state or district testing programs, new challenges have been raised to researchers, test developers, testing administrators, and psychometricians. In order to make sure testing results are both reliable and comparable, equating procedures are needed. Even with equating procedures, however, controlling budget and student testing time are still crucial in a large-scale performance assessment. This study shows that good equating results depend on well-done design. Especially for a large-scale performance assessment, it will make a big difference in budget and testing time if students can be assigned to write to one prompt rather than more than one prompt. Using the random-groups equating design, it is practical to equate prompts with each student having to respond to only one prompt; students' scores from different prompts can then be made comparable. Also, by calibrating for rater severity, it is practical to use only one rater to score each paper, especially if every-student testing is desired but resources are limited. This design, then, reduces the cost of equating, the amount of student testing time, and the amount and cost of teacher scoring time needed. Therefore, combining the FACETS model and the single-random-group design presented in this chapter, the equating method promises to offer broader applications for a wide variety of performance assessments.

REFERENCE

Engelhard, G., Jr. (1992). The measurement of writing ability with a many-faceted Rasch model. *Applied Measurement in Education*, *5*, 171–191.

6

AN EXAMINATION OF VARIATION IN RATER SEVERITY OVER TIME: A STUDY IN RATER DRIFT

Mark Wilson
Harry Case
University of California, Berkeley

INTRODUCTION

In recent years, there have been a number of attempts to incorporate essay questions and performance tasks into large-scale standardized examinations. For example, the Scholastic Aptitude Test (SAT) from the Educational Testing Service now includes an essay component, and Terra Nova from CTB-McGraw Hill includes a variety of performance-style assessments. There are several reasons why this

This research was carried out by the Berkeley Evaluation and Assessment Research (BEAR) Project, with funding from Los Angeles County Office of Education, for the California State Department of Education. A version of this paper was presented at the National Conference of the Council of Chief State School Officer's Association, Phoenix, Arizona, in June, 1996 (Wilson & Case, 1996). The opinions expressed, are, of course, solely those of the authors. We would like to thank staff of the California State Department of Education, especially Dale Carlson, Sue Bennett, Gerry Shelton, and Ellen Lee. We would also like to thank the staff of the Sacramento County Office of Education who organized the scoring sessions, especially Linda Murai and the raters at the two sites we visited, and the staff of CTB-McGraw Hill who provided mark-sense services at the sites. Programming for the project was carried out by Wen-chung Wang.

change has come about (see Khattri & Sweet, 1996 for a recent survey). Some advocates want to influence what is taught in schools; some want to remove the perceived negative influence of multiple-choice standardized tests. To the extent that assessment can drive instruction, some see that it is useful to include essay questions and performance tasks on assessments and required examinations. Another motivation is the belief that essay questions and performance tasks can measure cognitive skills that are, at best, indirectly measured (Resnick & Resnick, 1996) and, at worst, unmeasurable by multiple-choice or true-and-false questions. The intent is that a test yield a more complete assessment of the student's abilities by offering a broader range of item types (Hogan, 1981; Mislevy, 1991).

Essay questions and performance tasks are not without their drawbacks. They can be expensive and time consuming to take and to score, and introduce a certain amount of subjectivity into the scoring process. The subjectivity arises because raters need to make judgments about the student papers. Guidelines for such judgments cannot contain all possible contingencies and, therefore, the rater must make a judgment. It is this judgment that allows subjectivity to enter the scoring process. Contrariwise, it is this judgment that offers the possibility of a broader and deeper interpretation of test scores.

It is the discretion raters must use in assigning a score that raises the possibility that there are inconsistencies between raters. In particular, parents may be concerned that their child was scored by a rater who was consistently assigning lower scores and, in fact, earlier work (for example, Lunz, Wright, & Linacre, 1990; Wilson & Wang, 1995) has shown that there are instances where raters differ significantly (both statistically and substantively) in their severity. One of the goals of this chapter then, is to investigate ways to identify these inconsistencies between raters over the period of time while they are actively involved in the scoring process—so-called *rater drift*—so that timely corrective measures can be taken.

BACKGROUND

Failure to assign appropriate scores may the result of a variety of factors related to the rater, such as fatigue, failure to understand the intentions of those who created the scoring guidelines, distractions due to matters such as poor handwriting on the student's part, and distractions due to a prior student's responses. The types of errors that raters make have traditionally been classified into four different categories (Saal, Downey, & Lahey, 1980). They are (1) severity or leniency, (2) halo, (3) central tendency, and (4) restriction of range.

Rater *severity* or *leniency* is a consistent tendency on the part of the rater to give a score that is higher or lower than is appropriate, which is usually interpreted to mean higher or lower than the average of the other raters.

The *halo* effect may manifest itself in three distinct ways, all of which involve the rating of one response affecting the rating of another. The first may occur when a single response is scored on a number of subscales. Here, the rater scores the subscales based on some overall impression rather than looking at each subscale as a unique and independent domain. The second type of halo effect is between different responses from the same person. Here, the overall impression comes not from a single response but from several responses. Each response is therefore not scored on its own merit. The third manifestation of the halo effect is between persons. The response by a prior person may influence the scoring of a later response by a different person.

Central tendency and *restriction of range* are similar in that in both cases the rater is not making full use of the scoring range. However, in the case of central tendency, the rater rarely awards scores at the extremes (that is, the scores are [almost] always in the middle of the range), while in the case of restriction of range, the narrow band of scores awarded may be in any part of the range. Thus, central tendency is a special case of restriction of range. The causes of the problems may differ as well. A rater may adopt a rating strategy that looks like central tendency because staying in the middle of the range reduces the possibility of grossly misscoring any student (which means that discrepancy models used to check up on raters will not be very sensitive). However, a restriction of range, for example at the low end, may be due to a failure on the part of the rater to see distinctions between the different levels.

Several actions can be taken to promote more appropriate ratings. One action is to provide extensive training so that the rater more fully understands the intentions of the test designers and is aware of influences that can bias his or her judgment. This training may include instructions on how to interpret the scoring guidelines and opportunities to score sample items, and may be repeated during the scoring session.

A second action is to have double or triple readings of the student papers. Differences between scores by different raters can then be resolved in a number of ways. For example, they can be arbitrated by discussion and mutual agreement, by taking the average of the scores, by deferring to the more expert of the raters, or by referring them to another "expert." A variant on this strategy would have a more expert rater re-rate a sample from each rater's work. This strategy, termed "read-behinds," was used in the context described later in this chapter. Of course, the lighter the sampling, the less reliable this method will be in detecting discrepant raters.

A third approach uses information from one source of information about the rater to check for consistency with other pieces of information, such as other ratings. In the context to be described later, for example, information about a rater's ratings of students, along with the students' responses to multiple choice items, was combined to estimate a rater's severity. In this case, *all* ratings could be used, not just a sample. A rater would be considered severe if he or she tended to give scores that were lower than would be expected from other sources of information. If the rater tended to give higher scores, that would be considered leniency. Once detected, biased scores due to severity or leniency can be used as a basis for rater advisement, or the scores can then be adjusted to correct for the effect of the bias or both.

It is important to recognize that the use of a statistical model to check consistency has its limitations as a methodology. These models show their weakness with respect to the individual student. This is because, in general, mathematical models are premised on the belief that we expect raters, students, and items to act in a consistent fashion. However, raters may randomly assign a few scores that are too severe or too lenient even though, overall, they are doing a good job. Fit indices could be used to detect the most inconsistent patterns. Similarly, students who do well on most of an examination may make a mistake on a relatively easy question. In other words, we expect these models to be useful in identifying uniform severity problems and less useful with nonuniform severity problems. Consequently any adjustment to student scores may therefore improve the overall reliability and hence, improve the consistency of individual scores, but at the same time reduce the correctness of some (fewer) individual scores. Likewise, the use of mathematical models to help identify biased raters may be quite successful, but may not be so successful in helping us find inappropriate individual scores for student responses.

Nonetheless, there have been several studies that have demonstrated that mathematical models can be useful methods for analyzing rater performance. Several approaches have been used to calculate rater severity/leniency: analysis of variance (ANOVA; Guilford, 1954), ordinary least squares (OLS; Braun, 1988), or weighted least squares (WLS) regression (Raymond & Viswesvaran, 1993); and item response theory (IRT; Engelhard, 1994; Lunz, Wright, & Linacre, 1990; Wang & Wilson, 1996; Wilson & Wang, 1995).

A variety of ANOVA approaches exist, such as the investigation of rater *X* dimension designs and rater *X* ratee *X* dimension designs. All of these approaches attempt to establish the existence of a significant rater main effect. The OLS and WLS regression approaches use the following model.

$$Y_{ijk} = \alpha_{ik} + \beta_{jk} + \varepsilon_{ijk,} \qquad (1)$$

where:

Y_{ijk} is the rating given to candidate i by rater j on item k,
α_{ik} is the true rating for candidate i on item k,
β_{jk} is the leniency index for rater j on item k, and
ε_{ijk} is the normally distributed random error.

This model assumes that the error terms are normally distributed with an expected value of zero and that the variance of the errors across raters is equal. The OLS procedure provides an unbiased estimate of the vector of true ratings. If, however, the reliability of scoring varies from rater to rater, then the usual regression assumption of equal error variances across all raters and ratees will be violated and the use of WLS will be more appropriate.

Another approach to estimating rater severity utilizes IRT. A basic IRT model starts with the assumption that there are measurable latent characteristics called student ability and item difficulty and that these characteristics can be expressed in terms comparable to each other. The model then goes on to assume that the greater the student's ability relative to the item difficulty, the more likely the student is to answer the item correctly, or conversely, the lower the student's ability relative to the item difficulty, the more likely the student is to answer the item incorrectly. Note that a probabilistic model is used; that is, the student always has some probability of getting the answer right or wrong regardless of the ability of the student or the difficulty of the item. In the case of rated or judged items, one would think of rater severity as being the adjustment of item difficulty for a specific rater. For polytomous items, a further source of complication is the step from one category to the next, which we will refer to as a threshold.

The particular model that we will be using in this investigation is a polytomous Rasch-type model (Rasch, 1960/1980) with a linear model on the difficulties to allow for the effect of raters. Within this formulation, the log-odds of being in one score category, compared to the adjacent lower score category, is modeled as a linear function of different effects, in this case, student ability (θ_n), item difficulty (δ_i), rater severity (λ_j), and threshold (τ_k): If:

φ_{nijk} is the log odds that student n with a "true" proficiency of θ_n will receive from rater j a rating in category k as opposed to receiving a rating in category $k - 1$ for item i,
P_{nijk} is the probability of student n being rated k on item i by rater j, and
P_{nijk-1} is the probability of student n being rated $k - 1$ on item i by rater j,

then the log-linear model can be written:

$$\varphi_{nijk} = \ln[P_{nijk}/P_{nijk-1}] = \theta_n - \delta_i - \lambda_j - \tau_k. \tag{2}$$

We estimated the parameters using the random coefficients multinomial logit model (RCML; Adams & Wilson, 1996), and an adaptation of the ConQuest software (Wu, Adams, & Wilson, 1998), which uses a marginal maximum likelihood estimation procedure.

The IRT approach offers several advantages over regression analysis. First, the IRT model treats the student responses as ordered categorical responses, while the regression approaches assume that the item scores are measured on a linear scale. Second, the regression models estimate the effects for each item separately and ignore the fact that the item responses are actually repeated measures, each taken by one student. The IRT approach estimates the item parameters simultaneously and it directly models that the students answer the whole set of items. Third, the regression models assume the measured student's ability is a single fixed quantity and attribute all wrong answers on relatively easy questions as error. The IRT model, in contrast, views each student as capable of manifesting a range of responses according to the model, each with its own probability, and hence, incorrect responses to easy items are not necessarily a problem (although when there are many such, it may be considered misfit). Because of these advantages, we use the IRT approach in this study.

We will use the estimates of rater severity from the mathematical model as an aid in detecting severe or lenient raters. Previous work (Engelhard, 1994, 1996; Wilson & Wang, 1995) has shown that such effects can be detected by pooling data across rating sessions. In particular, the Wilson and Wang (1995) paper shows that strong rater effects can persist even in the presence of standard rater quality control efforts such as rater training and a 10% read-behind protocol (as is used in the context that follows). In this chapter, we also examine rater performance *within* scoring sessions. Scores from raters can be periodically analyzed while they are working and then a determination can be made as to how good a job the raters are doing. If this information is found to be useful, it could serve as part of a feedback mechanism by which poor raters are identified for counseling and, perhaps, retraining. The long-term goal of this research program is to investigate the usefulness of implementing a feedback mechanism to improve rater consistency.

There are several limitations to the following study that must be borne in mind when interpreting results. First, the follow-up interventions described here in were not under any sort of systematic control by us. Follow-up was left entirely to the discretion of the site administrators. In particular, we have documentation of only one follow up based on the IRT information (Case Study 10 later). Thus, the results must be seen as descriptive only. We are mainly concerned (1) to see if it is feasible to generate severity information within a scoring session, (2) to investigate the variation in rater severity within rating sessions, and (3) to examine the relationship (or lack thereof) of those IRT severity patterns to the expert re-ratings. We do have records of interventions based on these "read-behinds," but no comparable actions were taken based on the IRT information. We did have an opportunity to investigate some ancillary issues, and these will also be reported and discussed.

METHOD

Procedures at the Site

This study was conducted using the eighth-grade mathematics examination administered by the California Department of Education (CDE) for the Spring of 1994. CDE administered, statewide, a battery of tests designed to assess student ability in three grades: fourth, eighth, and 12th. In 1994 the exam included 8 multiple-choice items, 8 constructed response items (students are required to show their work even though the item is scored correct/incorrect), and 2 short performance tasks (of which only one was scored). There were 6 different forms, including common items and unique items. The exact format varies from year to year, for example, in 1993, the eighth grade mathematics examination included 7 multiple-choice items, and one performance task. The procedures were followed at two different scoring sites, which we will call sites A and B. At each scoring site, we had access to the multiple-choice data, but not the constructed response data, so we will ignore the constructed response data in the remainder of the analysis.

The procedures for the on-site study were as follows. Raters assembled in a large room for a 1-week scoring session during which they would score two different performance tasks (3 days for one and 2 days for the other). The raters were divided among seven tables of six under the supervision of a "table leader." The table leader was an experienced rater who had succeeded as a rater before. The whole scoring session was managed by a "chief reader." The raters first read the scoring rubric and then were shown examples of exemplar scores (answers that exemplified a particular scoring category within the rubric), followed by examples of borderline scores (answers, that would fall between scoring categories but for which a decision had to be made). After looking at the examples and discussing them with other raters and the table leaders, the raters were given 10 sample responses to score as a "calibration." The "correct" scores for these sample responses had been agreed upon beforehand by a team of experts, including the chief reader. In order to qualify as a rater, it was necessary that the score given by the rater matched the expert panel's score at least 8 out of 10 times. Raters who did not calibrate were retrained. If they failed to calibrate after a number of retrainings, they were sent home.

After the calibration test was passed, packets containing the responses from 20 students each were distributed to the raters who had calibrated. The responses within a packet were randomly assembled and the distribution of packets was also random. The raters would read the response and then assign a score between 1 and 4, by applying their understanding of the scoring rubric. After they had finished scoring the entire packet, they handed it to the table leader and selected another packet. The table leader would then select two responses from the scored packet (10% of the responses) and score them again. In theory, this would be done without looking at the initial score, but in practice, this did not seem to be an entirely correct assumption. These were the so-called "read-behinds." The table leaders maintained their own records of the results of the read-behinds and used CDE's preestablished performance standards to determine whether a rater needed additional attention. If the rater's score differed by more than one from the table leader's score on any performance task, the rater was immediately notified and the discrepancy was discussed and resolved. If, for any 10 consecutively sampled responses (across packets or within), the rater and table leader did not agree at least 80% of the time, retraining was administered. Retraining consisted of a discussion with the table leader about the general scoring rules and specific student responses that the table leader and rater had scored differently. Next, a calibration test was readministered. If the rater failed to achieve an 80% accuracy rate on the calibration test, then the retraining process was repeated. Depending on the extent of the problem, the rater might eventually have been dismissed.

After the table leader was finished with a packet, support staff would collect it and perform some basic quality control, such as checking to see if ID numbers matched. Then they would bring it to a scanning machine where the scores were scanned into a computer file. (This step and those that follow were not routine—they were carried out specifically for this study.) After the scores were scanned, they were combined with previously scored and stored multiple- choice responses, using the student's ID as the key for the merge. After a sufficient amount of data about raters had been collected, analyses to estimate the IRT model were conducted. The data for these analyses were not cumulative; that is, the analysis for the second period did not include the data from the first period. This decision was made because the goal was to identify intervals when the raters may have had changes in their performance. If the analyses had been cumulative, then results would be less sensitive to the most recent round of data available.

Analyses

One of our concerns was whether we would be able to produce results in a timely fashion. To that end, we tried to minimize the time needed for the analysis that was to be done by the computer. One of the best ways to speed up an IRT analysis is to reduce the number of parameters that need to be estimated and one way of doing this is to "anchor" some of the parameters. Anchoring means to fix parameters at a predetermined value so that they are not estimated during the analysis. We were able to calculate IRT parameter estimates for the multiple-choice item difficulties before going to the site because we were able to obtain a large sample of student responses in advance. This was possible because the multiple-choice items were machine scored and, therefore, the results were available well before the performance tasks were scored. The results of the IRT analysis using only the multiple-choice responses were then used as the anchoring values. Note that this means that the information available to link the data together consists of (1) the multiple-choice data, (2) the data from the rater's ratings, (3) the data from the table leader's re-ratings, and (4) the data from the rater's (successful) calibration sets.

During the course of our preliminary discussions with the organizers of the scoring site, a variety of different ways to present our results were put forward. Although it is standard in the research literature to present IRT results in the logit metric, it became clear that it would be more useful to present our results in the metric of the item; that is, the 1-to-4 scoring range use for mathematics performance tasks. This allowed the results to be compared directly to preexisting standards of rater quality. The actual results provided to the organizers of the scoring sessions were the effect a given rater had on an average student's score, relative to what the student would have received from the average rater.

The conversion process involved calculating the difference between the expected score for an average student with an average rater and the expected score when that average student is rated by the rater in question. This difference, D_j, is the severity effect in score units of the rater.

The formula for the expected score for an average rater (E) is as follows:

$$E = \sum_{k=1}^{4} kP_{ik}, \quad (3)$$

where P_{ik} is the probability of a student at the mean being rated k on item i by an average rater (calculated using the RCML model given in Equation 2). The formula for the expected score for rater j (E_j) is:

$$E_j = \sum_{k=1}^{4} kP_{ijk}, \quad (4)$$

where P_{ijk} is the probability of a student at the mean being rated k on item i by rater j. Thus, we can calculate $D_j = E - E_j$.

We had two opposing goals in trying to determine how often we should run analyses. On one hand, we wanted to produce analyses as often as possible, which would minimize the amount of data available. On the other hand we wanted to have accurate estimates, which means having as much data as possible. The issue here is the size of the standard errors. Without a sufficient amount of data, the standard errors of our estimates would be so large that it would be almost impossible to identify any level of poor performance.

Based on past experience and advice from the session organizers we estimated that a typical rater would be able to score about 60 responses per hour. We then estimated standard errors in the score metric, based on different numbers of responses, and used these as a basis for choosing approximately 180 responses as a reasonable target. This also is approximately the number of student papers scored by a typical rater in a half-day session. As a half-day is an administratively convenient unit of time, we chose that as the period of time analysis.

RESULTS

IRT Information on Rater Performance

Rater severities were calculated for each rater within each time period, along with a standard error for each severity. The size of the standard error is dependent on the number of responses scored by the rater and the amount of unexpected variation in the scores. To give an estimate of the confidence interval for the rater severity, we used a range of 1.96 standard errors on either side of the rater severity, which, under an assumption of normally distributed errors, corresponds to an approximate 95%

confidence interval. If this range did not include zero, then we concluded that the rater had a statistically significant bias, and the rater was flagged.

For each scoring period, we produced a chart that compared all the raters at the table and displayed a mean severity estimate and the approximate 95% confidence interval around the mean. Informal and anecdotal experience with charts that displayed the severities led us to believe that this was not the best way to communicate with the raters. Hence, we produced a second chart designed to make the information more interpretable. This chart displays the severity estimates in terms of expected counts in each score category. We counted the scores in each category for each rater, and compared it to the counts expected for the average rater when scoring the same set of students. We found that the expected counts for an average rater were extremely close to the observed average across all students and raters; in fact, it could not be distinguished when displayed on a graph. This happens because the student responses were randomly assigned to raters, so the distribution of students assigned to any one rater should be similar, over the long run, to that of all the students assigned to all the raters. If the number of papers scored by a single rater is large enough this is a reasonable assumption. Thus, instead of calculating the expected counts, we displayed the "room average" counts, which was a saving in computational time (and also easier to explain to the raters). Hence, the second chart shows the distribution of scores assigned by an individual rater and compares that to the distribution of scores assigned by all raters. Figure 1 shows

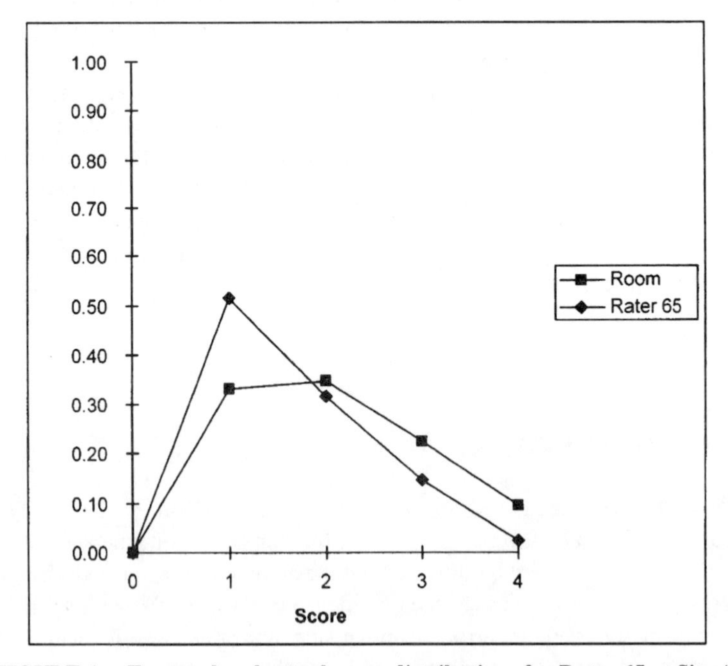

FIGURE 1. Expected and actual score distributions for Rater 65 at Site A.

the distribution of scores for Rater 65 in comparison to the distribution of all the raters combined. The severity of Rater 65 is reflected in the increased proportion of scores of 1, and the decrease in the proportions of scores 2, 3, and 4. The raters found this type of chart relatively easier to interpret—their interpretation was typically that Rater 65 had set the cut-off between a 1 and a 2 too high (which had also influenced the cut-off for the earlier scores). If this had been a feedback study, then we could have investigated the impact of such information. However, due to the exploratory nature of the study, we cannot report on systematic effects from such information (but, see Case Study 10 below).

Read-behind Information on Rater Performance

The read-behind approach provides the most direct method for checking rater performance, but it assumes (1) that the table leader is assigning scores correctly, and (2) that a 10% sample was adequate to detect inconsistencies or inadequacies. Although it would have been better to have *all* the responses scored twice, the figure of 10% was standard in CDE sessions because of time and cost constraints. We produced figures that contained the read-behind results for the table at which the rater was working, as well as for the individual rater. If a rater differed with the table leader more than 20% of the time, he or she was flagged. As an example, Figure 2 shows the performance of Rater 65 in comparison to Table 6 during peri-

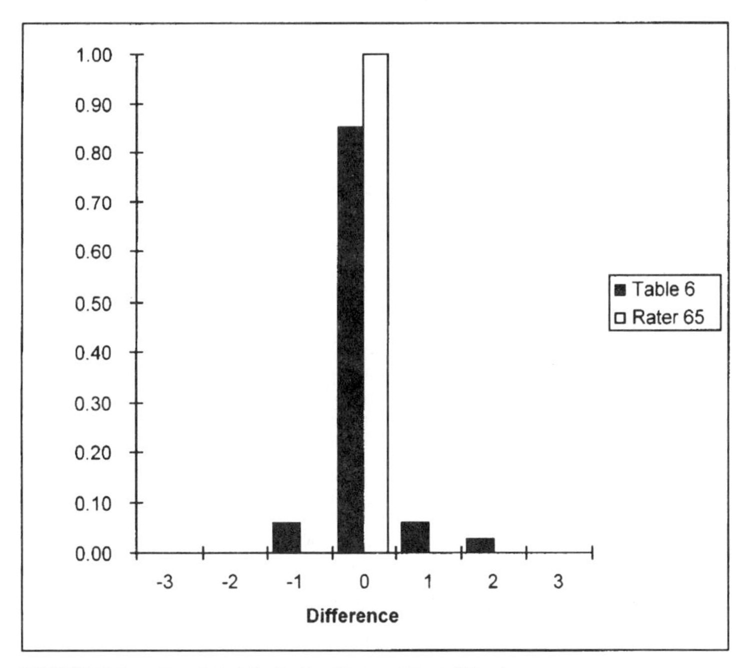

FIGURE 2. Read-behinds for Rater 65 at Site A.

od 1. During this period, Rater 65 matched the read-behind score of the table leader for *every* paper. Looking at Figure 2, we can see that during the same period, the table average for matches was about 85%. For approximately 4% of the read-behinds the raters at the table gave a score that was one point higher than the table leader (rater score – table leader score = 1). For approximately 4% of the read-behinds, the table leader gave a score that was one point higher than the raters (rater score – table leader score = –1). For approximately 2% of the read-behinds, the table leader gave a score that was one point lower than the raters (rater score – table leader score = 2). In this case, the results from the read-behinds for Rater 65 are quite contradictory to those from the IRT analysis. This we attribute to the light sampling employed in the read-behinds.

Case Studies of Rater Performance over Time

Severity estimates were calculated separately in the five periods for each of the raters. When examining the results of individual raters over time, a very wide range of scoring patterns can be observed. Some of these patterns will be described in the case studies listed below, which are illustrated in Figures 3 through 9. In these figures, at any time that a rater was flagged, the explanatory note is displayed below the x-axis. In addition, to the extent that we have documentation from table leader records that an intervention took place (for example, a discussion was held with the rater), a note is included above the flag notes. In these figures, the different severities are for the same rater over different periods.

The first two case studies are examples of raters who showed considerable drift in their severity.

Case Study 1

Rater 32 from Site A (Figure 3) was flagged in two consecutive periods but the IRT estimates were on opposite sides of the x-axis. This indicates that the rater had a major fluctuation in how he or she was scoring papers. In the first period, the rater was scoring student responses at about 1 score above what an average rater would give, for about 3 student responses out of 10. In the second period, it had become about 2 out of 10 *below*. It is interesting to note that the IRT estimates were very consistent from the second to the fifth periods although they were somewhat severe. In addition, the rater picked up a read-behind flag in the fifth period.

Case Study 2

Rater 54 from Site B (Figure 4) started out with an IRT flag in Period 1. The IRT estimate was below the x-axis. By Period 3, the rater was flagged again and this time the estimate was above the x-axis. This rater displays a steady drift upward in severity from Period 1 to Period 3 and then a steady drift downward in severity from Period 3 to Period 5. In the first period, the rater was scoring student responses at about 1 score below what an average rater would give on about 5 papers out of 10.

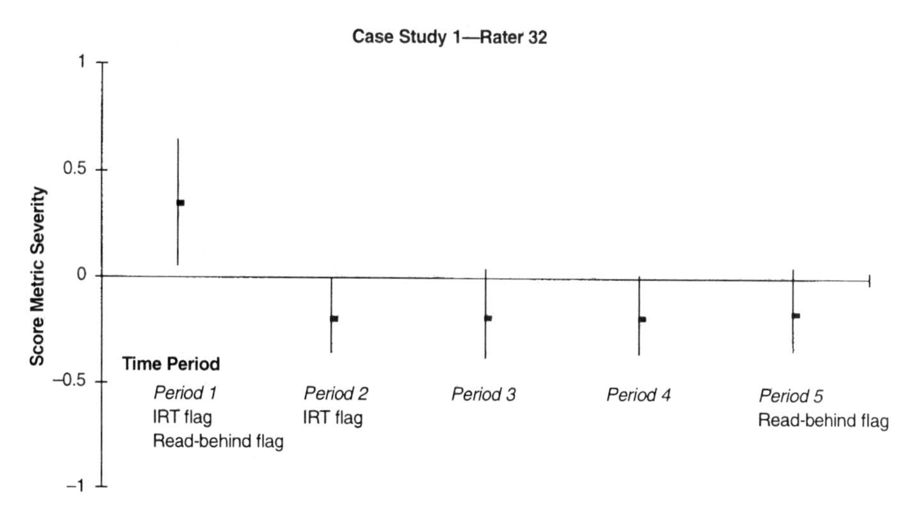

FIGURE 3. Case study 1.

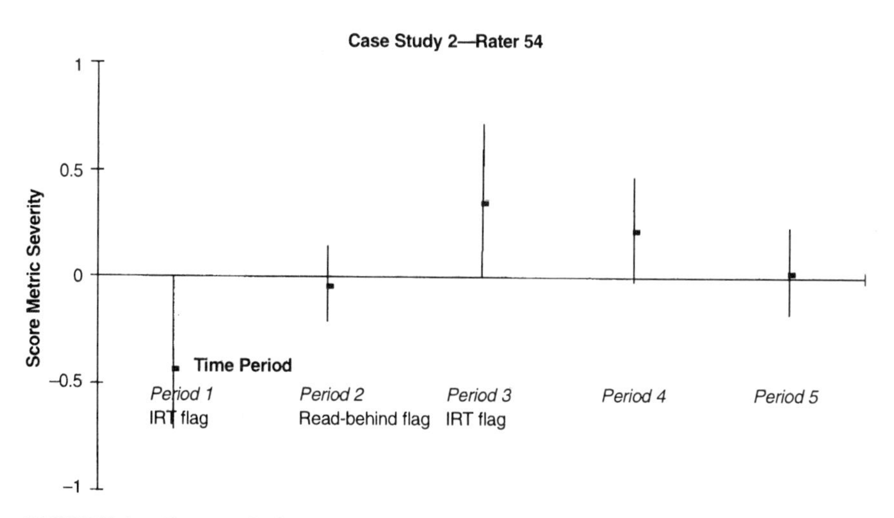

FIGURE 4. Case study 2.

In the second period, this had been reduced to less than 1 out of 10. In the third period, the rater switched to scoring papers at about 1 score above what an average rater would give on about 5 papers out of 10. Although there was no read-behind flag in either Period 1 or 3, there was a read-behind flag in Period 2. Again, it is an instance of a rater for whom the IRT flags and the read-behind flags are not in agreement.

In the next two cases, we show raters who showed very little drift in their IRT severity.

Case Study 3

Rater 14 from Site A (Figure 5) represents a rater whose severity does not drift above zero, although it is never significant in a statistical sense. The difference in scoring is, on average, about .1, which translates into a score of 1 more than an average rater would give on 1 out of 10 student responses. The rater receives three read-behind flags, although at no point was there an IRT flag. This is another indication that the two criteria report different types of information about rater performance, both of which may be important.

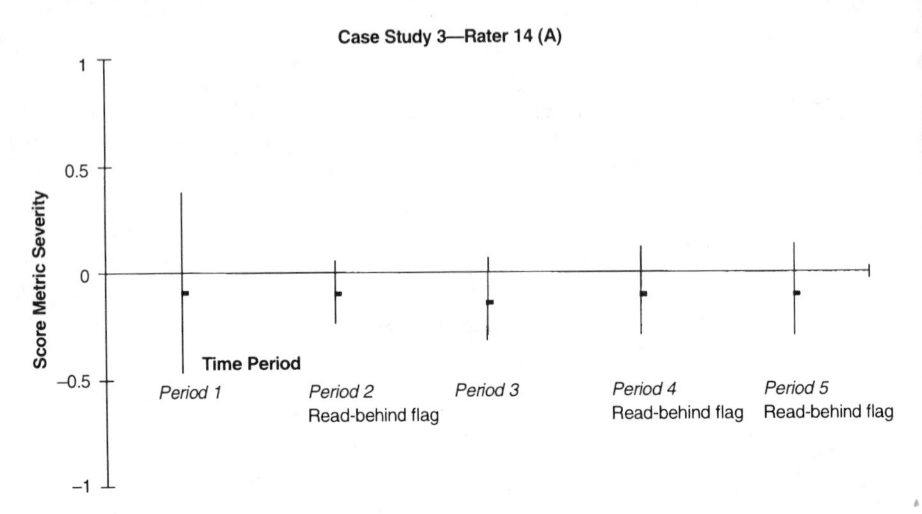

FIGURE 5. Case study 3.

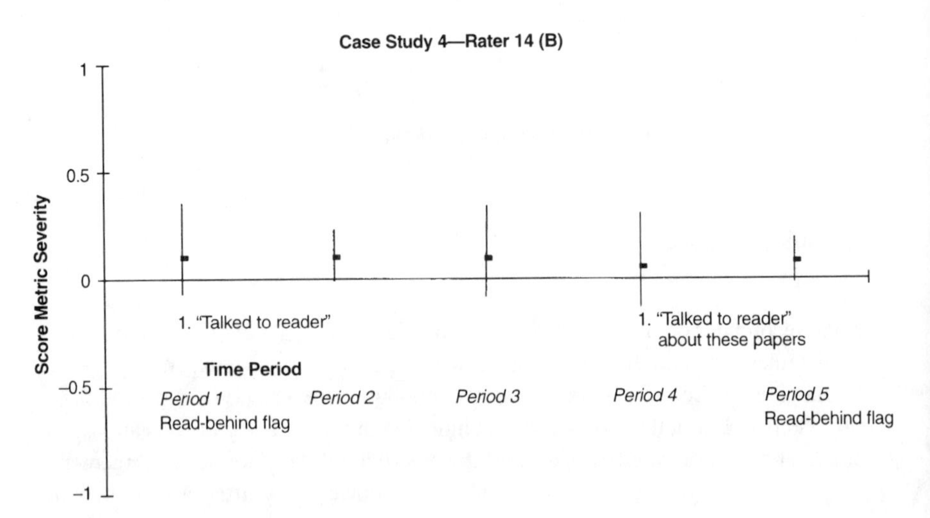

FIGURE 6. Case study 4.

Case Study 4

Rater 14 from Site B (Figure 6) is someone whose severity never drifts below zero although it never is statistically significant in its difference from zero. The difference in scoring is, on average, about .1, which translates into a score of 1 more than an average rater would give on 1 out of 10 student responses. This rater had multiple read-behind flags but at no time triggers an IRT flag. In addition, there were records of interventions but no indication that it had any effect on the severity of the rater's performance.

In the next case study, we show a rater for whom interventions seemed not to have had the desired effect on rater drift.

Case Study 5

Rater 65 from Site B (Figure 7) received a read-behind flag in Period 1 and an IRT flag in Period 3 but no intervention occurred until Period 4. In that period, the rater has both an IRT flag and a read-behind flag. However, after the intervention the rater's performance continues to drift down and the worst IRT estimate occurs in Period 5 along with a third read-behind flag. Across the five periods, the difference in scoring is, on average, about –.2, which translates into a score of 1 less than an average rater would give, on 2 out of 10 student responses.

In the next case study, we show a rater for whom at least some of the interventions seemed to have had a desirable effect on rater drift.

Case Study 6

Rater 52 from Site B (Figure 8) received interventions from the table leader in all of the first four periods but the IRT estimate only seems out of line in the first period according to the IRT drift. There may have been some other type of indi-

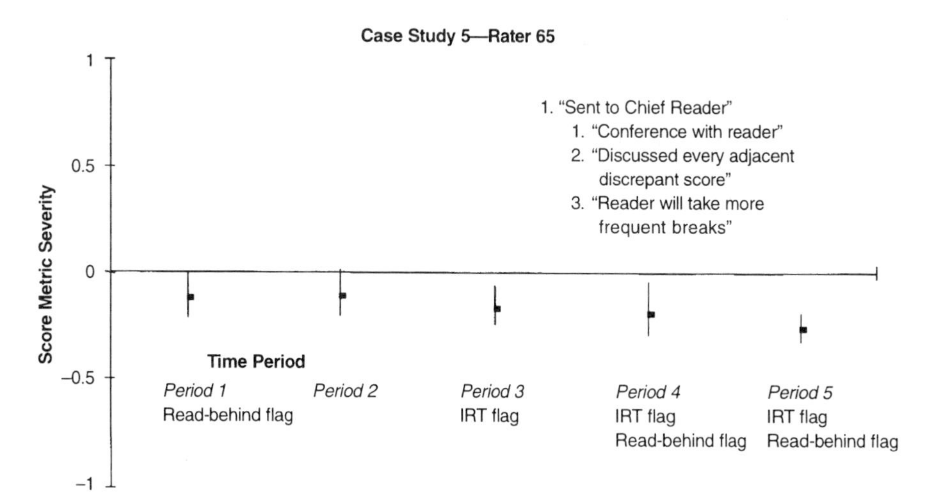

FIGURE 7. Case study 5.

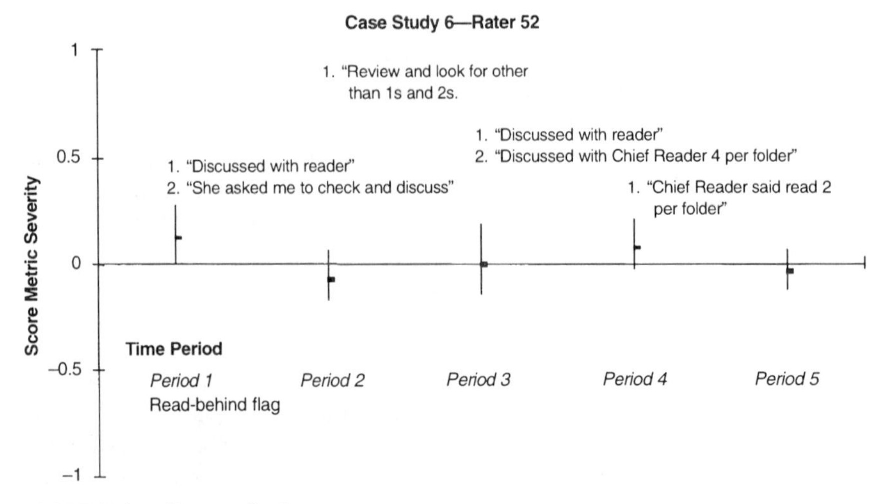

FIGURE 8. Case study 6.

cation that caused the table leader interventions, such as the rater talking too much at the table. In Periods 1 and 4, the difference in scoring is, on average, about .2 of a score, or 1 score more than an average rater would give for 2 out of 10 responses. In Periods 2 and 5 the difference in scoring is, on average, about –.1 of a score, or 1 score less than an average rater would give for 1 out of 10 student responses.

In the final case study, we show a rater for whom the intervention, which was based on the IRT information, seems to have succeeded.

Case Study 7

Rater 65 from Site A (Figure 9) received an IRT flag in Period 1 although the read-behind results showed all perfect matches (as is, mentioned earlier). During the conference after Period 1, it was decided that the chief reader herself would check the scoring on 20 of the student responses. The results of these special read-behinds were that out of 10 scores assigned a 1 by the rater, 50% of them should have been a 2. This information was conveyed to the rater, and also led to an increase in the read-behind rate by the table leader. The rater's severity drifted up, and for the next four periods, the rater's performance did not generate any more flags.

Comparison of IRT Information with Read-behind Information

The number of instances when the two criteria were in agreement for the regular raters (that is, not the table leaders) is detailed in Table 1. As might be inferred from the evidence provided by the case studies described previously, the two criteria are in agreement that there is an effect only rather rarely. But they are in

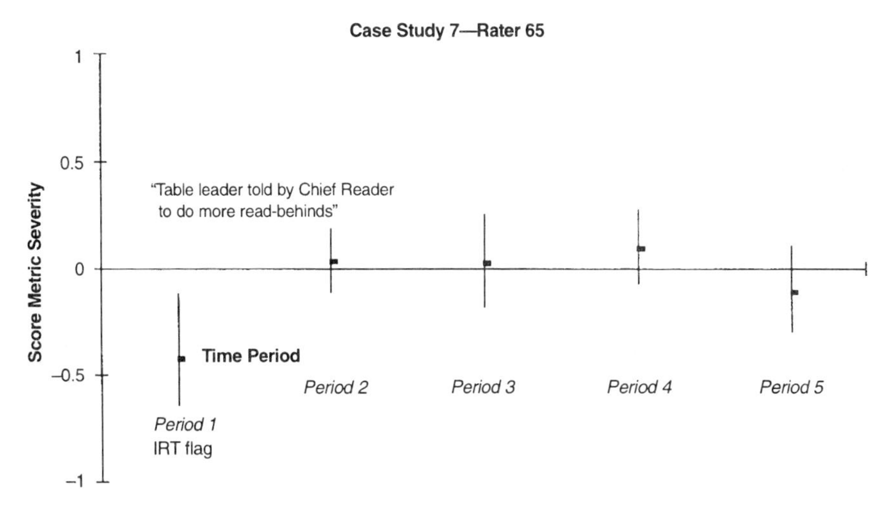

FIGURE 9. Case study 7.

agreement that there is no effect quite often. We cannot say which criterion is correct from this study, as there was no absolute criterion available to determine which flaggings were correct. What we can say is that the IRT information is definitely not redundant with respect to the traditional read-behind information.

Effect Size of Rater Severities

The preceding case studies and Table 1 give an impression that there is a discernible impact of rater severity. In order to obtain a measure of the overall impact of rater severity on the scores of students, we conducted a series of IRT analyses with the complete set of data from one of the sites. In these analyses, we calculated the effect sizes of (1) ignoring rater severities altogether, (2) considering rater severities as constant across periods, and (3) allowing rater severities to vary between periods. We did so by examining the total absolute residuals from each of these IRT analyses. The results are shown in Table 2: In the first column, we see

TABLE 1.
Number of Flaggings per site Per Criterion

	Agreement		Disagreement	
Site	No Flags	Both Flags	Only IRT Flag	Only R-B Flag
A	149	3	18	35
B	147	4	26	33

R-B, read-behind.

TABLE 2.
Effect Sizes of Rater Severities at Site B

Period	Ignore Rater	Constant Rater	Rater within Cycle	Number of Students
1	.10	.07	.03	4375
2	.12	.09	.04	3783
3	.14	.09	.05	3418
4	.09	.07	.03	4540
5	.12	.08	.04	4962

that typically, if we ignore rater severity altogether, that corresponds to an inaccuracy of between 9 and 14 score points in every 100 ratings. By adding in a constant rater effect, this is reduced to between 7 to 9 score points per 100. And allowing the rater severities to vary between periods reduces this to between 3 and 5 score points per 100 scorings. There is no absolute standard available here, but a reduction of one half to one third in the residuals is a clear indicator of the relative strength of the rater effects.

Influence of Table Leaders and Tables

At Site A, the table leaders received IRT flags 20% of all possible times while the regular raters were flagged at a rate of 10%. At Site B, the table leaders received IRT flags 20% of all possible times while the regular raters were flagged at a rate of 14%. That is, the table leaders, who had previously been identified as expert raters, were found to be less consistent over time than the regular raters. In interpreting this result, it is important to recall that the table leaders did not receive a random sample of the student work, as did the regular raters. Instead, their selection was (increasingly) determined by the performance of "poorer" raters, and often, their attention was drawn to specific score categories. Given this, it is not clear that their apparently less consistent behavior should be interpreted as indicating that they are not indeed good raters. However, given their status as leaders, it is reasonable to ask whether their rating severity or leniency was transmitted to the members of their tables.

An analysis was carried out to see if there was an effect from the raters being trained together and working together with a specific table leader. The analysis with the data from all periods combined was conducted with an additional parameter estimated for each table, but leaving out the data relating to table leaders. One possible explanation for a table effect would be that the raters absorbed the rating criteria of their table leader. If this was the case, then we would expect the correlation between the table leader severity (from the full data set) and the table effect (from this analysis) to be high. The results are given in Table 3. There are

TABLE 3.
Table Leader Severities and Table Effects

	Table Leader Severity		Table Effect	
Table	Site A	Site B	Site A	Site B
1	–0.455	–0.183	–0.098	–0.019
2	–0.258	0.021	0.088	–0.205
3	0.042	0.017	0.035	0.050
4	–0.263	0.158	0.039	–0.052
5	–0.244	–0.109	0.025	0.009
6	–0.375	0.162	0.016	0.157
7	–0.153	–0.166	–0.105	0.060

indeed statistically significant table leader effects, and statistically significant table effects, but these do not seem to be systematically related. At Site A, there was a weak and statistically nonsignificant correlation (r = 0.24, p = .60) between the table leader and the table, whereas at Site B there was a larger negative (but still nonsignificant) correlation (–0.51, p = .24). Thus, we cannot report a consistent pattern of relationship between table effects and table leader effects.

CONCLUSION

Procedures were implemented at two scoring sites and analyses were conducted on-site to determine IRT estimates of rater performance, and to examine patterns of rater drift in severity. It was found that we could obtain useful estimates from the ratings that accumulated over a half-day period. This information was communicated in ways designed to allow interpretation by raters and their table leaders. An array of case studies was displayed to show the patterns of drift in rater performance that was observed. From the different case studies we can see that (1) rater performance drifts significantly from period to period in both a statistical and substantive sense, and (2) the effect of interventions seems to vary from rater to rater. Examination of the case studies and tabulation of the results across the entire study show that the IRT and read-behind criteria are contributing different information about the raters.

Additional analyses were conducted to investigate two issues. First, the overall impact of the estimated rater severities was calculated using residuals from a series of IRT analyses. These showed that use of rater severities that varied between periods could reduce the amount of error considerably. Second, we observed that table leaders exhibited surprisingly large amounts of rater severity and leniency, but that these did not seem to be associated with tendencies by the raters they supervised to drift in a consistent way.

The analyses that we carried out at the sites, and those we conducted later, along with our reflections on these results lead us to a number of conclusions.

- The use of IRT models for on-site feedback is both feasible and potentially valuable.
- The use of training to eliminate rater severity effects is leaving many raters still with large severities.
- The use of expert re-ratings of samples of student work ("read-behinds") does not necessarily capture all the significant rater variation in severity.
- Raters' severities may drift significantly between scoring sessions.
- Adjusting student scores for rater severities can be used to reduce the effects of rater severity variation.
- The degree and direction of table leader severity does not necessarily determine the way that the severities of raters on that table will drift.

Moreover, our experiences at the sites also led us to a number of other conclusions that, although not directly supported by our data, are, we feel are important observations. First, the gap between ratings and feedback needs to be minimized in order to maximize the usefulness of the feedback. In the case of the CDE context, having direct entry of rater's ratings would make the feedback process both simpler and more effective. Second, having a simple mechanism to allow the identification and re-presentation of student scripts would also make the feedback much more useful. This is simply too difficult to organize in a paper-driven system—the CDE would need to move to a scanned image system to accomplish this. Third, although we have concentrated on rater severity effects in this study, we also noted interesting cases that looked more like central tendency effects. It would be a logical next step to investigate the usefulness of IRT models of central tendency.

REFERENCES

Adams, R. J., & Wilson, M. (1996). Formulating the Rasch model as a mixed coefficients multinomial logit In G. Engelhard & M. Wilson (Eds.), *Objective measurement: Theory into practice* (Vol. 3, pp. 143–166). Norwood, NJ: Ablex.

Braun, H. I. (1988). Understanding scoring reliability: Experiments in calibrating essay readers. *Journal of Educational Statistics, 13*, 1–18.

Engelhard, G., Jr. (1994) Examining rater errors in the assessment of written composition with a many-faceted Rasch Model. *Journal of Educational Measurement, 31*(2), 93–112.

Engelhard, G., Jr. (1996, April). *Models of judgment and Rasch measurement theory.* Paper presented at the American Educational Research Association, New York.

Guilford, J. P. (1954). *Psychometric methods.* New York: McGraw-Hill.

Hogan, T. P. (1981). *Relationship between free-response and choice-type tests of achievement: A review of the literature.* Green Bay, WI: University of Wisconsin. (Eric Document Reproduction Service No. ED 224 81).

Khattri, N., & Sweet, D. (1996). Assessment reform: Promises and challenges. In M. Kane & R. Mitchell (Eds.), *Implementing performance assessment: Promises, problems and challenges* (pp. 1–22). Mahwah, NJ: Erlbaum.

Lunz, M. E., Wright, B. D., & Linacre, J. M. (1990). Measuring the impact of judge severity on examination scores. *Applied Measurement in Education, 3*(4), 331–345.

Mislevy, R. J. (1991). A framework for studying differences between multiple-choice and free-response test items. In R. E. Bennet & W. C. Ward (Eds.), *Construction vs. choice in cognitive measurement* (pp. 75–106). Hillsdale, NJ: Erlbaum

Rasch, G. (1960). *Probabilistic models for some intelligence and attainment tests.* Copenhagen: Denmark's Paedagogistic Institut. (Reprinted 1980, University of Chicago Press.)

Raymond, M. R., & Viswesvaran, C. (1993). Least-squares models to correct for rater effects in performance assessment. *Journal of Education Measurement, 30*, 253–268.

Resnick, D., & Resnick, L. (1996). Performance assessment and the multiple functions of educational measurement. In M. Kane & R. Mitchell (Eds.), *Implementing performance assessment: Promises, problems and challenges* (pp. 23–38). Mahwah, NJ: Erlbaum.

Saal, F. E., Downey, R. G., & Lahey, M. A. (1980). Rating the ratings: Assessing the psychometric quality of rating data. *Psychological Bulletin, 88*(2), 413–428.

Wang, W., & Wilson, M. (1996). Comparing open-ended items and performance-based items using item response modeling. In G. Engelhard & M. Wilson (Eds.), *Objective measurement: Theory into practice* (Vol. 3, pp. 167–194). Norwood, NJ: Ablex.

Wilson, M., & Case, H. (1996, June). *An investigation of the feasibility and potential effects of rater feedback on rater errors.* Paper presented at the Council of Chief State School Officers National Conference, Phoenix, AZ.

Wilson, M., & Wang, W. (1995). Complex composites: Issues that arise in combining different modes of assessment. *Applied Psychological Measurement, 19*(1), 51–72.

Wu, M., Adams, R. J., & Wilson, M. (1998). ConQuest [Computer software]. Melbourne Australia: ACER Press.

7

A METHOD TO STUDY RATER SEVERITY ACROSS SEVERAL ADMINISTRATIONS

Thomas R. O'Neill
Mary E. Lunz
American Society of Clinical Pathologists

INTRODUCTION

Performance assessments are often thought to have greater validity than multiple-choice tests because the actual performance of the task is rated. However, the reproducibility of examination results derived from performance assessments is sometimes questioned because the performances must be graded by raters who often have different individual standards of excellence. Therefore, any given rating will be influenced not only by the examinee's ability and the item's difficulty, but also by a third facet, rater severity. In order for examination results to be meaningful, differences in raters must be accounted for, so that all results are expressed from the same frame of reference. The extension of the Rasch (1960/1980) model to the many facet Rasch model (MFRM; Linacre, 1989) has made accounting for rater severity possible, by placing rater severity in the same frame of reference as item difficulty and examinee ability. The MFRM estimates each rater's severity, each project's difficulty, and/or other such facets, and removes their influence before

A previous version of this manuscript was presented at the annual meeting of the American Educational Research Association in Chicago, Illinois, in March 1997.

computing an examinee's ability. In this way, similar examination results are expected for the same examinee even when different raters and projects are used.

Recent MFRM research supports that raters are able to maintain a consistent degree of severity even when rating examinees of very different ability levels (Lunz, Stahl, & Wright, 1996; O'Neill & Lunz, 1996). When the goal is to carry forward the same scale, a linking strategy must be employed so that the severity of new raters is expressed in the same frame of reference as that of the original raters. Using common raters to link together two test administrations requires that the raters maintain their same degree of severity in the second administration as in the first. For this reason, studies regarding the stability of rater severity across administrations are important. To this end, Lunz and colleagues (1996) compared the severity of 11 raters across two test administrations that were 6 months apart. They found that, generally, raters do not change their severity across administrations.

This chapter describes a retrospective method that can be applied to data spanning many administrations for the purpose of assessing rater stability over time. The retrospective multiadministration method is demonstrated on data from a histotechnology performance assessment that spans a 10-year period. How to prospectively use the retrospective information is also discussed.

Administration-to-Administration Equating

As part of the equating process, rater stability is verified from administration to administration. This is done by comparing the severity of several common "anchor" raters on the current administration with their degree of severity from the prior administration and then checking that their severity on the current administration places them in the same relative position as in the past. If their relative positions hold, it is reasonable to conclude that their severity has not changed. In cases where only one or two of the anchor raters have changed positions, it is reasonable to conclude that those one or two raters have changed their degree of severity and should be treated as new raters, but the rest of the anchor raters can be used to link the new raters to the established scale. Yet, if several raters change places and the number of anchor raters is few, it becomes more complicated to determine which of the anchor raters changed their severity and which remained the same. To prevent this from happening, psychometricians try to employ as many *stable* precalibrated raters as possible, so that any anomalous raters will stand out more clearly.

Multiadministration Analysis

The retrospective multiadministration method begins by taking advantage of the larger number of ratings available by pooling the ratings from all administrations and computing ability measures for all examinees and computing calibrations for all of the projects (and other facets that represent agents of measurement). Because the number of ratings per project is greater for the pooled analysis than for any individual analysis, the pooled analysis produces the most precise calibrations for projects. Because examinees do not overlap administrations (examinees that overlap failed the first time and hopefully improved before retaking the test), the pooled analysis and individual administration analyses provided the same number of observations for each examinee. While the number of observations are the same, the examinee ability estimates from the pooled analysis will contain less error because they are based on the more precise project calibrations.

Next, each administration is reanalyzed individually, but with each examinee's ability and each project's difficulty set to its value from the pooled analysis. In this way, rater severity becomes the object of measurement and the test's other facets (examinee ability and project difficulty) are held constant. Next, the individual rater's severity estimates are collected for each administration, and then plotted (± 2 *SE*s) to illustrate the intrarater cross-administration changes.

METHODS

The Rasch Model

The Rasch (1960/1980) model is a logistic latent trait model of probabilities which analyzes items and persons independently, and then expresses both the item difficulties and the person abilities on a single continuum. The MFRM extends the Rasch model to account for other differences in context, such as particular items, projects, raters, tasks, session, and so on, so that the results generalize beyond the specific occasion in which the data were collected. In this way, the actual examinee ability level is expressed so that the particular items or raters are of no importance.

In this study, the focus is the degree of rater severity across administrations. Because MFRM accounts for differences in the particular examinees or projects rated, it is possible to focus on changes in rater severity across administrations.

Examinees

The examinees were candidates for histotechnician certification. Each of the 4,683 examinees submitted a work-sample for evaluation. The participants in this study represent the examinees from 17 different administrations that span 10 years. The number of examinees per administration ranged from 168 to 385, with 275 examinees being the average. Failing examinees were permitted to submit another examination, but were treated as independent cases because (it was to be hoped) the examinees had taken steps to improve their ability before retesting.

Raters

The raters were experts in their field. While only 11 to 20 raters were needed to grade any single administration, a total of 57 different raters were used over the 10 years. Most raters graded in more than one administration and, on average, raters graded in 6 administrations. Prior to each grading session, the raters attended a 3-hour orientation session to refamiliarize them with the scoring criteria and the particular projects under consideration. The performance of the nine raters who graded in 10 or more administrations is reported in detail in this chapter. Other raters showed comparable patterns, but graded in only one to nine administrations.

Instruments

The examination requires examinees to submit 15 projects (histology slides) made according to prespecified requirements for type of tissue and stain. The projects varied across administrations, but there was sufficient overlap to equate the different versions of the exam. Each project is rated on three different tasks; thus, each exam is evaluated on the basis of 45 ratings. The three tasks and the rating scales for the three tasks remained the same during the period of this study. Ratings on the examination were modeled as being governed by four facets: (1) examinee ability, (2) project difficulty, (3) task difficulty, and (4) rater severity, as follows:

$$\log \left(\frac{P_{nitjx}}{P_{nitjx-1}} \right) = B_n - D_i - C_t - S_j - F_x , \quad (1)$$

This equation specifies the relationship between the probability of a response and those facets where:

P_{nitjx} is the probability of the examinee receiving score x by rater j on project i on task t,
$P_{nitjx-1}$ is the probability of examinee receiving score $x - 1$ by rater j on project i on task t,
Bn is the ability of examinee n (Facet 1),
Di is the difficulty of project i (Facet 2),
Ct is the difficulty of task t (Facet 3),
Sj is the severity of rater j (Facet 4), and
Fx is the difficulty of achieving score x relative to score $x - 1$ on the rating scale.

Procedures

FACETS (Linacre, 1994), an MFRM computer program, was used to calibrate candidates, raters, projects, and tasks. The initial pooled data analysis included all

17 administrations that established a benchmark scale. Because the number of observations per task and per project was greater for the pooled analysis than for any individual administration, the pooled analysis produced the most precise calibrations for projects and tasks. Although the pooled analysis and individual administration analyses provided the same number of observations per examinee, the pooled analysis examinee ability estimates contain less error because the project and task difficulty calibrations have been estimated more precisely.

Next, the 17 administrations were reanalyzed individually, but each examinee's ability, each project's difficulty, and each task's difficulty was set to its value from the pooled analysis. In this way, only rater severity calibrations were permitted to vary because the test's other facets (examinee ability, project difficulty, and task difficulty) were held constant. Thus, the examinees, projects, and tasks become agents used to assess the raters, so the raters are defined as the object of measurement. The rater severity calibrations from each administration were collected and summarized.

Analysis

First, the separation reliability for each facet from the pooled analysis was computed. Next, descriptive statistics for each exam administration were computed from the individual analyses. In order to get a clear picture of rater severity over time, only the raters who graded in 10 or more administrations had their severity (± 2 *SE*s) plotted across administrations. The raters who graded in fewer than 10 administrations ($N = 47$) were calibrated, but their severities were not plotted for this study.

RESULTS

Pooled Analysis

The pooled analysis demonstrated that the test adequately discriminated among examinees (separation reliability = .81). The slide-projects were significantly different in difficulty (separation reliability = .99), and the severity estimates of raters were significantly different from each other (separation reliability = .97). The tasks were also very different in difficulty from each other (separation reliability > .995). The errors of measurement were very small owing to the large number of observations used in the pooled analysis.

Fit of Pooled Data to the Model

The overall fit of the pooled data to the model was examined using infit and outfit mean square statistics. Mean square fit statistics can range in value from zero to infinity, but the expected value is 1.0. For the purpose of identifying misfit in the data, adequately fitting data were defined as having fit statistics ranging from .6 to 1.5. This range seemed to make the most sense given the nature of the examination. Although some examinees, projects, and raters did not fit the model as well as expected, all were kept in the analysis because subsequent analyses would be based upon subsets of the pooled data.

Most of the 4,683 examinees fit the model reasonably well. Excluding two examinees with maximum scores, the average examinee mean square was 1.03 (*SD* 0.28) for infit and 1.00 (*SD* 0.34) for outfit. However, some examinees manifested more stochasticity than the model predicted. Of the examinees, 5.5% exceeded the 1.5 limit on infit, as did 6.1% on outfit. Conversely, some examinees showed less stochasticity than the model predicted. Of the examinees, 4.2% were below the 0.6 limit on infit, as were 4.1% on outfit. Given the large number of examinees, it is not surprising that a small percentage of the examinees would not fit the model well. The average project Infit was 1.08 (SD 0.19) and average project outfit was 1.01 (*SD* 0.22). Three projects did not fit as expected: Two had an outfit of 1.6 and 1.7, and the third had an infit of 1.6. The task infit and task outfit statistics ranged from 0.9 to 1.1, indicating that the three tasks fit the model well. The average rater infit was 1.04 (*SD* 0.11) and average rater outfit was 1.00 (*SD* 0.13). The rater fit statistics ranged from 0.7 to 1.3 indicating that all 57 raters adequately fit the model.

Individual Administration Analysis

Table 1 shows the mean project difficulties, rater severities, and examinee ability estimates for the pooled analysis and each administration. Although there was some variability across administrations, the mean project difficulty and rater severity remained reasonably comparable, indicating that, overall, the administrations were of similar difficulty. Because the same tasks were used across all administrations, the mean task difficulty was identical. Examinee ability estimates also showed some variability across administrations, but overall the examinee pool was comparable.

Of the 57 raters, nine graded in at least 10 administrations during this 10-year period. The severity of these nine raters is listed in Table 2 and was plotted (± 2 *SE*s) across administrations (Figures 1 through 3). Upon inspection of these plots, three things can be seen. First, it is clear that different raters have different levels of severity. Second, raters are usually able to maintain a self-consistent level of severity across administrations. Third, some raters are more consistent than others over time.

A comparison of severity estimates for Rater 1 and Rater 2 (Figure 1) illustrates that different raters maintain different personal levels of severity. Rater 2 has an overall severity of –1.90 (*SE* = ±0.02) and Rater 1 has a severity of –1.17 (*SE* = ±0.02). The side-by-side comparison of the their plotted severities makes this dif-

TABLE 1.
Individual Administration Summary Statistics for Raters, Projects, and Examinees

	Rater Severity			Project Difficulty			Examinee Ability		
Administration	Mean	*SD*	*N*	Mean	*SD*	*N*	Mean	*SD*	*N*
1	−1.41	(.20)	13	−.07	(.37)	15	.27	(.72)	226
2		—			—			—	
3	−1.58	(.23)	14	−.11	(.37)	15	−.03	(.68)	210
4	−1.51	(.26)	18	−.12	(.32)	15	−.17	(.67)	300
5	−1.65	(.43)	18	−.07	(.37)	15	−.09	(.72)	217
6	−1.45	(.25)	17	.01	(.31)	15	.06	(.64)	324
7	−1.51	(.18)	15	.02	(.40)	15	−.03	(.59)	260
8	−1.53	(.20)	19	−.04	(.33)	15	.00	(.73)	321
9		—			—			—	
10	−1.60	(.21)	18	−.04	(.33)	15	.04	(.66)	385
11	−1.52	(.29)	16	−.10	(.34)	15	.04	(.69)	271
12	−1.60	(.39)	18	−.10	(.30)	15	−.03	(.65)	381
13	−1.39	(.40)	12	−.02	(.36)	15	−.02	(.57)	265
14	−1.57	(.29)	18	−.07	(.34)	15	.09	(.71)	364
15	−1.62	(.27)	19	.02	(.38)	15	−.12	(.62)	192
16	−1.54	(.29)	18	.02	(.38)	15	−.10	(.57)	339
17	−1.57	(.38)	11	−.04	(.36)	15	−.08	(.59)	168
18	−1.61	(.33)	16	−.04	(.36)	15	.10	(.67)	281
19	−1.58	(.32)	12	−.07	(.31)	15	.10	(.67)	179
Pooled	−1.55	(.35)	57	0.00	(.35)	53	0.00	(.66)	4,683

Presented in logits.

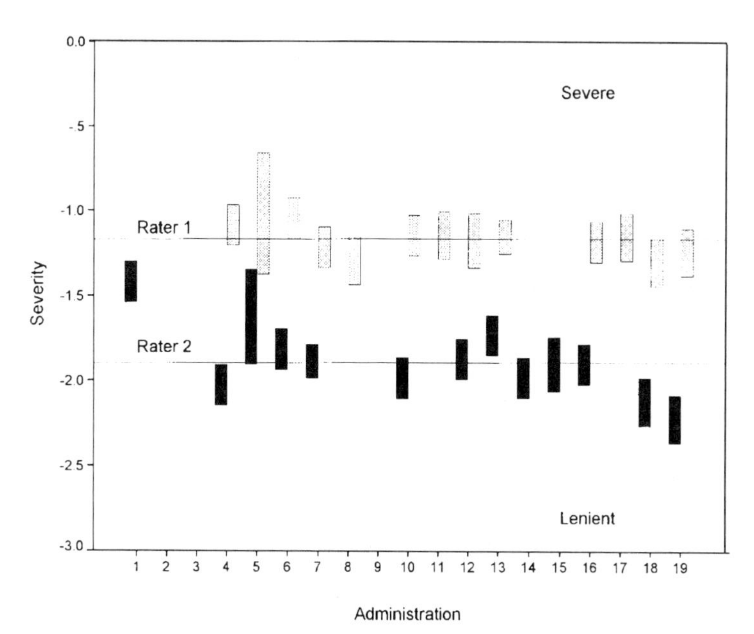

FIGURE 1. Raters maintain different degrees of severity.

TABLE 2.
Raters' Severity Across Administrations (For Nine Selected Raters)

Rater	Administration																			
	1	2	3	4	5	6	7	8	9	10	11	12	13	14	15	16	17	18	19	Pooled
1	—	—	—	−1.09	−1.02	−1.05	−1.22	−1.30	—	−1.20	−1.15	−1.18	−1.20	—	—	−1.19	−1.16	−1.31	−1.25	−1.17
2	−1.42	—	—	−2.03	−1.63	−1.82	—	−1.99	—	−2.00	—	−1.88	−1.70	−1.99	−1.91	−1.91	—	−2.13	−2.33	−1.90
3	—	—	−1.70	−1.56	−1.96	−1.89	—	−1.58	—	—	−1.75	−2.04	—	−1.90	−1.55	−1.83	−2.23	−1.90	−1.55	−1.80
4	−1.01	—	−1.70	−1.53	−1.91	−1.60	−1.18	−1.50	—	−1.70	−1.54	−1.71	—	−1.57	−1.61	−1.68	—	−1.58	−1.73	−1.59
5	−1.07	—	−1.60	−1.53	−1.84	−1.39	−1.60	−1.69	—	−1.70	—	—	−1.60	—	—	—	−1.69	—	−1.56	−1.58
6	—	—	—	—	−2.05	−1.56	−1.59	−1.30	—	−1.70	−1.39	—	−1.70	−1.57	−1.37	−1.58	−1.61	−1.54	—	−1.57
7	−1.45	—	−1.70	−1.71	−1.70	−1.48	−1.55	−1.99	—	—	−1.76	−1.65	—	−1.75	−1.96	−1.87	—	−1.92	—	−1.72
8	−1.48	—	—	−1.39	−1.40	−1.04	−1.54	−1.42	—	−1.50	−1.20	—	−1.40	−1.31	−1.47	−1.63	−1.50	−1.50	—	−1.41
9	—	—	—	−1.30	−0.95	−1.09	−1.34	−1.40	—	−1.30	—	−1.29	—	−1.58	−1.51	−1.51	—	—	—	−1.36

[a] The unit of measurement is logits (log-odds units); an "—" indicates that the rater did not grade during that session.

ference obvious. The raw score impact of this difference in rater severities can be substantial. For example, an examinee who receives 57 out of a possible 75 points from Rater 2 would likely receive only 47 points from Rater 1.

Usually raters can maintain a self-consistent level of severity across administrations. To illustrate, these nine raters, as a group were consistent 67% of the time; that is, a rater's severity was within two standard errors of the rater's overall degree of severity. A higher percentage can be expected if frequently inconsistent raters are screened out in advance.

Another finding is that some raters are more consistent than others. This is illustrated by comparing Rater 1 to Rater 3 (Figure 2). Rater 1 exhibits a very consistent degree of severity across a span of 8 years whereas Rater 3 exhibits a noticeably less consistent degree of severity across a similar span of time. While severity for Rater 3 showed some variation across administrations, the degree of severity *within* administrations was relatively uniform (infit $MS = 1.1$, outfit $MS = 1.1$). Another, less striking, detail of the data is that for five of the nine raters, the first severity estimate is not in line with the other severity estimates.[1] This is probably the result of a new rater learning how to apply the rating scale. It seems that most raters settle on a uniform degree of severity that they can apply consistently after one or two administrations.

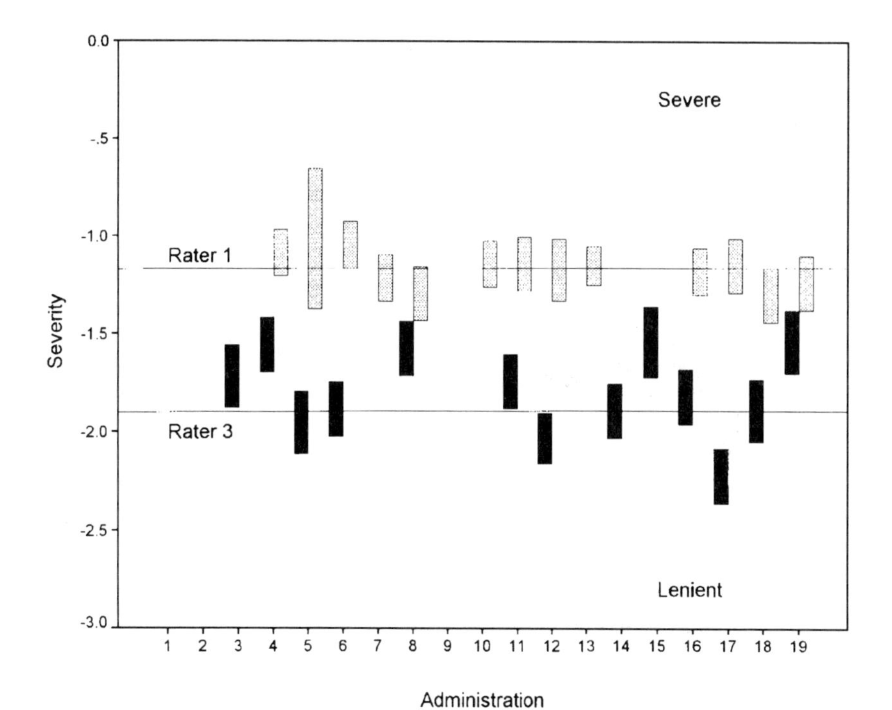

FIGURE 2. Some raters are more consistent than others.

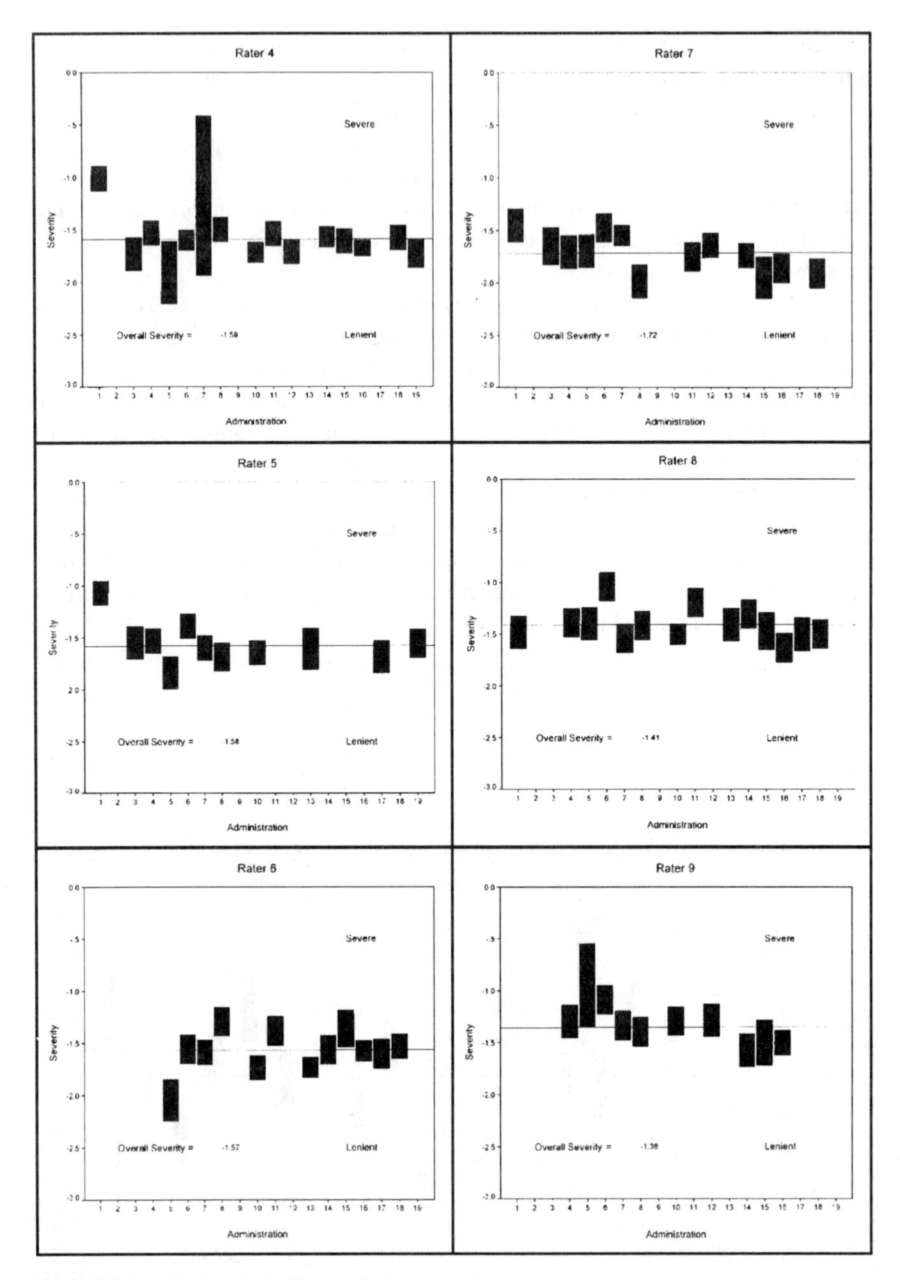

FIGURE 3. Comparison chart of rater severity.

DISCUSSION AND CONCLUSIONS

Generally, raters have their own unique internal standard which they apply fairly consistently. The results of this study confirm that raters' perceptions of excellence are not interchangeable, but are usually self-consistent. However, some raters maintain a standard more consistently than others, and even very consistent raters can vary occasionally. It can never be known in advance exactly how severe a particular rater will be on any given occasion. Yet, a rater's past performance often suggests how he or she will rate in the future. This information can be helpful to psychometricians who are organizing or equating performance assessments across administrations.

The method for analyzing rater severity proposed in this chapter is not a replacement for an ongoing equating procedure, but it can aid developers of more established exams, those with historical data, in making decisions about raters. For example, a psychometrician may select a few raters to participate in several consecutive administrations for the purpose of maintaining the same frame of reference for rater severity. Common raters should be selected on the basis of their documented ability to maintain a uniform level of severity. Armed with historical information, psychometricians can seek out stable raters such as Rater 1 (Figure 2) for this purpose. Others, such as Rater 3 (Figure 2) can still be used across administrations because their degree of severity is consistent within administrations. However, knowing that their across-administration degree of severity has more variance, the psychometrician would not want to use them as a link back to the initial scale. They should be thought of as new raters each time they grade.

Additionally, viewing rater severity in this manner can generate hypotheses regarding how individual raters behave over time. For example, Figure 3 suggests that Rater 7 is becoming slightly more lenient with experience. A similar tendency could be suggested for Rater 9 based on the last three administrations, but the data is less persuasive. If the psychometrician thinks that there has been a shift in severity and that the new level severity is likely to be stable, he or she may want to consider updating the rater calibration bank with the new severity calibration. Another hypothesis suggested by this data set is that Rater 5 initially had occasional problems maintaining a uniform degree of severity, but had internalized a standard by administration 7 and thereafter was very stable. Perhaps, as a means of anticipating the future performance of Rater 5, administrations 1 through 6 should not be considered.

The primary importance of this chapter is the methodology used to investigate rater severity. In practice, one would only use this methodology to analyze several administrations because it is rather labor intensive. An advantage of this method is that the plotted calibrations with their error bands (±2 *SE*s) provide a useful description of rater behavior over time. This picture permits the psychometrician to verify that things are going well or to identify problem areas.

The most obvious information noticeable from these charts is which raters are consistent and which are erratic across administrations. As stated earlier, this information can be used to select anchor raters, but it can also be used after the data have been collected. Suppose that out of 14 raters, only 4 raters had a known degree of severity. Further, suppose that two of these four anchor raters were more lenient by approximately the same amount on the current administration than in earlier administrations. How would the psychometrician know if the two raters who really became more lenient were more lenient? It would seem equally plausible that the two raters that really remained the same had become more severe. A potential answer is to review the historical performance of the four raters. It seems *probable* that the historically more stable raters would be less likely to be the ones who changed.

To prevent the preceding scenario, enough common raters should be employed so that if a small percentage of raters change in severity, it will be easy to identify which raters changed. Reviewing the historical data can allow the psychometrician to make a good guess, that given the available pool of anchor raters (with known severity and cross-administration stability), (1) which raters should be selected, (2) how many of the raters are expected to change severity during this administration, and (3) how many raters will be needed to clearly identify those that have changed severity.

These findings are also important because they address a primary concern about the reliability of performance examinations. To achieve any reproducibility in pass/fail or placement decisions, differences among raters must be accounted for both within and across administrations. These data confirm again that raters do have their own unique perceptions of excellence and that perception sometimes changes over time.

NOTE

1. Rater 2 in Figure 2; Rater 3 in Figure 3; Raters 4, 5, 6, and 7 in Figure 4.

REFERENCES

Linacre, J. M. (1994). *FACETS, a computer program for analysis of examinations with multiple facets.* Chicago: Mesa Press.

Linacre, J. M. (1989). *Many-facet Rasch measurement.* Chicago: MESA Press.

Lunz, M. E., Stahl, J. A., & Wright, B. D. (1996). The invariance of rater severity calibrations. In G. Engelhard & M. Wilson (Eds.), *Objective measurement: Theory into practice* (Vol. 3, pp. 99–112). Norwood, NJ: Ablex.

O'Neill, T. R., & Lunz, M. E. (1996). *Examining the invariance of rater and project calibrations using a multi–facet Rasch model.* Paper presented at the annual meeting of the American Educational Research Association, New York.

Rasch, G. (1980). *Probabilistic models for some intelligence and attainment tests.* Chicago: MESA Press. (Original work published 1960)

8

DETECTING RATER EFFECTS IN SIMULATED DATA WITH A MULTIFACETED RASCH RATING SCALE MODEL

Edward W. Wolfe
Michigan State University

Chris W. T. Chiu
University of Pittsburgh

Carol M. Myford
Educational Testing Service

INTRODUCTION

As performance-based and portfolio assessments become more popular as a means of linking large-scale educational testing to classroom instruction, more attention has been directed toward understanding how the use of raters influences the validity and reliability of test scores. Raters may introduce error into examinee scores

Portions of this research were supported by a postdoctoral fellowship at Educational Testing Service. A previous version of this manuscript was presented at the Annual Meeting of the National Council on Measurement in Education, March 1997, in Chicago, Illinois.

in a variety of ways—unfamiliarity with or inadequate training in the use of the rating scale, fatigue or lapses in attention, deficiencies in some areas of content knowledge that are relevant to making scoring decisions, or personal beliefs that conflict with the values espoused by the scoring rubric. In any case, when raters exhibit problematic rating behaviors, it may be possible to identify unique patterns in the data that correspond to specific types of rater errors. For example, raters who make errors because of fatigue are likely to make more random errors as time progresses. On the other hand, raters who are unable to differentiate between the various categories in a rating scale are likely to assign a disproportionate number of ratings in the middle of the rating scale.

The purpose of this chapter is to illustrate how common patterns of rater errors may be detected in a large-scale performance assessment setting. In the following sections, we identify several rater effects and describe a scaling method that can be used to detect these effects in operational data sets. We also present the results of analyses of several simulated data sets that are generated to exhibit each of these rater effects. In addition, we provide an example of how these procedures can be applied to operational data.

RATER EFFECTS

Previous research concerning rater effects has identified several ways that raters may introduce error into examinee scores (Saal, Downey, & Lahey, 1980, provide a good review of much of this work). Our work focuses on three continua that depict some of the more commonly cited of these effects. In this section, we describe how rater *accuracy/indiscrimination*, *leniency/harshness*, and *centrality/extremism* manifest themselves in the ratings assigned by raters. We also describe how a multifaceted Rasch rating scale model can be used to analyze and detect various aberrant patterns in rating data.

Accuracy/Indiscrimination

A common concern of those who interpret ratings is the extent to which raters make seemingly random errors. We represent this concern with an *accuracy/indiscrimination* continuum. At one end of this continuum, *accuracy*, the ratings assigned by a rater contain no error. Such ratings are accurate representations of the examinee's proficiency. On the other end of the continuum, a rater's ratings are not representative of the examinee's proficiency at all. In fact, the ratings may seem random in nature. This situation might occur if a rater had little or no understanding of a scoring rubric and used it indiscriminately, the rater uses a set of criteria that are completely different than those contained in the scoring rubric, or the rater does not possess the content knowledge necessary to make the fine discriminations depicted by the scoring rubric. When raters differ in their position on the accuracy/indiscrimination continuum, we cannot be sure how representative any individual rating is of the examinee's proficiency.

The accuracy/indiscrimination continuum can be depicted as the difference between a particular rater's estimate of an examinee's proficiency (θ_r) and that examinee's actual proficiency (θ). If a rater is completely accurate, the difference between the rater's rating and the examinee's actual proficiency ($\theta_r - \theta$) would be zero. On the other hand, we would expect $|\theta_r - \theta|$ to be quite large (sometimes positive and sometimes negative) across the range of examinee proficiency for a rater who uses the scoring rubric indiscriminately. Most raters, however, will fall somewhere between these two extremes, and a "reasonable" amount of random variation will be established for evaluating the performance of individual raters.

Leniency/Harshness

We represent the most commonly investigated rater effects on a continuum labeled *leniency/harshness*. This continuum describes the extent to which a rater assigns systematically higher or lower ratings than do other raters. In operational settings, this effect is likely to arise when a rater uses standards to evaluate examinee performance that differ substantially from the standards used by other raters. If a rater exhibits harshness, then the ratings assigned by that rater will tend to underestimate the examinee's proficiency across the proficiency continuum (that is, $\theta_r - \theta$ is negative). On the other hand, if a rater exhibits leniency, then the ratings assigned by that rater will tend to overestimate the examinee's proficiency across the proficiency continuum (that is, $\theta_r - \theta$ is positive). When raters differ in their positions on the leniency/harshness continuum, we do not know whether an examinee's score is truly a measure of examinee proficiency or is a reflection of a rater's idiosyncratic set of standards.

Centrality/Extremism

We represent another common rater effect on a continuum labeled *centrality/extremism*. This continuum depicts the extent to which raters under- or overutilize the categories contained in the rating scale. If a rater exhibits centrality, then the ratings assigned by that rater tend to cluster in the center of the rating scale. This situation might occur if a rater were concerned about being singled out for assigning ratings that disagree with ratings assigned by other raters. The rater may "play it safe," overusing the middle categories of the rating scale while avoiding the extreme categories. Centrality results in accurate rating in the central range of the ability continuum, but overestimates of examinee proficiency for nonproficient examinees (that is, $\theta_r - \theta$ is positive) and underestimates of examinee proficiency for highly proficient examinees (that is, $\theta_r - \theta$ is negative). By contrast, extremism occurs when raters tend to cluster ratings in the extreme categories of the rating scale. For these raters, there is no middle ground. Examinees do well or poorly, but not in between. This results in accurate rating in the tails of the proficiency distribution, but large amounts of error associated with the ratings assigned to examinees with average proficiency. When raters differ in their positions on the centrality/extremism continuum, we can never be sure whether the ratings assigned by a particular rater will be accurate and when they will be biased by the examinee's location on the underlying scale.

Multifaceted Rasch Rating Scale Model

Rasch measurement theory provides one way of examining the rater effects described above. The multifaceted Rasch rating scale model (MFRRSM; Linacre, 1989a) describes the probability that a specific examinee (n) will be rated with a specific rating scale step (x) by a specific rater (k) on a specific task (i). This probability (Equation 1) depends on four parameters: the examinee's *proficiency* (θ_n), the rater's *leniency/harshness* (λ_k), the task's *difficulty* (δ_i), and the difficulty of each scale step (that is, the *threshold* between two adjacent rating scale categories, τ_x). Calibration of rating data produces a separate parameter estimate and a standard error for that estimate for each examinee, rater, task, and scale step in the measurement context. This model assumes that a common rating scale structure applies to each task (that is, that τ_j is constant across tasks).[1] The model also assumes that the data conform to the predictions of the MFRRSM. Departures in the data from model-generated expected values indicate potentially mismeasured examinees, raters, or tasks.

$$P(x|\theta,\lambda,\delta,\tau) = \frac{\exp \sum_{j=0}^{x} \left[\theta_n - \lambda_k - \delta_i - \tau_j\right]}{\sum_{x=0}^{m} \exp \sum_{j=0}^{x} \left[\theta_n - \lambda_k - \delta_i - \tau_j\right]}, \qquad x = 0, 1 \ldots, m, \qquad (1)$$

where $P(x|\theta,\lambda,\delta,\tau)$ is the probability that the response of examinee n to task i is assigned rating scale category x by rater k when the rating scale has $m + 1$ categories.

To evaluate the degree to which the response patterns associated with individual elements of the measurement context are inconsistent with the MFRRSM, two fit statistics are generated for each parameter estimate (Wright & Masters, 1982). Both of these fit statistics are based on the mean of the squared standardized residuals of the observed scores from their expected scores. The *outfit* statistic is simply the mean of these standardized residuals (Equation 2). Outfit statistics are sensitive to departures in the data in the extreme rating categories. The *infit* statistic, on the other hand, weights each standardized residual by its variance (Equation 3). As a result, infit statistics are more sensitive to unexpected ratings that fall near the center of the rating scale. The infit and outfit statistics have an expected value of 1.00 and can range from 0.00 to ∞. A 0.1 increase in a fit statistics is associated with a 10% increase in unmodeled error. In general, elements with fit statistic values ranging from 0.6 to 1.5 are considered to show adequate fit to the model (Wright & Linacre, 1994).

$$outfit_{\lambda} = \frac{\sum_{n=1}^{N} \sum_{i=1}^{I} z_{rni}^{2}}{NI}, \quad (2)$$

where, N is the number of examinees, I is the number of items, and $z_{rni}{}^{2}$ is the standardized score residual.

$$infit_{\lambda} = \frac{\sum_{n=1}^{N} \sum_{i=1}^{I} z_{rni}^{2} W_{rni}}{\sum_{n}^{N=1} \sum_{i}^{I=1} W_{rni}}, \quad (3)$$

where W_{rni} is the variance of the score residual.

DATA SIMULATION AND ANALYSIS

Next we describe a series of simulations that demonstrate how each rater effect manifests itself in the raw scores and the rater calibrations and fit statistics that are produced by FACETS (software that performs MFRRSM scaling; Linacre, 1989b). In addition, we investigate the accuracy of decisions that are made when one attempts to identify rater effects in individual raters within a decision theoretical framework. The following sections describe how data generation was conducted and how the data sets were analyzed using FACETS.

Data Generation

To demonstrate how various rater effects manifest themselves in MFRRSM analyses, data were simulated to represent six types of rater effects: (1) no effect, (2) harshness (3) leniency, (4) centrality (5) extremism, and (6) indiscrimination. For the *no effect* condition, 10 data sets were generated. For each of the remaining conditions, 10 data sets were generated for each of four effect sizes. This resulted in a total of 210 data sets (that is, 10 *no effect* data sets + [10 replications × 5 rater effects × 4 effect sizes]).

Data generation was performed using the guidelines described by Harwell, Stone, Hsu, and Krisci (1996). More specifically, for each data set, unique identifiers were created for 100 examinees and 100 raters. Each examinee responded to a single test item, and each of the 100 raters assigned ratings on a five-point rating scale to each of the 100 examinees' responses. The 10 replications within an effect type-by-effect size combination differed only through the use of different seed values used for number generation. Ratings were generated by first generating parameters to be substituted into Equation 4 for each examinee-by-rater pairing. Details of parameter generation follow. Next, probability distributions were generated for each examinee-by-rater-by-rating scale step possibility. Then, an "observed" probability was generated for each examinee-by-rater pairing from a U(0,1) distribution. Finally, the observed probability was compared to the probability distribution for each examinee-by-rater pairing to determine the raw score. These raw scores were subjected to FACETS analyses as described in the following sections.

Parameter Generation

A two-parameter logistic rater model (2PLRM; Wolfe, 1997) was employed for the purpose of data generation. This model is useful in our simulations because it depicts how both rater leniency/harshness and centrality/extremism manifest themselves in polytomously scored test data (Equation 4). When data are generated under the 2PLRM, location parameters are specified for each examinee, each item, each rater, and each step threshold of the polytomous rating scale, and a slope parameter is specified for each rater. As with the MFRRSM, θ depicts the examinee's ability, δ depicts the item's difficulty, λ depicts the rater's leniency or harshness, and τ depicts the scale step's difficulty relative to the next lower scale step. In addition, the 2PLRM depicts the rater's centrality as γ. The rater centrality parameter describes the extent to which individual raters exhibit more or less variability in their distributions of ratings.

$$P(x|\theta,\lambda,\gamma,\delta,\tau) = \frac{\exp \sum_{j=0}^{x} \gamma_k(\theta_n - \delta_i - \lambda_k - \tau_j)}{\sum_{x=0}^{m_i} \exp \sum_{j=0}^{k} \gamma_k(\theta_n - \delta_i - \lambda_k - \tau_j)}, \qquad x = 0, 1, 2, \ldots, m, \tag{4}$$

where, $P(x|\theta,\lambda,\gamma,\delta,\tau)$ is the probability that the response of examinee n to task i is assigned rating scale category x by rater k when the rating scale has $m + 1$ categories.

For the *no effect* condition, data were generated using Equation 4 with examinee ability (θ) sampled from a N(0,1) distribution; item difficulty (δ) set equal to 0.00; rating scale step difficulties (τ) set to –3.00, –1.00, 1.00, and 3.00 for the first, second, third, and fourth thresholds, respectively; rater leniency/harshness (λ) set equal to 0.00; and rater centrality/extremism (γ) set equal to 1.00. The 10 *no effect* data sets served as the basis for generating the remaining data sets. For the rater *leniency*, *harshness*, *centrality*, and *extremism* conditions, ratings associated with 10%of the raters were altered by adding (*leniency* and *harshness*) or multiplying (*centrality* and *extremism*) a predetermined constant in Equation 4 during data generation. By using the same seed values used for the *no effect* condition for each of the four effect sizes under each rater effect, we produced four parallel sets of 10 data files, each based on one of the 10 *no effect* data sets. The constant values used for each effect type-by-effect size combination are shown in Table 1.

TABLE 1.
Parameter Values for Rater Effects Used During Data Generation

Effect Type	Effect Size	Parameter	Value
Leniency	Small	λ	–0.50
	Medium		–1.00
	Large		–1.50
	Very large		–2.00
Harshness	Small	λ	0.50
	Medium		1.00
	Large		1.50
	Very large		2.00
Centrality	Small	γ	0.75
	Medium		0.50
	Large		0.25
	Very large		0.10
Extremism	Small	γ	1.50
	Medium		2.00
	Large		2.50
	Very Large		3.00

For the *indiscrimination* condition, the ratings of 10% of the raters were altered by substituting values in the *no effect* data sets with values sampled randomly from a symmetrical, unimodal, five-point distribution (roughly normal in shape). Four effect sizes were defined by using different degrees of random value replacement for the 10% of the raters chosen to exhibit indiscrimination. These effect sizes were 5% (small), 15% (medium), 30% (large), and 60% (very large); so that larger degrees of indiscrimination were associated with a larger proportion of the total ratings assigned by an individual rater being random in nature.

Analyses

The goals of our analyses were twofold: (1) to identify how various rater effects manifest themselves in the raw scores and the rater calibrations and fit statistics produced by FACETS, and (2) to determine the accuracy of decisions when one categorizes raters as being aberrant or non-aberrant based on these statistics.

Rater Effect Manifestation

To investigate how rater effects manifest themselves, we examined the raw score distributions and the rater calibrations and fit statistics produced by FACETS. More specifically, we examined the descriptive statistics (mean, standard deviation, skewness, and kurtosis) of raters' assigned raw scores, rater calibrations ($\hat{\lambda}$), and rater fit statistics (mean-square infit$_\lambda$ and outfit$_\lambda$) by comparing aberrant and non-aberrant raters' descriptive statistics under each of the rater effects to the *no effect* statistics. For each of the 210 data sets, descriptive statistics were generated for the assigned raw scores, rater calibrations, and rater fit statistics; and the mean and standard deviation of each statistic was computed for the 10 replications within each rater effect. These statistics were also generated for the aberrant raters separately. Each of the 210 data sets were also scaled using FACETS (Linacre, 1989b). For each data set, we defined an MFRRSM that contained three facets: (1) *examinee*, (2) *rater*, and (3) *scale step*. The *examinee* facet was scaled so that more able examinees received higher logit values (that is, positively oriented). The *rater* facet was oriented so that more harsh raters received higher logit values (that is, negatively oriented). The *rater* facet was centered at zero, and the *examinee* facet was noncentered. We examined the average and the standard deviation of the rater calibrations and the rater mean square infit and outfit statistics (across the 10 replications within each rater effect) for all raters and the aberrant separately, comparing the values for each rater effect to those of the *no effect* condition.

Decision Theory Analysis

It is clear from prior research concerning the MFRRSM that rater leniency and harshness manifest themselves in the rater calibrations (Engelhard, 1994; Lunz, Wright, & Linacre, 1990). However, it is not clear from previous research how rater centrality, extremism, and indiscrimination should manifest themselves. Previous work in this area (Engelhard & Stone, 1998; Smith, 1996; Wolfe & Chiu, 1997) suggests that rater extremism and indiscrimination should manifest themselves in increased rater fit statistics. However, the picture is not so clear with respect to the influence of rater centrality on fit statistics. Although rater fit would typically decrease with the introduction of a centrality effect (Engelhard & Stone, 1998; Smith, 1996; Wolfe & Chiu, 1997), under some conditions (that is, widely spaced rating categories or large examinee variability) rater fit indices may actually increase (J. M. Linacre, personal communication, May 4, 1998).

In addition, each of these conditions should influence the raw score variability associated with aberrant raters differently. For example, we would hypothesize that centrality should result in smaller score variance for aberrant raters because most scores assigned by these raters fall only in the middle-most rating scale categories. Extremism should have the opposite effect. The score variance should be larger for aberrant raters, we would hypothesize, because most scores are assigned only in the extreme rating scale categories. On the other hand, we would hypothesize that indiscrimination should have only a small influence on the raw score variance, assuming that the aberrant raters assign scores in each rating scale category in proportions similar to those assigned by nonaberrant raters. Of course, if indiscriminate raters assigned scores uniformly across the rating scale categories (that is, with equal probability in each scoring category), the score variance would increase. However, the upper limit of the score variance under such a situation should still be smaller than it would be under the extremism condition.

We evaluated these hypotheses through our rater effect manifestation analyses. We then conducted decision analyses (Mooney, 1997; Shavelson, 1988), using appropriate indicator statistics to determine the accuracy of diagnoses of rater aberrance for each of the five rater effects. Indicator statistics and selection criteria for each rater effect are summarized in Table 2. For each rater effect, raters were rank ordered[2] according to the appropriate indicator statistic(s), and raters were diagnosed based on their rankings on these indicator statistics.[3] In an operational setting, rater evaluators would have criteria for identifying raters who are suspect of exhibiting some type of aberrance (for example, a certain percent of the most extreme rater calibrations or rater infit or outfit that exceeds a specific value). In our decision analyses, our criteria for "diagnosing" aberrance consisted of selecting the 10% of the raters with the highest indicator statistic rankings. For example, under our harshness condition, the 10 (out of 100) raters who had the highest rater calibration rankings were diagnosed as being aberrant. All other raters were diagnosed as being nonaberrant.

Thus, we had two pieces of information about each rater: (1) the rater's "diagnosis" based on our diagnosis criteria, and (2) the rater's "true" status based on our data simulation procedures. For the purpose of evaluating the accuracy of our decisions about individual raters, the average number of correct diagnoses was determined across the 10 replications within an effect type-by-effect size combination by comparing diagnosis status to true status. Evaluation of accuracy rates focused on the average proportion of correct and incorrect diagnoses for aberrant and nonaberrant raters within each effect type-by-effect size combination. *Accurate diagnoses* indicated that the rater's true group membership (aberrant or nonaberrant as specified during data generation) was correctly diagnosed based on the diagnosis

TABLE 2.
Indicator Statistics and Diagnosis Criteria for Decision Analyses

Rater Effect	Indicator Statistic(s)	Diagnosis Criteria
Leniency	λ	Lowest 10% nominated as aberrant
Harshness	λ	Highest 10% nominated as aberrant
Centrality	$\frac{infit_\lambda + outfit_\lambda}{2}$, SD_x	Highest 10% (average ranking) nominated as aberrant
Extremism	$\frac{infit_\lambda + outfit_\lambda}{2}$, SD_x	Highest 10% (average ranking) nominated as aberrant
Indiscrimination	$\frac{infit_\lambda + outfit_\lambda}{2}$	Highest 10% nominated as aberrant

Note. For each rater effect, raters were rank ordered based on the magnitude of the indicator statistic(s). The nomination criteria define how raters were diagnosed as "aberrant" for the decision analyses. Average rater fit was ranked from highest to lowest for both centrality and extremism. SD_x was ranked from lowest to highest for centrality and from highest to lowest for extremism. It should be noted that rater fit typically *decreases* with the introduction of rater centrality. However, under the conditions simulated in this demonstration, fit statistics increased.

criteria (that is, aberrant or nonaberrant as dictated by indicator statistics). *Inaccurate diagnoses* indicated that the rater's group membership was incorrectly diagnosed based on the indictor statistics.

RESULTS

The results of our simulations suggest that rater effects are identifiable in the MFRRSM rater statistics and the raw score standard deviations associated with individual raters. More specifically, rater harshness and leniency manifest themselves in the rater calibrations, rater centrality and extremism manifest themselves in the rater fit indices and the standard deviations of rater raw scores, and rater indiscrimination manifests itself in the rater fit statistics. In addition, attempts to use these indicators to diagnose rater harshness and leniency in individual raters results in accurate diagnosis regardless of the size of the effect. Diagnosis of rater centrality and extremism, on the other hand, was accurate only when the effect size was large or very large. Rater indiscrimination diagnosis was not accurate for medium, large, and very large effect sizes.

Condition Manifestation

Table 3 summarizes the descriptive statistics for the raw scores associated with individual raters by showing the average mean, standard deviation, skewness, and kurtosis across the 10 replications within each rater effect type-by-effect size combination. As shown by the statistics for the *no effect* condition, the average raw score was 1.97 on the five-point scale with a standard deviation of 0.85. Note that aberrant raters exhibited the same mean and standard deviation. The raw score distributions were nearly symmetrical and were slightly platykurtic. In general, each of the rater effects influenced either the mean or the standard deviation of the raw scores. More specifically, the *leniency* condition resulted in lower average assigned scores for the group of aberrant raters. The *harshness* condition, on the other hand, resulted in higher average assigned scores. The *centrality* condition resulted in smaller standard deviations of assigned raw scores for the aberrant raters. This shrinkage caused the distribution of raw scores to be slightly more peaked as evidenced by the kurtosis statistics. The *extremism* condition resulted in considerably larger raw assigned score standard deviations for the aberrant raters, while the *indiscrimination* condition resulted in only modest increases in raw assigned score standard deviations.

Table 4 summarizes the descriptive statistics of the rater facet from the FACETS output. That is, the table contains the average rater calibrations (in logits), standard error, mean square infit statistic, and mean square outfit statistic for each effect type-by-effect size combination. The average rater calibration in the *no effect* condition was 0.00, and the average standard error of these calibrations was

TABLE 3.
Average Descriptive Statistics for Simulated Raw Scores Associated with All and Aberrant Raters

Rater Effect	Effect Size	Group	Mean	*SD*	Skewness	Kurtosis
No Effect		All	1.97	0.85	0.01	–0.19
		Aberrant	1.97	0.85		
Leniency	Small	All	1.94	0.85	0.02	–0.20
		Aberrant	1.75	0.85		
	Medium	All	1.92	0.86	0.02	–0.21
		Aberrant	1.53	0.83		
	Large	All	1.90	0.87	0.00	–0.23
		Aberrant	1.32	0.82		
	Very large	All	1.89	0.88	–0.01	–0.25
		Aberrant	1.12	0.79		
Harshness	Small	All	1.99	0.85	0.01	–0.20
		Aberrant	2.19	0.85		
	Medium	All	2.01	0.86	0.01	–0.21
		Aberrant	2.41	0.83		
	Large	All	2.03	0.87	0.03	–0.22
		Aberrant	2.62	0.83		
	Very large	All	2.05	0.89	0.05	–0.25
		Aberrant	2.83	0.80		
Centrality	Small	All	1.97	0.84	0.01	–0.18
		Aberrant	1.98	0.79		
	Medium	All	1.97	0.84	0.01	–0.17
		Aberrant	1.99	0.75		
	Large	All	1.97	0.84	0.01	–0.16
		Aberrant	1.99	0.72		
	Very large	All	1.97	0.83	0.01	–0.16
		Aberrant	2.00	0.72		
Extremism	Small	All	1.97	0.86	0.02	–0.21
		Aberrant	1.95	0.97		
	Medium	All	1.96	0.87	0.02	–0.21
		Aberrant	1.95	1.10		
	Large	All	1.96	0.89	0.02	–0.21
		Aberrant	1.93	1.20		
	Very large	All	1.97	0.84	0.01	–0.18
		Aberrant	1.93	1.29		
Indiscrimination	Small	All	1.97	0.85	0.01	–0.18
		Aberrant	1.97	0.88		
	Medium	All	1.97	0.86	0.01	–0.18
		Aberrant	1.96	0.93		
	Large	All	1.97	0.87	0.01	–0.17
		Aberrant	1.98	1.01		
	Very large	All	1.97	0.88	0.01	–0.16
		Aberrant	1.99	1.10		

Note. The values shown are the average of each descriptive statistic across 10 replications within each effect type-by-effect size combination.

TABLE 4.
Average Descriptive Statistics of FACETS Output Associated with All and Aberrant Simulated Raters

Rater Effect	Effect Size	Group	Rater Calibration	*SE*	Rater Infit	Rater Outfit
No Effect		All	0.00	0.14	1.00	1.00
		Aberrant	0.00	0.14	1.00	1.00
Leniency	Small	All	0.00	0.14	1.00	1.00
		Aberrant	–0.42	0.14	1.00	1.00
	Medium	All	0.00	0.14	1.00	1.00
		Aberrant	–0.83	0.15	0.98	0.98
	Large	All	0.00	0.14	1.00	1.00
		Aberrant	–1.24	0.15	0.99	0.99
	Very large	All	0.00	0.14	1.00	1.00
		Aberrant	–1.66	0.15	1.00	1.00
Harshness	Small	All	0.00	0.14	1.00	1.00
		Aberrant	0.41	0.14	1.00	1.00
	Medium	All	0.00	0.14	1.00	1.00
		Aberrant	0.83	0.15	0.99	0.99
	Large	All	0.00	0.14	1.00	1.00
		Aberrant	1.24	0.15	1.00	1.00
	Very large	All	0.00	0.14	1.00	1.00
		Aberrant	1.66	0.15	1.00	1.00
Centrality	Small	All	0.03	0.14	1.00	1.00
		Aberrant	0.04	0.14	1.01	1.01
	Medium	All	0.06	0.14	1.00	1.00
		Aberrant	0.02	0.14	1.07	1.07
	Large	All	0.05	0.14	1.00	1.00
		Aberrant	0.00	0.14	1.16	1.16
	Very large	All	0.05	0.14	1.00	1.00
		Aberrant	0.00	0.14	1.23	1.23
Extremism	Small	All	0.00	0.14	1.00	1.00
		Aberrant	0.02	0.14	1.04	1.04
	Medium	All	0.00	0.14	1.00	1.00
		Aberrant	0.03	0.14	1.17	1.18
	Large	All	0.00	0.14	1.00	1.00
		Aberrant	0.06	0.14	1.34	1.34
	Very large	All	0.00	0.14	1.00	1.00
		Aberrant	0.07	0.14	1.50	1.50
Indiscrimination	Small	All	0.00	0.14	1.00	1.00
		Aberrant	0.02	0.14	1.17	1.17
	Medium	All	0.00	0.14	1.00	1.00
		Aberrant	0.01	0.14	1.48	1.48
	Large	All	0.00	0.14	1.00	1.00
		Aberrant	–0.02	0.14	1.91	1.91
	Very large	All	0.00	0.13	1.00	1.00
		Aberrant	–0.04	0.13	2.39	2.40

Note. The values shown are the average value for each statistic across 10 replications within each effect type-by-effect size combination.

0.14. Both the mean square infit and outfit statistics had an average of 1.00. As we hypothesized, the *leniency* condition resulted in average rater calibrations for the aberrant raters that were smaller than the average rater calibration for all raters. By contrast, the *harshness* condition resulted in average rater calibrations for aberrant raters that were larger than the average rater calibration for all raters. These figures show that, in the measurement situation we modeled, each increase of rater harshness or leniency of a one-half true score variance unit resulted in about a 0.40 logit unit change in aberrant raters' calibrations.

The addition of rater centrality to our simulated data resulted in increases in both rater infit and outfit statistics. As shown, reducing the variance of scores assigned by aberrant raters to 75% or 50% of the variance of scores assigned by other raters resulted in only small increases in rater fit statistics. However, a reduction of variance to 25% or 10% of that exhibited by nonaberrant raters resulted in average increases in rater fit for aberrant raters to 1.16 and 1.23, respectively. Rater extremism resulted in more apparent changes in rater fit statistics. With an increase of variance in aberrant raters' scores to 200%, 250%, and 300% of that contained in nonaberrant raters' scores, average rater fit statistics increased to 1.18, 1.34, and 1.50, respectively. The addition of random scores under the indiscrimination condition had an even more profound influence on rater fit statistics. As shown, even an addition of 5% random ratings resulted in an average rater fit statistic of 1.17. Larger percentages of random ratings resulted in proportional increases in rater fit.

Table 5 summarizes the trends that are revealed by the rater effect manifestation analyses. Our simulations suggest that, as would be expected, leniency and harshness manifest themselves as lower and higher assigned raw scores and rater calibrations. Centrality, extremism, and indiscrimination all manifest themselves in the assigned raw score standard deviations and rater fit statistics. More specifically, centrality manifests itself as smaller raw score standard deviations and larger rater fit statistics (at least in this simulation). Extremism manifests itself as larger raw score standard deviations and rater fit statistics. Rater indiscrimination is more difficult to depict with such a general statement. In the current study, the introduction of rater indiscrimination resulted in small increases in raw score standard deviations and large increases in rater fit indices. However, different operationalizations of rater indiscrimination would likely influence these statistics in different ways. In our case, we replaced a proportion of the scores assigned by aberrant raters with a random score drawn from a distribution that had a shape similar to that of the raw scores assigned by other raters. If, on the other hand, we were to replace the scores of aberrant raters with a random score drawn from a distribution that is much flatter (say, a uniform distribution—a condition that begins to approximate extremism), we might expect a much larger increase in the raw score standard deviation. Hence, the differentiation between rater extremism and indiscrimination is difficult to make because these two rater effects are not completely distinct from one another.

TABLE 5.
Rater Effect Manifestations

Rater Effect	Raw Score	Raw SD	Rater Logit	Rater Fit
Leniency	Decreases	—	Decreases	—
Harshness	Increases	—	Increases	—
Centrality	—	Decreases	—	Increases
Extremism	—	Increases	—	Increases
Indiscrimination	—	Small increase	—	Large increase

TABLE 6.
Decision Analysis Results for Simulated Raters

		Aberrant		Nonaberrant	
Rater Effect	Effect Size	Accurate Diagnosis (%)	Inaccurate Diagnosis (%)	Accurate Diagnosis (%)	Inaccurate Diagnosis (%)
Leniency	Small	83	17	99	1
	Medium	90	10	99	1
	Large	90	10	99	1
	Very large	90	10	99	1
Harshness	Small	83	17	99	1
	Medium	90	10	99	1
	Large	90	10	99	1
	Very large	90	10	99	1
Centrality	Small	36	64	93	7
	Medium	68	32	96	4
	Large	82	18	98	2
	Very large	87	13	99	1
Extremism	Small	35	65	93	7
	Medium	66	34	96	4
	Large	84	16	90	1
	Very large	88	12	99	1
Indiscrimination	Small	47	53	94	6
	Medium	89	11	99	1
	Large	100	0	100	0
	Very large	100	0	100	0

Decision Analyses

Table 6 shows the percent of accurate and inaccurate diagnoses for aberrant and nonaberrant raters under each effect type-by-effect size combination. As would be predicted, diagnostic accuracy increased as effect size increased under each of the rater effects. Under the harshness and leniency rater effects, diagnostic accuracy was high, even for small effect sizes. Recall that a small leniency/harshness effect size was defined as a 0.5 logit difference in aberrant raters' ratings. In our simulations, that 0.5 logit difference resulted in a raw score difference of only 0.20 raw score points on the six-point rating scale (4% of the maximum possible range). Under the centrality and extremism rater effects, diagnostic accuracy was somewhat lower, particularly for small and medium effect sizes. Recall that small and medium effect sizes were defined as a 25% and 50% reduction in true score variance for the centrality rater effect and an increase in true score variance to 1.5 and 2 times its original size for the extremism rater effect. Only when the centrality and extremism effect sizes were large or very large (that is, a 75% or 90% reduction in true score variance under the centrality rater effect and an increase in true score variance to 2.5 to 3 times its original size for the extremism rater effect), did the percentage of correctly identified aberrant raters exceed 80%. The accuracy of diagnosis for aberrant raters under the indiscrimination condition was very good when effect size was medium, large, and very large (that is, 15%, 30%, and 60% random ratings, respectively).

DISCUSSION AND CONCLUSIONS

Through our simulations, we have shown how rater effects can be detected using rater calibrations, rater fit indices, and raw score standard deviations. Specifically, our simulations show that rater harshness and leniency are associated with larger and smaller average rater calibrations than are typical of other raters. Rater centrality is associated with assigned raw score standard deviations that are smaller and rater fit indices that are larger than those of nonaberrant raters. However, it should be noted that this finding is contrary to those found in operational settings (see, for example, Engelhard, 1994). This is probably owing to the fact that our distribution of ratings was slightly platykurtic whereas many operational distributions for performance assessments are leptokurtic.

We also found that rater extremism is associated with assigned raw score standard deviations that are smaller and rater fit indices that are larger than those of nonaberrant raters. In addition, rater indiscrimination results in inflation of rater fit indices and small increases in assigned raw score standard deviations. With respect to the accuracy of decisions concerning whether individual raters are manifesting each of these rater effects, we found positive results. For the harshness and leniency conditions, the MFRRSM produced very accurate indicators of rater aberrance, regardless of the effect size of the rater effect. The accuracy of diagnoses for the centrality and extremism conditions were not adequate when the effect sizes were small or medium, but diagnostic accuracy was sufficient with large centrality and extremism effect sizes. Indiscrimination, on the other hand, exhibited adequate diagnostic accuracy as long as the effect size was not small.

Limitations

Of course, because these analyses are based on simulated data, there are several limitations to the interpretation of our results. One limitation of this study is the fact that we did not investigate how the distribution shape might influence the indicators that we examined. For example, it is likely that the shape of the assigned raw score distribution (for example, skew and kurtosis) might influence not only the behavior of the indicator statistics that we examined but also the diagnostic accuracy of those indicator statistics. The same can be said for the underlying distributions upon which our data were generated. For example, by increasing the variability between raters (relative to the variability between examinees), we would certainly alter the rater calibrations.

Another limitation of our study is that our investigation included several features that may not be common in operational data. We only investigated a situation in which examinees respond to a single item. In some operational settings, such an approach to rater monitoring might be suitable—a rater monitor might be interested in evaluating raters as they rate examinee responses on an item-by-item basis. However, in some operational settings, a single rater rates examinee responses to several items at one time. Our simulations do not address rating procedures such as these. Also, we simulated a situation in which all raters assign ratings to all examinees. Because such rating procedures are costly, they are seldom implemented operationally.

A third limitation of our study is the fact that we investigated only a limited set of criteria for identifying aberrant raters. We could have investigated the usefulness of other potential statistical indicators of rater aberrance, but we chose to examine only the ones that are the most common focus of rater evaluation efforts. We also diagnosed raters as being aberrant in the same proportion that we introduced rater aberrance into the data. In operational settings, one does not know how many raters are truly aberrant. As a result, one might choose to select a certain proportion of raters that is much larger or smaller than the proportion of raters who are truly aberrant. Our study minimized this problem. An alternative method for identifying aberrant raters would be to diagnose raters as being aberrant only if their indicator statistics exceed a specific value. Our study did not investigate this alternative.

Future Research

These limitations point to several potential areas for further study. Additional simulation work is necessary to identify how assumptions about underlying distributions, different designs for distributing responses among raters, and selection of identification criteria influence diagnosis of rater aberrance. In addition, these methods could be extended to the examination of other rater effects. For example, Engelhard (1994) suggests that halo effects might be detectable using fit statistics. Wolfe and Myford (1997), on the other hand, describe an equating method that might be useful for identifying rater drift over time. Also, other models might be used for examining some of these rater effects. Congdon (1997) describes the use of the partial credit model for examining changes in raters' distributions of ratings over time. Such a method could easily be extended to the examination of rater centrality and extremism. Wolfe (1997) describes an extension of the generalized partial credit model that may be useful for identifying both rater leniency/harshness and centrality/extremism simultaneously. And, of course, the most important extension of our work is to apply these methods to actual rating data to determine the degree to which the rater effects that we seek to identify manifest themselves in operational settings.

NOTES

1. It should be noted that it is possible to define a multifaceted Rasch model that allows the rating scale structure to vary from one task to another (or even from one rater to another). We refer to this model as a multifaceted Rasch partial credit model (MFRPCM). Although our results only use the MFRRSM, the methods we describe in this chapter can be used with the MFRPCM as well.

2. Rankings were averaged using Blom's (1958) rank order method, in which ranks are based on proportion estimates using the following formula:

$$\frac{r - \left(\frac{3}{8}\right)}{w + \left(\frac{1}{4}\right)}$$

where r is the rank order of the element and w is the sum of the case weights.

3. Note that our diagnosis of raters as being aberrant is based on normative information (that is, the highest 10% of the indicator statistics rankings). Diagnostic decisions can also be made in an absolute framework by choosing a cut value for the indicator statistics and diagnosing raters as being aberrant if their indicator statistics exceed that cut value. For example, Wright and Masters (1982) suggest that a reasonable range for fit indices extends from 0.80 to 1.20. For the sake of consistency across both rater effects and indicator statistics, we chose to use normative criteria only in our decision analyses.

REFERENCES

Blom, G. (1958). *Statistical estimates and transformed beta variables.* New York: Wiley.

Congdon, P. J. (1997, March). *Some changes in rater characteristics over twelve months.* Paper presented at the Ninth International Objective Measurement Workshop, Chicago.

Engelhard, G. J., Jr., (1994). Examining rater errors in the assessment of written composition with a many-faceted Rasch model. *Journal of Educational Measurement, 31*, 93–112.

Engelhard, G. J., Jr., & Stone, G. E. (1998). Evaluating the quality of ratings obtained from standard-setting judges. *Educational and Psychological Measurement, 58*, 179–196.

Harwell, M., Stone, C. A., Hsu, T. C, & Krisci, L. (1996). Monte Carlo studies in item response theory. *Applied Psychological Measurement, 20,* 101–125.

Linacre, J. M. (1989a). *Many-facet Rasch measurement.* Chicago: MESA Press.

Linacre, J. M. (1989b). *A user's guide to FACETS: Rasch measurement computer program.* Chicago: MESA Press.

Lunz, M. E., Wright, B, D., & Linacre, J. M. (1990). Measuring the impact of judge severity on examination scores. *Applied Measurement in Education, 3*, 331–345.

Mooney, C. Z. (1997). *Monte Carlo simulation*. Thousand Oaks, CA: Sage.

Saal, F. E., Downey, R. G., & Lahey, M. A. (1980). Rating the ratings: Assessing the psychometric quality of rating data. *Psychological Bulletin, 88*, 413–428.

Shavelson, R. J. (1988). Decision, error, and power. In *Statistical Reasoning for the Behavioral Sciences* (2nd ed., pp. 282–312). Boston: Allyn and Bacon.

Smith, R.M. (1996). Polytomous mean-square fit statistics. *Rasch Measurement Transactions, 10*, 516–517.

Wolfe, E. W. (1997, October). *A two-parameter logistic rater model.* Paper presented at the annual meeting of the Florida Educational Research Association, Orlando, FL.

Wolfe, E. W., & Chiu, C. W. T. (1997, March). *Detecting rater effects with a multi-faceted rating scale model.* Paper presented at the annual meeting of the National Council on Measurement in Education, Chicago.

Wolfe, E. W., & Myford, C. M. (1997, March). *Detecting order effects with a multi-faceted Rasch model.* Paper presented at the annual meeting of the National Council on Measurement in Education, Chicago.

Wright, B., & Linacre, M. (1994). Reasonable mean-square fit values. *Rasch Measurement Transactions, 8,* 370.

Wright, B. D., & Masters, G. N. (1982). *Rating scale analysis.* Chicago: MESA Press.

9

UNMODELED RATER DISCRIMINATION ERROR

Peter J. Congdon
Joy McQueen
Australian Council for Educational Research

INTRODUCTION

This chapter explores ways in which to describe the amount of error that can exist when a measurement model is used that assumes that all raters apply a common understanding of the rating scale. A common understanding of the rating scale is defined where all raters require the same amount of ability or achievement change to move between particular score levels. That is, all raters are discriminating similarly.

A number of sources of error have been discussed in the literature on performance assessment (see, for example, Cantor & Hoover, 1986; Engelhard, 1992; Engelhard, Gordon, & Gabrielson, 1991; Gabrielson, Gordon, & Engelhard, 1995; McNamara, 1996; Ruth & Murphy, 1988). Prominent among these sources is the variance associated with raters. This is a reflection of the concern that, no matter how carefully constructed, the reliability of a rating scale is critically dependent on the raters who operate it (Overall & Magee, 1992). As Dunbar, Koretz, and Hoover (1991, p. 291) put it, "fallible raters can wreak havoc on the trustworthiness of scores and add a term to the reliability equations that does not exist in tests that can be scored objectively."

A previous version of this paper was presented at the International Conference on Latent Trait Theory: Rasch Measurement, Perth, Australia, in January 1998.

As few assessment programs can afford the time and expense of having every piece of work assessed by every rater, steps must be taken to ensure that no candidate is disadvantaged by the chance allocation of his or her work to a particular rater, however self-consistent that rater may be. As Webb, Raymond, and Houston (1990, p. 16) put it, whenever an incomplete rating design is used, investigation of rater effects is an ethical obligation. Any discovery of significant rater effects should inform quality control procedures during the rating period, and indicate where adjustment of candidates' scores is needed to compensate for these effects.

Rater discrimination, which can be measured and controlled for using multifaceted Rasch models, is a rater characteristic that should be monitored closely before reporting person measures when rater-independent measures are desired. If a rater spreads performances of a given ability along a continuum more or less than other raters, then the rater is in disagreement with the other raters on the location of some of the score categories on that continuum, even after adjusting for average rater severity differences. Variation in rater discrimination has been described as differences in raters' resolving powers, where resolving power is reflected in the effect of a fixed increment in ability on the increment in assigned rating (Cason & Cason, 1984).

Raters of writing performances have shown a significant interaction with the score level difficulties. When person abilities from a single rating were compared to measures derived from four raters, significant differences, due to variation in rater discrimination, were found (McNamara & Adams, 1994). In a study of the stability of rater characteristics of two raters over a 12-month period (Congdon, 1997), one of the two raters studied showed a significantly different level of discrimination over the period when re-rating the same performances.

In measurement systems that produce person measures that are intended to be generalizable or comparable to person measures from different raters or from different occasions, it may be conceptually more simple to use a fixed or common scale step structure than a scale step structure that is unique to individual raters or single occasions. That is, the distance between score categories is kept the same for all raters and occasions and no interactions are parameterized. However, where interactions do occur and are not accounted for, the measurement error increases and the data fit to the model becomes problematic.

With multifaceted Rasch measurement models such as Andrich's rating scale model, which uses a common scale step structure across raters, or Masters' partial credit model, which uses a unique scale step structure for each rater, the total information supplied by each rater, or the area under the information curve, is exactly the same. What differs between these models is the amount of information at any location on the continuum when all raters have the same average severity. The total test score remains a sufficient statistic in both models, which is not the case with models that contain a slope parameter. With two-parameter measurement models that employ a slope parameter based on the discrimination index of each rater, not only can the amount of information at any location on the continuum vary for each rater, but also the total amount of information supplied by each rater can vary. This condition no longer allows the total test score to be a sufficient statistic owing to the weighting effect of the slope parameter. Although both Rasch models and models with a slope parameter attempt to parameterize differences among raters, they do this in different ways. The models that use a slope parameter for each rater can weight any rater's responses either up or down, which has the consequence of statistically determining that one rater's responses are worth more or less than another rater's responses. As there is no differential weighting of the value of raters responses with the Rasch models, an important difference exists in the infered validity of each rater's contribution to the measurement process.

DATA FIT TO THE MODEL

Variation in rater discrimination can become a problem when a measurement model is used that assumes the difference in ability between two adjacent scores is applied equally by all raters and that descriptions of person performance are independent of the characteristics of the raters. The fit of the data to such models is one way to monitor the validity of this assumption. Part of the problem of assessing data fit to the model is determining what amount of data misfit will compromise the utility of the model estimates.

Assessing data fit to the model can be done in a number of ways. The methods used in the study described here include the use of a deviance statistic for comparing alternative models, comparison of the summary fit statistics for each facet and the elements within those facets, and comparison of parameter estimates where the parameters are common to both models. The fit statistics used here were developed for marginal maximum likelihood estimation procedures for generalized item response models (Wu, 1997) and provide a measure of the relative consistency of each rater's performance. Mean square fit statistics greater than 1.0 indicate that there is more variation in the observed responses than expected and, conversely, values less than 1.0 indicate less variation than expected. The level at which these values become problematic is arbitrary and perhaps dependent on the impact that person measures will have on individuals or the allowable tolerance at particular boundaries or cut scores. Acceptable ranges suggested by other researchers include 0.5 to 1.5 (Lunz, Stahl, & Wright, 1996) and 0.8 to 1.2 (Linacre, 1989).

The fit statistics for individual raters can be used to identify which raters are most different from the group of raters. For example, when using a rating scale model, the fit of the rater severity estimate, if high, may indicate that the rater is more discriminating than the other raters, or if low, could indicate that the rater is less discriminating than other raters. After identifying raters with possible misfitting response styles, the problem of determining their effect on the usefulness of the modeled estimates remains.

The study described in this chapter attempts to identify and quantify some of the potential problems with models that assume equal rater discrimination and, in doing so, to provide some ideas for quality control procedures when applying models that incorporate rater response characteristics. Two different multifaceted Rasch measurement models are used within the context of writing performance data. One model uses a fixed rating scale (RS) structure for all raters, while the other uses a partial credit (PC) structure for each rater.

METHOD

The Test Background

The test discussed here is one component of a large, state-wide program of literacy and numeracy testing, the purpose of which was to provide teachers and parents with information on individual student performances. This study focuses on the writing assessment component.

In 1996 almost 47,000 students participated. They were allowed 30 minutes to write up to two pages in response to a single prompt designed to elicit a newspaper report of a recent event. Each performance was scored by two raters. Scoring followed a criterion-referenced rating scale based on the curriculum framework in use. There were six described levels within the scale, for each of two performance dimensions. This study deals with rater severities on one performance dimension only.

A half-day training session was conducted in the week prior to the start of rating. Following the training, the raters scored a set of performances, and multifaceted Rasch analysis was carried out so that rater severity and fit to a common rating scale step structure could be estimated. Raters with unacceptable fit (that is, those who were grading inconsistently relative to all raters combined) were excluded from the rest of the program. Raters who were not using the whole scale, or who were significantly more lenient or more severe than the other raters, were counselled.

Raters worked at designated tables of about eight, each with its table leader. The role of the table leader was to monitor the quality of scores given by raters at their table, and to counsel raters whose ratings differed by more than one grade from their own. The table leader also acted as a point of reference and offered guidance in dealing with problems.

The design of the study linked the 16 raters within each of six working days in the rating period. During the rating period, the 16 raters graded 8,285 performances, and each rater averaged 1,036.

The 16 raters were randomly selected from the pool of raters who were prepared to work in both the morning and the afternoon sessions each day. Twelve of the 16 had previously participated in at least one rating program of this kind.

Measurement Model

The models used to analyze the data from this study were multifaceted versions of the Rasch model, as implemented by the ConQuest software (Wu, Adams, & Wilson, 1996). The ConQuest software produces measures of all facets and elements in a common metric (logits) and a variety of commonly used fit statistics. The label facet is used to describe a group of components, for example, raters. The label element is used to identify components of the facet, for example, Rater 1, Rater 2, Rater 3, and so on.

The first model (RS) used the facets of *person ability*, *rater severity,* and a rating scale made up of five ordered *steps* representing increasing ability. It was intended that the structure of the rating scale remain constant across these facets. The RS model was defined as:

$$\ln\left(\frac{P_{n,i,j}}{P_{n,i,j-1}}\right) = B_n - (R_i + S_j), \qquad (1)$$

where:

$P_{n,i,j}$ is the probability of person n being rated j by rater i,
$P_{n,i,j-1}$ is the probability of person n being rated $j - 1$ by rater i,
B_n is the writing ability of person n,
R_i is the severity of rater i, and
S_j is the difficulty of scoring step j relative to step $j - 1$.

The second model (PC) used the facets of *person ability*, *rater severity* and a rating scale made up of five ordered *steps* representing increasing ability. It was intended that the structure of the rating scale could vary with rater. This allowed each rater to define a unique rating scale based on their own response pattern. The PC model was defined as:

$$\ln\left(\frac{P_{n,i,j}}{P_{n,i,j-1}}\right) = B_n - (R_i + S_{ij}), \qquad (2)$$

where:

$P_{n,i,j}$ is the probability of person n being rated j by rater i,
$P_{n,i,j-1}$ is the probability of person n being rated $j - 1$ by rater i,
B_n is the writing ability of person n,
R_i is the severity of rater i, and
S_{ij} is the difficulty of scoring step j relative to step $j - 1$ for rater i.

Both applications of these models had rater nominated as the single noncentered facet and the same data were used for both analyses. The estimate standard errors were calculated using the "quick" option in ConQuest. This option ignores the covariances between response model parameters. Parameter covariance was not considered to be a problem with the following results as the rater facet was not constrained.

RESULTS

Data Fit to the Models

The data fit to the two modeled conditions are shown in Table 1. The deviance statistic is –2 times the value of the log-likelihood function evaluated at the maximum. The smaller the deviance, the better the fit of the data to the model. To compare two models, the deviances associated with each of the models are computed and the test statistic is the absolute value of the difference between the two deviances. This statistic has a chi-square distribution with the degrees of freedom equal to the difference in the number of response parameters modeled.

The results show a significant improvement in the data fit to the PC model compared to the RS model. This indicates that the additional facet, *rater* by *step*, is accounting for some of the variation in the data that is in addition to rater severity. That is, by using a model that acknowledges each rater's unique interpretation of the rating scale, a statistical improvement has been made in describing the response patterns of the raters.

Tables 2 and 3 show some of the output produced by the ConQuest software, on the rater facet under both models. The tables contain the rater labels (VARIABLES), severity estimate (ESTIMATE), standard error of the estimate (ERROR), four different fit statistics, unweighted mean square residual (UNWGHTED MNSQ), unweighted T (UNWGHTED T), weighted mean square residual (WGHTED MNSQ) and weighted T (WGHTED T), parameter separation reliability and a chi-square test of estimate differences.

TABLE 1.
Comparison of Data Fit to Model.

Model	Deviance	Parameters	Deviance change	Parameter change	chi-square p-value
RS	38778.0	22			
PC	37918.9	82	859.1	60	0.000

TABLE 2.
Rating Scale Model, Rater Facet Output.

TERM 1 Rater

Rater VARIABLES	ESTIMATE	ERROR	UNWGHTED FIT MNSQ	UNWGHTED FIT T	WGHTED FIT MNSQ	WGHTED FIT T
5	–1.398	0.050	1.07	1.60	1.05	1.10
12	–0.452	0.052	0.85	–3.60	0.89	–2.50
16	–0.407	0.054	0.88	–2.80	0.94	–1.40
10	–0.371	0.051	0.84	–4.10	0.81	–4.70
14	–0.332	0.049	0.96	–0.90	0.96	–0.90
11	–0.258	0.046	1.20	4.80	1.15	3.50
9	–0.213	0.053	1.01	0.30	1.02	0.40
4	–0.168	0.060	1.46	8.00	1.44	7.50
15	–0.107	0.062	0.90	–1.90	0.91	–1.80
3	–0.062	0.055	0.96	–0.90	0.96	–0.90
13	–0.017	0.051	0.83	–4.30	0.87	–3.00
7	0.069	0.059	0.82	–3.90	0.85	–3.10
1	0.152	0.054	0.88	–2.90	0.89	–2.50
8	0.315	0.052	0.96	–0.90	0.99	–0.30
6	0.577	0.054	1.20	4.30	1.17	3.60
2	0.613	0.054	0.89	–2.70	0.89	–2.50
Mean	–0.129		0.98	–0.62	0.99	–0.47
SD	0.469		0.17	3.61	0.16	3.12

Separation reliability = 0.987.
Chi-square test of parameter equality = 1359.555; df = 16; sig level = 0.000.

Rater Fit

The weighted and unweighted T statistics reported here were devised to give a normalized equivalent of the corresponding mean square residual statistics (Wright & Masters, 1982). With the normalized property of the T statistic, it is convenient to establish values of greater than absolute 2 as being significantly different at the 5% level or misfitting. The average number of observations per rater with these data was 1,036. The number of raters with misfitting weighted and unweighted T statistics under the RS model was 9 and 10, respectively, compared to 2 and 2 under the PC model.

The difference in the fit of the step facet and the rater facet under the two modeled conditions are summarized in Table 4. The standard deviation (*SD*) of the T value is expected to be close to 1.0 and the mean close to zero when the data fit the model. The average mean square residual value is expected to be close to 1.0 and the *SD* close to zero (Wright & Masters, 1982; Wu, 1997). By noting the change in these statistics between models, the impact of model difference can be gauged. The mean changes most for the weighted T step facet, with the PC model

TABLE 3.
Partial Credit Model, Rater Facet Output

TERM 1 Rater

Rater VARIABLES	ESTIMATE	ERROR	UNWGHTED FIT MNSQ	UNWGHTED FIT T	WGHTED FIT MNSQ	WGHTED FIT T
5	-1.105	0.051	1.02	0.50	1.01	0.20
16	-0.707	0.054	0.94	-1.50	0.97	-0.50
12	-0.493	0.057	0.98	-0.50	1.03	0.50
14	-0.317	0.048	0.95	-1.30	0.96	-1.10
4	-0.295	0.052	0.98	-0.40	0.97	-0.50
13	-0.186	0.055	0.95	-1.30	0.98	-0.30
11	-0.174	0.044	1.01	0.40	1.01	0.40
15	-0.100	0.065	0.97	-0.50	0.99	-0.30
7	-0.057	0.066	1.01	0.30	1.03	0.50
9	-0.039	0.053	0.92	-2.00	0.92	-1.70
3	-0.036	0.057	1.02	0.40	1.02	0.40
1	0.136	0.055	0.86	-3.30	0.87	-3.00
10	0.173	0.054	0.91	-2.10	0.91	-2.10
8	0.357	0.052	0.95	-1.20	0.95	-1.00
6	0.397	0.051	1.04	0.90	1.04	0.90
2	0.898	0.060	0.95	-1.20	0.94	-1.20
Mean	-0.097		0.97	-0.80	0.98	-0.55
SD	0.464		0.05	1.15	0.05	1.08

Separation reliability = 0.986.
Chi-square test of parameter equality = 1171.396; *df* = 16; sig level = 0.000.

producing a statistic that is closer to the value expected when the data fit the model. This may not be unexpected, given the differences in the response parameters being modeled; however, it is interesting to note the amount of change in the rater fit statistics. There was a reduction of both rater fit standard deviations by approximately 70% from the RS model to the PC model, which indicates an improvement in the estimation of the rater parameter.

Rater Severity

When there is a significant difference between the fit of the data to the different models, it is possible that some of the parameter estimates will differ between models. The rater severity estimates produced from the alternate models were examined for differences using a standardized difference and chi-square test. The results (Table 5) show that 7 of the 16 raters' severity estimates had changed significantly at the 5% level. The largest change was 0.544 logits (Rater 10) and the smallest significant change was 0.169 logits (Rater 13). In terms of the effect these differences

TABLE 4.
Facet Fit Summary

	Weighted Mean Square Residual				Weighted T			
	Step		Rater		Step		Rater	
Model	Mean	*SD*	Mean	*SD*	Mean	*SD*	Mean	*SD*
RS	1.043	0.040	0.987	0.157	1.285	0.881	–0.469	3.116
PC	0.989	0.081	0.975	0.049	–0.209	0.880	–0.550	1.079

TABLE 5.
Rater Severity Comparisons

	PC Model Severity		RS Model Severity					
Rater	Logits	*SE*	Logits	*SE*	Difference	Standardized Difference	Chi-square	*p* Value
1	0.136	0.055	0.152	0.054	–0.016	0.208	0.043	0.836
2	0.898	0.060	0.613	0.054	0.285	–3.531	12.465	0.000
3	–0.036	0.057	–0.062	0.055	0.026	–0.328	0.108	0.743
4	–0.295	0.052	–0.168	0.060	–0.127	1.600	2.559	0.110
5	–1.105	0.051	–1.398	0.050	0.293	–4.102	16.830	0.000
6	0.397	0.051	0.577	0.054	–0.180	2.423	5.873	0.015
7	–0.057	0.066	0.069	0.059	–0.126	1.423	2.026	0.155
8	0.357	0.052	0.315	0.052	0.042	–0.571	0.326	0.568
9	–0.039	0.053	–0.213	0.053	0.174	–2.321	5.389	0.020
10	0.173	0.054	–0.371	0.051	0.544	–7.324	53.641	0.000
11	–0.174	0.044	–0.258	0.046	0.084	–1.320	1.741	0.187
12	–0.493	0.057	–0.452	0.052	–0.041	0.531	0.282	0.595
13	–0.186	0.055	–0.017	0.051	–0.169	2.253	5.077	0.024
14	–0.317	0.048	–0.332	0.049	0.015	–0.219	0.048	0.827
15	–0.100	0.065	–0.107	0.062	0.007	–0.078	0.006	0.938
16	–0.707	0.054	–0.407	0.054	–0.300	3.928	15.432	0.000
Average	–0.097		–0.129			Chi-square	121.846	
Difference	0.032					*p* value 15 *df*		0.000

will have on the person measures, 0.544 and 0.169 logits are equivalent to 29% and 9% respectively, of a standard deviation in the person measure distribution.

Rater Function

The rater by step facet in the PC model has produced estimates that enable each rater's interpretation of the rating scale to be measured and compared to the common scale step structure, which is defined under the RS model. Each rater's modeled severity and unique use of the scale step structure from the PC model has been plotted as a rater function curve in Figure 1. This figure shows the modeled relationship between ability on the x-axis and expected raw score on the y-axis. The amount of ability required to move between these raw scores can vary from rater to rater. These curves show each rater's modeled severity at any point along the ability continuum.

The rater functions defined by the PC model have then been compared to the rater functions defined by the RS model to demonstrate the differences between the models (Figure 2). Figure 2 shows in expected raw scores the amount and locations of rater function differences. The main amount of disagreement between the models appears to be toward the extremes. These differences are composed of differences in rater severity and rater discrimination. The error component of these functions increases toward the extremes, as approximately 80% of the data are

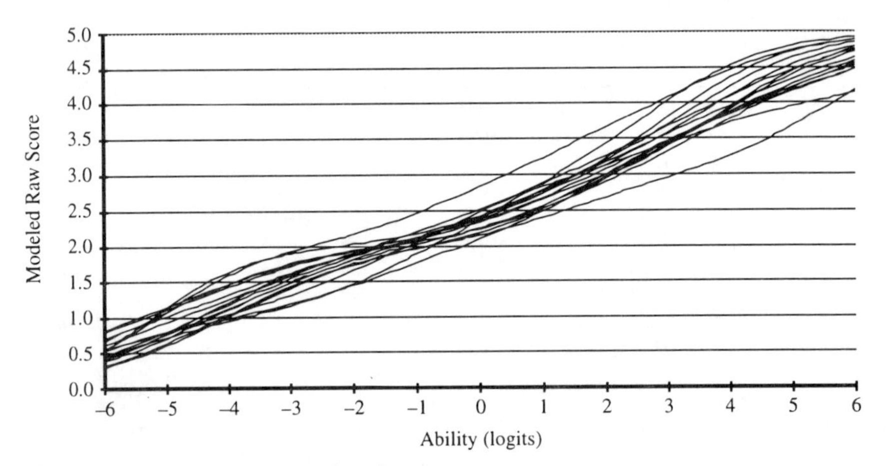

FIGURE 1. Partial credit rater function curves.

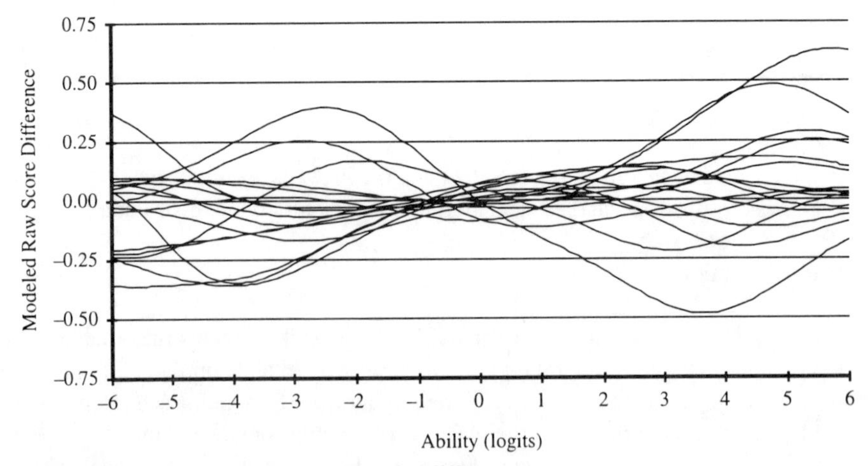

FIGURE 2. Between model rater function differences.

located within the range of –3.0 to 3.0 logits. Positive differences show that the resulting person measures would have been higher and negative differences lower under the RS model compared to the PC model. These differences show that under the PC model, some raters tend to give relatively high scores to people of low ability and relatively low scores to people of high ability, where for some other raters the pattern is the reverse.

Rater Discrimination

The modeled locations of the score step difficulties, for four raters, are shown with and compared to the RS model score step difficulties in Table 6. These score step difficulties are the thresholds described by Masters (1988, pp. 11–29), where the values represent the ability level required for an individual to have a 50% chance of passing that step. The four raters used here are the two with the most positive RS model weighted T fit statistics (Raters 4 and 6) and the two with the most negative RS model weighted T fit statistics (Raters 10 and 13). For these individual raters, the locations of the score level difficulties were defined under the PC model. To compare only rater discrimination with this method, rater severity was controlled for using the same severity estimate (average from the RS model) in each case. That is, the origin for each of the four raters was kept the same as the RS model origin. The result is a comparison of the discrimination of these four raters to the average discrimination of all raters.

For each of the four raters, there was a significant difference from the RS model score level difficulty estimates on at least three of the five score levels.

Relative Unmodeled Error

As the two models used here produce different parameter estimates, it is useful to try to measure the amount of error that would exist in the person measures if the model that explained less of the variation in responses (RS model) was used in preference to the PC model. The differences in average rater severity have already

TABLE 6.
Score Step Thresholds with Standard Errors

	Score Level Difficulty				
Score	RS model	Rater 4	Rater 6	Rater 13	Rater 10
1	–6.0(0.06)	–5.7(0.18)	–6.7(0.25)*	–7.0(0.31)*	–6.7(0.38)
2	–3.1(0.03)	–1.7(0.10)*	–2.4(0.04)*	–4.0(0.16)*	–3.7(0.12)*
3	0.4(0.02)	0.3(0.09)	0.6(0.09)	1.0(0.08)*	–0.4(0.08)*
4	2.9(0.03)	2.3(0.12)*	2.4(0.12)*	3.0(0.12)	2.9(0.12)
5	4.9(0.06)	3.7(0.20)*	5.1(0.20)	6.0(0.20)*	7.0(0.20)*

* Indicates statistical difference from RS model ($p < .05$).

been demonstrated along with the differences in rater discrimination, and rater function as the combination of rater severity with rater discrimination. Perhaps more importantly, there remain the person measures that can also be affected by such a choice in models. To demonstrate the effect of the model differences on person measures, the amount of relative unmodeled error has been calculated and reported here.

The amount of unmodeled error in the RS model relative to the PC model for person measures (logits) has been calculated as the root mean square error (RMSE) using Formula 3 to give the error component associated with a single scoring of each performance;

$$y = \sqrt{\sum_{r=1}^{N} \frac{(su_r - sc_r)^2}{N}}, \tag{3}$$

where;

su_r is the PC model expected score for rater r,
sc_r is the RS model expected score for rater r, and
N is the number of raters used in calibration.

Formula 4 was used to calculate the error component that reflects the number of raters used to score each performance.

$$RMSE = \frac{N - t}{N - 1} * y, \tag{4}$$

where t is the number of raters scoring each performance.

These calculations were applied at multiple points along the person ability (logits) continuum to produce the results shown in Figure 3. The range associated with each point is the RS model expected score ± 2 RMSE. Figure 3 shows the relationship of the average amount of unmodeled error in expected scores with person ability when the RS model is used in preference to the PC model. This figure shows that the choice of model can account for at least 0.5 of an expected score outside the ability range of –2.0 to 2.0 logits.

While the focus so far has been on the relative error between the two measurement models, there remains the point that in most measurement processes the person measures produced are estimates with an error component. This error component is an important component of the process of determining significant differences in performance. Although substantive differences in performance levels, based on the marking guide, may be a desirable method for categorizing performance levels (that is, a score of two is substantively different from a score of

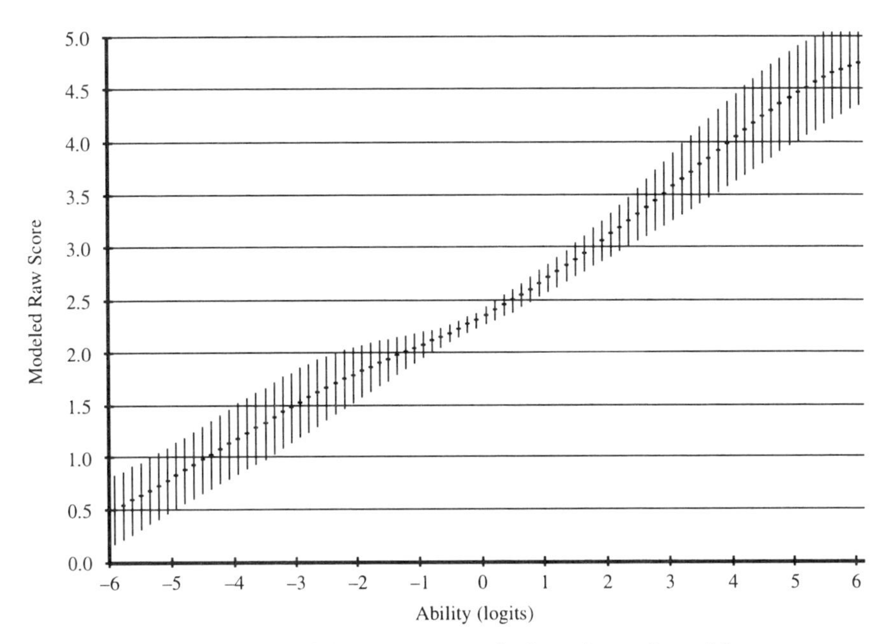

FIGURE 3. Relative unmodeled raw score error in the rating scale model.

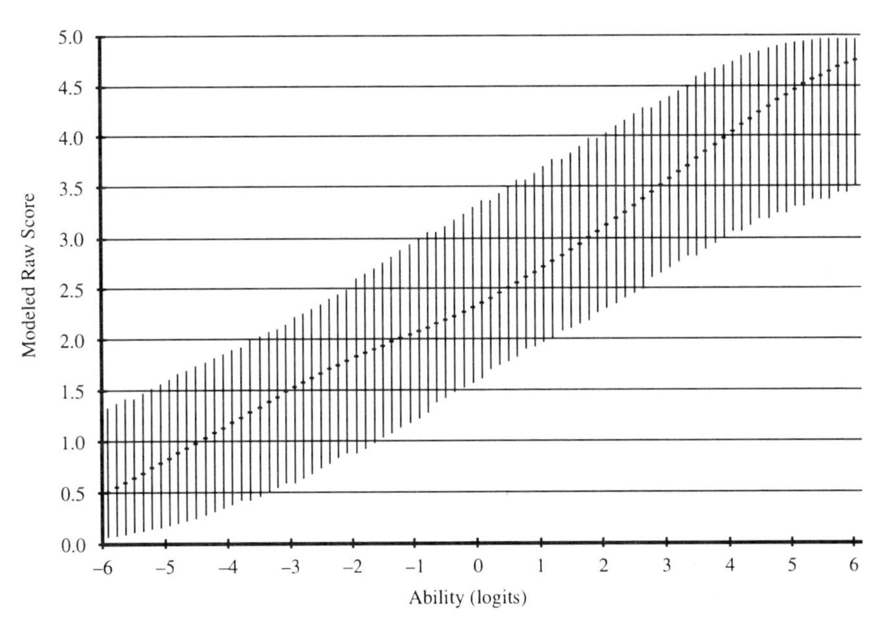

FIGURE 4. Person measurement error in rating scale model.

one), the amount of error in the complete measurement process should be considered before claiming that these scores are statistically different from each other.

In Figure 4, maximum likelihood estimates of person ability with ± 2 standard errors are shown with the expected scores based on the RS model. The error range shown represents the amount of person measurement error. These results suggest that an ability range of 4 to 5 logits represents significant difference in performance with these data. In addition to this error range is the error shown in Figure 3, which when added to the RS model person measures increases the range required before these measures become statistically different.

CONCLUSIONS

When the intention of the measurement process is to estimate the magnitude of a trait and then use that estimate to decide on a particular outcome with some degree of certainty, the level of certainty or precision required can be compared with the amount of error in the estimates being used. If the measurement error of the estimates is small enough not to interfere with the outcome for which these estimates are being used, then the measurement instrument has demonstrated utility. When the amount of error associated with the estimate becomes large enough to compromise the validity of the outcomes claimed, then the measurement process has problems.

When standards of performance are monitored over time it can be convenient to categorize persons into descriptive levels and report the percentage of persons in those levels. Depending on how well the data fit the measurement model used, consideration should be given not only to the standard error of the person measures but also to the amount of misclassification owing to data to model fit before any comparisons are made.

The importance of the differences found in the rater functions between the models is suggested to be related to the number of person measures affected by them and the impact that these person measures have. For instance if a substantial number of persons were located where major differences in rater function were being observed or a boundary or cut score was located in a similar position, the amount of error in the person measures or the precision of the boundary or cut score may need to be adjusted to avoid mismeasurement or misclassification.

Measurement models that apply a common scale step structure across all raters have the potential to maintain the same scale step structure over different occasions and with different raters, when the data fit the model. The distance between adjacent score steps can be kept the same and their location on the continuum is determined by the relative severity of the corresponding rater. When individual raters change their discrimination or new raters are introduced with different discrimination properties, the data fit to the measurement model and the objectivity of the person measures produced can both become questionable. If the majority of the raters become more or less discriminating than on previous occasions the range of person measures may shrink or expand respectively. This type of variation in rater discrimination has the potential to compromise the objectivity of person measure comparisons.

It is possible to use the PC model not only as a method for monitoring raters' performances under the RS model, but to produce comparable measures of person performance that are free of the raters used to score individual performances and different scoring occasions. Achieving this with the PC model would require a form of calibration or equating that held a reference discrimination. This reference discrimination could, for example, be based on a set of exemplar performances. It would be desirable that these performances be sufficient in both range of scores and number of performances at each score category to help produce stable estimates of all possible step parameters.

The successful operation of a rater-based measurement has some dependence on monitoring rater performance. When the fit statistics suggest that the response styles of raters may not be congruent with the model in operational use, part of the process of monitoring rater performance can include using more complex models that parameterize likely areas deviation. This process can quantify the amount of mismeasurement that is occurring under the operational model and allows diagnostic feedback to raters.

REFERENCES

Cantor, N. K., & Hoover, H. D. (1986, April). *The reliability and validity of writing assessment: An investigation of rater, prompt within mode, and prompt between mode sources of error.* Paper presented at the annual meeting of the American Educational Research Association, San Francisco.

Cason, G. J., & Cason, C. L. (1984). A deterministic theory of clinical performance rating. *Evaluation and the Health Professions, 7,* 221–241.

Congdon, P. J. (1997). *The stability of rater characteristics over a 12-month period.* Paper presented at the Ninth International Objective Measurement Workshop, Chicago.

Dunbar, S. B., Koretz, D. M., & Hoover, H. D. (1991). Quality control in the development and use of performance assessments. *Applied Measurement in Education, 4*(4), 289–303.

Engelhard, G., Jr. (1992). The measurement of writing ability with a many-faceted Rasch model. *Applied Measurement in Education, 5*(3), 171–191.

Engelhard, G., Jr., Gordon, B., & Gabrielson, S. (1991). The influences of mode of discourse, experiential demand, and gender on the quality of student writing. *Research into the Teaching of English, 26*(3), 315–336.

Gabrielson, S., Gordon, B., & Engelhard, G., Jr. (1995). The effects of task choice on the quality of writing obtained in a statewide assessment. *Applied Measurement in Education, 8*(4), 273–290.

Linacre, J. M. (1989). *Many-facet Rasch measurement.* Chicago: MESA Press.

Lunz, M. E., Stahl, J. A., & Wright, B. D. (1996). The invariance of judge severity calibrations. In M. R. Wilson & G. Engelhard. Jr (Eds.), *Objective measurement: Theory into practice* (Vol. 3, pp. 99–112). Norwood, NJ: Ablex.

Masters, G. N. (1988). Measurement models for ordered response categories. In R. Langeheine & J. Rost (Eds.), *Latent trait and latent class models* (pp. 11–29). New York: Plenum.

McNamara, T. F. (1996). *Measuring second language performance.* New York: Addison Wesley Longman.

McNamara, T. F., & Adams, R. J. (1994). Exploring rater behaviour with Rasch techniques. In *Selected Papers of the 13th L.T.R.C.* Princeton, NJ: Educational Testing Service.

Overall, J. E., & Magee, K. N. (1992). Estimating individual rater reliabilities. *Applied psychological Measurement, 16*(1), 77–85.

Ruth, L., & Murphy, S. (1988). *Designing writing tasks for the assessment of writing.* Norwood, NJ: Ablex.

Webb, L. C., Raymond, M. R., & Houston, W. M. (1990, April). *Rater stringency and onsistency in performance assessment.* Paper presented at the annual meeting of the American Educational Research Association and the National Council on Measurement in Education, Boston.

Wright, B. D., & Masters, G. N. (1982). *Rating scale analysis.* Chicago: MESA Press.

Wu, M. L. (1997). *The development and application of a fit test for use with marginal maximum likelihood estimation and generalised item response models.* Melbourne, Australia: Australian Council for Educational Research.

Wu, M., Adams, R. J., & Wilson, M. (1996). *ConQuest: Generalised item response modelling software.* Melbourne, Australia: Australian Council for Educational Research.

10

SETTING STANDARDS ON PERFORMANCE EXAMINATIONS

Mary E. Lunz
Measurement Research Associates

INTRODUCTION

Wisdom and experience define expectations for examination standards and give the criterion standard meaning and value within its context. If there is no value associated with passing a test or achieving a credential, it is unlikely that anyone would attempt to do so. When measuring educational achievement, the standard provides criteria for distinguishing those candidates who can demonstrate the required knowledge, skill, and/or professional behavior, from those who cannot. Standards are assumed to be well established and familiar, and to have value for those who implement them, and for the candidates whose performance is evaluated with them. Decisions made using established standards are assumed to have meaning beyond the particular candidates or judges or cases or skills represented on a specific test administration.

A previous version of this paper was presented at the Ninth International Objective Measurement Workshop, Chicago, Illinois, in March 1997.

BRIEF REVIEW OF STANDARD SETTING METHODS

The first issue in setting a standard is to determine if it will be based on norm-referenced or criterion-referenced approaches. Norm-referenced standards are determined by the performance of a candidate reference group. The standard is set based on reference group performance, and then applied to all candidates. This process usually occurs after the examination administration. A criterion-referenced standard is established prior to the administration of the examination based on an assessment by a group of experts, with regard to the level of knowledge and skill that represents satisfactory or passing performance. Methods for establishing criterion-referenced standards for multiple choice examinations have been studied in some detail.

An early methodology for setting criterion-referenced standards was to decide arbitrarily the percentage of items a competent candidate should be able to answer correctly, without reference to the difficulty of the items or the knowledge and skill the items were meant to test. The percent correct related neither to the domain of knowledge and skill nor to the measurement system. The standard was an ideal in the minds of the experts formed from intuitive beliefs about acceptable test performance (75% correct is a popular expectation), regardless of the nature or difficulty of the test.

In an effort to add some validity to standard setting methodologies for multiple-choice examinations, judgmental standard setting methods such as those of Nedelsky (1954) or Angoff (1971) were developed and implemented. Experts attempted to predict the response patterns or performance of candidates who are "minimally competent." Modifications to this basic concept include presentation of empirical item data and feedback to judges on their consistency (Norcini, 1990) with provision to change item ratings. A more objective approach was developed by Grosse and Wright (1987), who established standards using a selected group of criterion items and sometimes a criterion sample of candidates, in conjunction with Rasch analysis techniques, for standard setting purposes. Stone (1995) also used objective techniques that asked subject matter experts to identify the items with "essential content" and then to identify the mastery level that represented minimal competence on those items. Each of these approaches requires the input of subject matter experts who can agree on the level of performance that must be demonstrated to pass a multiple-choice examination.

Standards for performance examinations are more difficult to establish for several reasons. First, there are three or more facets (for example, examiners, projects, tasks) to account for in the examination process. The interactions among these facets determine how the scoring system is set up and the meaning associated with passing the examination. Second, some performance examinations, such as oral examinations, have minimal written documentation, and acceptability of candidate performance is defined by the perception of each individual judge. Third, performance examinations have a temporal quality owing to the situation-specific interaction of candidates, judges, and/or examination materials. Fourth, performance examinations are essentially different, parallel examination forms for each candidate. For example, candidates may be evaluated by different judges, at different times, on the same cases or problems; candidates may submit different portfolios or projects; or judges may choose to ask different questions or stress different areas in an oral examination. All of these factors, plus numerous others, cause performance examinations to be unique parallel examination forms. Thus, when setting a standard, all forms and variations of the performance examination must be considered in the standard setting process. Because candidates take different forms of the performance examination, the standard must be appropriate for all examination forms or differences among the forms must be accounted for before the standard is used to make pass/fail decisions about candidates. This is a considerable challenge, and few have attempted to establish norm- or criterion-referenced standards for performance examinations. Thus, the process of making pass or fail decisions has historically been given to individual judges, and the reliability of the results has often been considered to be suspect.

Whenever a judge is allowed to make a pass/fail or placement decision, the judge is implementing an individual standard, which also becomes the standard for passing or failing for the particular candidate being assessed. Even when specific criteria have been established for such judgments, judges differ in their interpretations of criteria because of their individual experience, education, and personal expectations (Tubs & Moss, 1974), despite extensive training as a rater. It is difficult to alter 30 years of experience in a 3-hour training session. When individual judges score and classify a candidate's performance, the decision is essentially unreproducible (Lunz, Stahl, & Wright, 1990), because there are probably no two judges who view the candidate's performance in exactly the same way. Some judges may make similar decisions based on the information available, or training, or both, but there is no guarantee that the same decision will be made by two judges, or that the individual judge will make the same decision about the same candidate twice. This is a primary reason why performance examinations have historically been called unreliable.

Interjudge reliability was used as a method to argue reliability for individual judge standards for performance examinations. In this approach to standard setting, judges are trained to implement the grading criteria so that the ratings of judges are presumed to be consistent. High interjudge reliability was thought to be evidence that candidates would earn comparable ratings regardless of the judge who did the rating. The following methods have been used to train judges to be consistent in their grading of performance examinations:

1. Defining, in detail, the meaning of each category on the rating scale, is likely to focus judges' perceptions so they use similar criteria for rating candidates.
2. Defining the meaning of the points of the rating scale may help judges use the rating scale consistently.
3. Providing workshops on the content of the cases, expectations, and outcomes may help to focus judge expectations.

4. Providing experience in rating candidates and then comparing the ratings of two judges for the same candidate may provide insight into similarities and differences among judges.
5. Providing feedback concerning rating patterns during a specific test administration may also be helpful.

When interjudge reliability is the basis for implementing a standard, the argument is that candidates have a comparable probability of success regardless of the judge to whom they are randomly assigned. Most of the perception literature indicates that people have unique perceptions, and interjudge reliability studies often result in only slightly better than chance agreement (Kortez, 1993). This suggests that other methods of establishing standards for performance examinations may be advisable.

Engelhard and Gordon, in the first chapter of this volume, evaluated the quality of judgments in the context of setting criterion standards for an examination of writing ability using the multifacet Rasch model. The components were the judge's view of writing competence (for example, personal standard) and the predetermined quality of the packet. The operational judges were asked to identify the packets that should pass and those that should fail. Even though there were significant differences in judge severity, the packets fell into two distinct groups so that it was fairly straightforward to establish a criterion standard.

Jaeger (1996) defined an approach for setting standards for performance examinations called iterative judgmental policy capturing, which used distinct multidimensional exercises to elicit responses from judges. Judges were trained and then asked to apply the principles to teacher profiles. As might be expected, results indicated the panelists could make these judgements; however, the algorithm for translating the results into a standard was unclear.

Glass (1978) argues that underlying the concept of achievement measurement there must be a continuum of knowledge and proficiency which is tied to a repertory of behaviors ranging from incompetent to highly competent. This is commensurate with the principles of latent trait theory (Rasch 1960/1980). However, with performance examinations the number of facets that influence the score of a candidate increase from two to three, five, or more. Understanding the influence of each facet on candidate scores is almost impossible without the assistance of computerized analysis such as the multifacet Rasch model (see Linacre, 1989).

Error in Standard Setting

Two sources of error are unavoidable when setting standards. The first is judgment error. Each judge has a unique perspective that influences his or her requirements for competence. How the judge feels that day and the dynamics of the group standard setting process influence the standard established by the group. Judgment error is usually not systematically measured, even though it affects the meaning and interpretation of the standard.

The second type of error in setting standards is measurement error. Error occurs in the estimation of task difficulties, candidate abilities, and examiner severities, and/or any other facet. This means that an item difficulty or a candidate ability estimate is not really a point, but a region of points on the scale. We have the most confidence in the accuracy of scores located outside the error of measurement that surrounds the pass point. Some candidates, however, are really minimally competent or minimally incompetent so that clear decisions for them are impossible.

Standards for Performance Examinations

The first step when setting a standard for a performance examination is to establish the *philosophy* under which the standard will be established, *norm* or *criterion* referencing. The next step is to *construct a pass point* on a scale using that philosophy. After the initial pass point has been identified, the third step is to *refine* the pass point, using the error of measurement, and the desired level of confidence in the decision. Candidates in the error of measurement surrounding the *pass* point *are* always the most difficult to place accurately.

RELATIVE OR NORM-REFERENCED STANDARDS FOR PERFORMANCE EXAMINATIONS

If a performance scale with raw scores or ability estimates is constructed, it is possible to set a norm-referenced standard. Of course, raw scores do not account for differences in the examination forms taken by candidates, but at least individual judges are not directly responsible for making pass/fail decisions about candidates.

To establish a norm referenced standard, an appropriate reference group is identified. Typically, candidate total test scores are scaled and summary statistics calculated using total scores. Scores at 1.00 or 1.5 standard deviations below the reference group mean locate the standard for acceptable performance. The rationale for this approach is that candidates who pass should be able to perform at a designated level, compared to an appropriate norm group. The ordering of the candidates determines those who pass and those who fail. The norm-referenced standard can be established for each examination administration, or examinations can, theoretically, be equated so that the same standard from the same norming group is used across administrations.

A norm-referenced standard requires a distribution of candidate scores. The more observations collected for each candidate performance, the lower the error of measurement and the more reliable the score and standard is likely to be. If there is a sufficient number of observations of each candidate's ability to calculate a mean and standard deviation, then this method can be used for setting standards for performance examinations. When this approach was used by a medical specialty board, the first-time U.S.-educated candidates served as the reference group.

The standard was established at one standard deviation below the mean of the reference group and applied to all candidates. A more specific example is unnecessary because the method has been used for many years and types of examinations.

CRITERION STANDARDS: DEFINITION AND SCALING

A criterion standard, by its nature, is determined from collective input of experts, and theoretically, encompasses the perspectives of a group of experts who define what a successful candidate should know or be able to do (Cizek, 1996). Thus, the criterion standard derives from collective judgment, rather than individual judgment. It represents the quality of performance that the candidate should be able to demonstrate within the context of the examination. The interpretation of the standard derives from the complexity and difficulty of the content being tested relative to the ability of the candidate. The higher a candidate's ability, the more knowledge and skill mastered, and the more likely he or she can meet the standard. This is a basic assumption of the Rasch model (1960/1980).

The criterion standard is translated to a pass point, the logical point on a benchmark scale that represents the level of mastery required for acceptable or competent performance. To pass, the candidate must be able to demonstrate a level of knowledge and skill that is commensurate with the criterion standard, regardless of the cases, tasks, or judges included on the performance examination.

Because performance examinations require the intervention of a judge, the differences in judges and any other facets (for example, tasks) that are part of the performance examination, must be acknowledged and accounted for when total performance scores are calculated (Linacre, 1989). Considering judges or tasks to be invariant (see Fagot, 1991) is unrealistic and a primary cause of the unreliability typically associated with performance examinations. Judges distinguish between acceptable and unacceptable candidate performances using a rating scale. The ratings are then summed across cases, tasks, and judges, and so on, to determine a score. The first problem is how to establish a criterion standard when multiple facets must be accounted for in the process. Another problem is to ensure that the complexity of the rating is also represented in the summary score. The input of each judge must be preserved such that it can be explained in the interpretation of the results.

When a criterion standard, represented as a pass point on a benchmark scale, is implemented, any candidate who can demonstrate the required level of performance passes, regardless of the particular judges or cases encountered. Candidates pass or fail based on the established criterion standard, rather than based on individual judge standards. This leads to decision reproducibility because the contextual bias produced by judge severity and case difficulty is removed, thus improving overall examination reliability (Lunz, Stahl, & Wright, 1990).

The four criterion-referenced standard setting methods described in this chapter depend on the use of the multifacet Rasch model (Rasch, 1960/1980; Linacre, 1989) and the FACETS computer program for analysis of examinations with more than two facets (Linacre, 1996). The FACETS program constructs a variable on which candidate ability estimates are located, after the contextual bias caused by the particular examination form presented to the candidate is accounted for. Thus, candidate ability estimates are located on the benchmark scale when these methods of criterion-referenced standard settings are used.

The standard setting methods have been pilot tested with data from several medical specialty and allied health certification boards. The context of the standard is very important. When there is concrete evidence of candidate ability, such as an essay, practical examination, or portfolio, the judges rate the quality of the evidence without interaction with the candidate. When oral examinations, or other types of examinations that require direct interaction with the candidate are used, the examination is more temporal. For example, there is no way to document how the examiner reacted to the appearance or attitude of the candidate. Another contextual aspect is the design of the examination. Perhaps, standardized cases are used, so a case difficulty can be calculated. The number of tasks or skills on which the candidates, projects, or portfolios are rated also influences the implementation of a standard. Some of these issues are outlined in Lunz and Bergstrom (1998). The standard setting methods presented have been pilot tested in contexts that were appropriate for their use.

CRITERION STANDARDS BASED ON COMPARISONS OF GLOBAL ASSESSMENTS AND ANALYTICAL SCORES

This method is based on the belief that judges can identify globally satisfactory performance and provide a global assessment (pass or fail; competent or incompetent), as well as detailed analytical ratings on defined tasks that can be used to construct a scale. Analytical scores are compared to the more subjective global assessments to determine the location of the region of passing on the scale. It is a criterion-referenced standard because the analytical scoring is associated with specifically defined acceptable behaviors, and the subjective global assessments confirm the analytical scoring. Once the standard is established, all candidates are required to achieve that standard to pass.

The sample of "criterion examinations" must be carefully selected. The standard setting examinations should represent minimal acceptable or unacceptable performance, so they will be close to where the pass point should be located on the scale. Criterion examinations are scored twice, first using the detailed analytical methods, and again with a global assessment of overall performance. The analytical and global assessments are compared to identify a region where the analytical score of the *lowest* globally acceptable criterion examination and the *highest* globally unacceptable criterion examination resides. A pass point is then established through consensus of experts, within this region. The steps are shown in Figure 1. The criterion standard is based on explicit examples of performance of the required tasks.

The performance examination used to demonstrate this method of standard setting uses data from an allied health certification board and evaluates the skills of histotechnicians. Each candidate produces a set of laboratory slides. Thus, a concrete manifestation of the candidate's ability is present, only the judgment of the quality of that performance is temporal. This criterion standard is constructed from a combination of detailed analytic ratings and global classifications of examinee performances. The detailed analytical ratings of each "criterion examination" are included in the total sample of examinations. The ability estimates of the criterion examination indicate the region of the criterion or standard on the benchmark scale. The global classifications locate the transition from satisfactory to unsatisfactory. The intersection of these indicators should be close to the pass point. The resulting pass point is based on actual performance of tasks considered to be necessary for competence. This pass point, once established, can serve as an absolute and detailed representation of the difference between passing and failing performances. Figure 2 shows the graph of the region of the pass point.

Setting criterion standards from criterion examinations connects the standard and actual performances to the detailed professional expectations of the judges who are evaluating the performances. The specific "criterion" examinations used to set the standard can be preserved as explicit documentation of borderline com-

1. Identify a set of 3 or 4 criterion examinations that appear to demonstrate slightly above or below satisfactory performance. These criterion examinations will be graded using both analytical and global ratings by a group of experienced judges.
2. Include the criterion examinations in with the sample of examinations for analytical grading. They are graded using the analytical rules.
3. Examiners make an additional summary statement on the overall acceptability of the candidate's performance with a global rating. This may take the form of a pass/fail decision or other appropriate assessment. The comparison of the analytical scoring and the global judgment of acceptability relate the global decision to an ability measure, and provide the criteria for establishing a criterion standard.
4. Identify the region of the criterion standard on the examination scale by comparing the analytical scores and the global ratings. This will identify the region on the scale that represents minimally acceptable performance.
5. The committee of experts makes the final decision about the exact location of the pass point on the scale.
6. Implement the pass point for all candidate performances.

FIGURE 1. Global Versus Analytical Standard Setting.

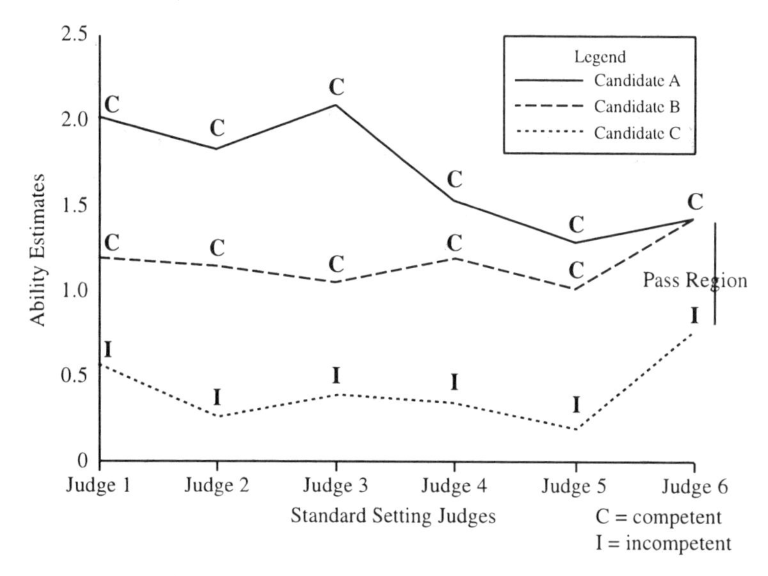

Analytic Scoring for Criterion Examinations

	A		B		C		Global Ratings		
Judge	Meas	(*SE*)	Meas	(*SE*)	Meas	(*SE*)	A	B	C
3	2.02	(.34)	1.18	(.25)	.56	(.23)	C	C	I
4	1.83	(.36)	1.14	(.27)	.26	(.23)	C	C	I
34	2.09	(.40)	1.04	(.26)	.38	(.23)	C	C	I
35	1.53	(.34)	1.19	(.29)	.34	(.23)	C	C	I
36	1.28	(.28)	1.01	(.25)	.18	(.22)	C	C	I
50	1.43	(.30)	1.42	(.28)	.75	(.24)	C	C	I

C, competent; I, Incompetent.

Note. Three criterion examinations (A, B, C,) are graded by six judges (3, 4, 34, 35, 36, 50) using detailed analytical methods and global judgments. The pattern of analytical scores that represents competent and incompetent is defined.

FIGURE 2. Benchmark performances: Analytical versus global ratings.

petence. The standard can be reviewed, verified, and revised with additional standard setting studies as needed. These new studies can include a regrading of the original criterion examinations as a check of the stability of the judging.

This approach to standard setting is best for performance examinations for which candidates produce samples such as essays, work samples, or specifically defined projects, and/or different rating scales are used to grade the tasks (for example, 0/1 for tasks A and B versus 0, 1, 2, 3 for task C). It is less effective for examinations that depend on interviews. The example is pretty clear cut in that the analytical and global ratings support each other; however, sometimes judges are not as consistent in their global ratings, making the pass region more difficult to identify. There is always the error of measurement to consider when setting a standard. Adjustments within the error of measurement can be made as deemed appropriate.

FAIR AVERAGE APPROACH

The fair average score is calculated as part of the FACETS program (Linacre, 1996). It is the average score earned by the candidate after the characteristics of the examination form are accounted for by the multifacet Rasch analysis. This accounts for the contextual bias caused by the severity of the examiners and the difficulty of the tasks or subject areas, or both.

The data used for this example are from an oral examination of a medical speciality board. Candidates were rated on three skills—diagnosis, treatment, and outcome—by each of two examiners. The first examiner used two standardized protocols and the second examiner used two additional standardized protocols. Therefore, each candidate earned a total of 24 ratings; a rating of 3 (excellent), 2

Rating Scale Developed to Indicate Performance

Example: 3 = Excellent
*2 = Satisfactory (Criterion for passing)
1 = Unsatisfactory

Fair average formula accounts for the characteristics of the examination form taken by a candidate, then translates back from a logit ability estimate to an average score using the following formula:

$$\text{Fair Average} = RSR * (e^{LPC} / 1 - e^{LPC})$$

where:

RSR = rating scale range (number of categories)
LPC = logit percent correct
B_o = observed ability
B_n = ability estimate corrected for contextual bias
m = rating scale centering [depends on number of categories]
x = rating scale slope [intercept]

$$B_o = m + x * \log (R/W)$$

$$B_n = B_o - D_i - C_j - T_m$$

$$LPC = (B_n - m) / x$$

Note. The fair average score accounts for the characteristics of the examination taken by a candidate and, therefore can be interpreted as an absolute requirement to pass.

FIGURE 3. Fair average approach.

CANDIDATES	Obsvd Score	Obsvd Ratings	Obsvd Average	**Fair Avrge**	Ability Measure	S.E.
1122	58	24	2.4	**2.3**	.93	.36
1133	58	24	2.4	**2.3**	.91	.36
1126	58	24	2.4	**2.3**	.88	.36
1145	58	24	2.4	**2.3**	.80	.35
1096	54	24	2.3	**2.3**	.77	.35
1101	54	24	2.3	**2.3**	.75	.35
1129	58	24	2.4	**2.3**	.72	.37
1073	54	24	2.3	**2.3**	.71	.35
1089	54	24	2.3	**2.3**	.69	.35
1113	56	24	2.3	**2.3**	.68	.35
1119	56	24	2.3	**2.2**	.60	.35
1057	52	24	2.2	**2.2**	.57	.35
1001	46	24	1.9	**2.2**	.56	.35
1077	52	24	2.2	**2.2**	.56	.35
1131	56	24	2.3	**2.2**	.46	.36
1114	54	24	2.3	**2.2**	.44	.35
1146	54	24	2.3	**2.1**	.33	.34
1140	54	24	2.3	**2.1**	.28	.37
1170	52	24	2.2	**2.1**	.25	.35
1141	52	24	2.2	**2.1**	.23	.34
1071	50	24	2.1	**2.1**	.20	.34
1072	50	24	2.1	**2.1**	.20	.34
1153	52	24	2.2	**2.0**	.13	.34
1148	52	24	2.2	**2.0**	.11	.33
1162	52	24	2.2	**2.0**	.11	.34
1058	48	24	2.0	**2.0**	.08	.35
1070	48	24	2.0	**2.0**	-.03	.34
1121	50	24	2.1	**2.0**	-.03	.34
1144	34	18	1.9	**2.0**	-.10	.38
1147	50	24	2.1	**2.0**	-.11	.33
1163	50	24	2.1	**2.0**	-.13	.34
FAIR AVERAGE OF 2.0 TO INDICATE SATISFACTORY PERFORMANCE = PASS POINT						
1081	50	24	2.1	**1.9**	-.14	.34
1116	48	24	2.0	**1.9**	-.26	.34
1132	50	24	2.1	**1.9**	-.27	.34
1161	48	24	2.0	**1.9**	-.36	.34
1180	46	24	1.9	**1.8**	-.45	.33
1143	46	24	1.9	**1.8**	-.47	.34
1123	46	24	1.9	**1.8**	-.49	.34
1124	44	24	1.8	**1.7**	-.72	.34
1179	42	24	1.8	**1.7**	-.90	.34
1025	38	24	1.6	**1.6**	-.96	.40
1152	42	24	1.8	**1.6**	-.96	.35
1169	42	24	1.8	**1.6**	-.98	.36
1177	40	24	1.7	**1.6**	-1.13	.34
1172	38	24	1.6	**1.5**	-1.52	.37

CANDIDATES	Obsvd Score	Obsvd Ratings	Obsvd Average	**Fair Avrge**	Ability Measure	S.E.
MEAN	59.6	23.9	2.5	**2.5**	1.64	.42
SD	8.0	0.8	0.3	**0.4**	1.25	.10

Reliability .88

Scale = 3 = Excellent; 2 = Satisfactory; 1 = Unsatisfactory

FIGURE 4. Pass point based on fair average.

(satisfactory), or 1 (unsatisfactory) on each skill for each protocol. This rating scale suggests that a candidate should earn an average score of 2.00, or satisfactory, to be considered competent to pass the examination. The problem is that an observed average score does not account for the severity of the examiners who rated the candidate, or the difficulty of the standardized protocols (rotated among examination sessions) encountered by the candidate, but the fair average does.

The first step is to identify the logit ability estimate for a *fair average of 2.00,* which is the fair average score that represents *satisfactory.* The logit ability estimate also accounts for the actual severity of the judges and difficulty of the protocols encountered by a candidate. The logit ability estimate, then, stands as the pass point or may be adjusted by the error of measurement to avoid Type 1 or Type 2 errors in passing or failing candidates. This process combines the ratings given by the examiners, the rating scale category definitions and interpretations, and the score earned by the candidate to establish a criterion standard. Pass/fail decisions are directly linked to the standard for satisfactory performance on the rating scale.

Figure 3 shows the fair average formula as implemented in the FACETS program, and Figure 4 shows the pass point on the examination scale. The fair average of 2.00 translates to –.13 logits on the benchmark scale. By adding or subtracting the error of measurement, the level of confidence in the pass/fail decision is established.

SYNTHETIC CANDIDATE APPROACH

Another alternative for setting a criterion standard is to develop a synthetic candidate. First, a group experts meet, and by consensus, determine the ratings that indicate acceptable performance for the barely passing candidate for each task on each project or case or task. Theoretically, the expected ratings take into account the perceived difficulty of the project. Ratings representing performance at the expected levels, are entered into the data file as a "synthetic candidate." The synthetic candidate is then analyzed with the other candidate performances. Theoretically, this synthetic candidate performance should be close to the pass point or at the pass point, as it is designed to represent the standard setting committee's conception of minimally acceptable performance. The construction of the minimally competent candidate performance requires some careful consideration of the difficulty of the projects and the expectation for task performance within cases or projects by the standard setting committee. Experts must agree on the appropriate rating for each so that the judge severity for the synthetic candidate is, essentially .00. Figure 5 shows the steps in developing a synthetic candidate data record.

This method is useful whether one rating scale or more than one rating scale with different categories is used in the examination. It also requires that the candidate submit work samples or projects to be evaluated. Figure 6 shows the scale and pass point

1. Standard setting group discusses the expected performance of the minimally competent candidate on each task on each project/case.
2. Assign ratings by group consensus, on each task for each case. These ratings vary based on the perceived difficulty of completing the task within each case.
3. Complete a data sheet for the synthetic candidate and enter into the computer.
4. Analyze the synthetic candidate data with those of the actual candidates.
5. Note the estimated ability measure of the synthetic candidate on the scale.
6. Verify the logic of setting a pass point at that level with the standard setting committee.
7. If necessary adjust the pass point by the SEM.
8. The data sheet for the synthetic candidate serves as evidence of the expectations for passing performance. Implement the standard for scoring candidates.

Sample Synthetic Candidate Record

Project Rating Scale	Task 1 0–3	Task 2 0–1	Task 3 0–1
Project 1	2	1	1
Project 2	2	1	0
Project 3	2	1	1

Task 1: 0 = Unacceptable, 1 = Marginal, 2 = Satisfactory, 3 = Excellent.
Tasks 2 and 3: 0 = Unacceptable, 1 = Acceptable.

Rating scales differ among tasks in this example. A minimum of 2 is expected for Task 1 across projects, and full credit is expected across projects on Task 2. Task 3 is considered to be more difficult on Project 2, so a lower rating is judged appropriate for minimal competence in the overall scheme of the examination. Passing or Failing the examination is based on the total score. In the example, the synthetic candidate earned an ability estimate of 1.21. This was used as the criterion standard.

FIGURE 5. Synthetic Candidate Data Record Construction.

for an examination in which candidates submitted defined work sample projects. The projects were graded on three tasks, each with a unique rating scale.

The synthetic candidate performance acts as the criterion to define a pass point. It represents input and consensus about barely acceptable performance from a group of qualified experts. The ability measure represented by the synthetic candidate can be carried forward through equating processes or rescaled with the data for each examination administration. Adjustment for level of confidence can be made to the criterion standard by adjusting for the error of measurement.

Exam Nu	Obsvd Score	Obsvd Count	Obsvd Average	Ability Measure	S.E.
7	84	47	1.8	2.60	.40
50	82	47	1.7	2.59	.36
40	84	47	1.8	2.51	.40
48	84	47	1.8	2.47	.40
24	79	47	1.7	2.26	.32
4	81	47	1.7	2.08	.34
9	79	47	1.7	2.08	.32
43	82	47	1.7	2.05	.36
46	76	47	1.6	2.05	.29
45	76	47	1.6	2.01	.29
42	80	47	1.7	2.00	.33
2	80	47	1.7	1.94	.33
23	77	47	1.6	1.87	.30
49	74	47	1.6	1.81	.28
3	76	47	1.6	1.78	.29
41	75	47	1.6	1.75	.28
15	73	47	1.6	1.74	.27
38	75	47	1.6	1.72	.28
21	75	47	1.6	1.65	.28
35	74	47	1.6	1.52	.28
32	72	47	1.5	1.48	.27
14	70	47	1.5	1.45	.26
27	70	47	1.5	1.41	.26
17	73	47	1.6	1.38	.27
37	66	47	1.4	1.38	.24
53	66	47	1.4	1.38	.24
11	65	47	1.4	1.36	.24
52	70	47	1.5	1.36	.26
1	70	47	1.5	1.31	.26
34	66	47	1.4	1.29	.24
26	65	47	1.4	1.26	.24
39	71	47	1.5	1.23	.26
47	68	47	1.4	1.23	.25
51	64	47	1.4	1.21	.24
*******	**69**	**47**	**1.5**	**1.21**	**.25**
PASS POINT BASED ON THE SYNTHETIC CANDIDATE					
8	66	47	1.4	1.17	.24
5	63	47	1.3	1.15	.24
10	64	47	1.4	1.10	.23
13	63	47	1.3	1.07	.23
16	66	47	1.4	1.00	.24
22	59	47	1.3	.98	.23
33	59	47	1.3	.98	.23
12	60	47	1.3	.96	.23
19	63	47	1.3	.94	.23
6	58	47	1.2	.82	.22
31	55	47	1.2	.77	.23
18	62	47	1.3	.69	.23
25	62	47	1.3	.64	.23
30	57	47	1.2	.62	.22
29	63	47	1.3	.59	.23
44	63	47	1.3	.59	.23
20	51	47	1.1	.48	.22

*****Synthetic Candidate**
(ability estimate = 1.21)

Nu	Obsvd Score	Obsvd Count	Obsvd Average	Measure	Model S.E.
Mean	69.4	47.0	1.5	1.43	.27
S.D	8.2	0.0	0.2	.54	.05

Reliability .74

See Figure 5 for explanation of the construction of the synthetic candidate

FIGURE 6. Standard based on a synthetic candidate.

ASSESSMENT OF PROJECT OR CASE DIFFICULTY

This approach to setting criterion standards for performance examinations is most analogous to standard setting methods for multiple-choice examinations and involves assessing the difficulty of each case or standardized protocol, based on expected performance from the minimally competent (satisfactory) candidate. The general process is for the group of experts to discuss the assets and liabilities of the each case to get a feeling for how easy or difficult the case will be for the candidate. The interpretations of the rating scale categories must be clearly defined. Each expert may assess each case independently, or a group of experts may assess case difficulty by consensus. Each case assessment answers the question, "What rating should a barely satisfactory or competent candidate receive for each task associated with the case?" The ratings are then summed and divided by the number of experts or ratings, or both, to construct a *predicted average* for the case or project and a *predicted average* for the total examination (all cases). This *predicted average* can then be used for standard setting purposes.

Theoretically, the predicted average for each case represents the level of performance deemed acceptable to pass. This process serves several useful purposes. First, it forces the standard setting committee to review each case from a new perspective—namely, scoring—and to determine the meaning of different ratings within the context of each case. Second, it provides subjective expert assessment of the difficulty of each case which documents the meaning of the standard. Third, it confirms for examiners that cases or protocols usually differ in difficulty. Some cases represent problems that are more difficult for the candidate than others. This can later be confirmed with actual test results.

The overall predicted average is the basis for identifying the region where the criterion standard should be established. A committee of experts then makes a final decision concerning the pass point. The steps in the process are presented in Figure 7.

1. Identify the standardized cases being used in an examination.
2. Standard setting committee provides ratings for each skill for each case individually or by consensus. Ratings should reflect the perceived difficulty of scoring a high rating on the skills within the case for the barely passing candidate.
3. Sum or average the predicted ratings across skills and cases to construct the *predicted average* for the overall examination.
4. Administer the test.
5. Analyze the data using multifacet methods.
6. Compare predicted average to actual performance of candidates (observed average and fair average) on the skills, case by case.
7. Determine the criterion region (mean rating ± *SEM*) using the predicted and perhaps fair average.
8. Set the pass point using input plus judgment of the standard setting committee.
9. All candidates who meet the standard pass, others fail.

FIGURE 7. Steps in standard setting using standardized cases.

The factor that is not accounted for in this process is the severity of the experts with regard to expectations for acceptable/competent performance of each case. Consequently, when sums or averages are calculated, all judges are treated as if they are comparable in their expectations for passing performance. The reality is that all judges are not comparable; however, differences among them average in the standard setting process. Another validation is to compare the predicated average to the fair average calculated by the FACETS program (Linacre, 1996). The predicted average is located on the scale, and all candidates can be required to meet that level of performance. Adjustments for the error of measurement can be made to determine the level of confidence in the decision.

The pilot study for this technique was carried out by a medical specialty board. The group of experts analyzed and gave ratings for the skills using all of the standardized cases planned for the oral examination. After discussion of the case, experts made independent ratings of the expected performance for the barely passing candidate in each of the skill areas for each of the standardized cases. The rating scale was 3 (excellent), 2 (satisfactory), and 1 (unsatisfactory). Figure 8 shows the results. The observed and predicted average for each standardized case are shown. Generally, the predicted average for each case was lower than the observed average for each case. Cases 10 and 25 had the greatest difference between predicted and observed. Cases 9 and 17 were the only cases in which predicted performance was higher than actual performance. The criterion standard in this example would be a predicted average of 2.02, and would be located on the scale using the observed average or the fair average, which also accounts for the impact of examiners. This would be transcribed to a logit ability estimate and located on the scale to identify the region where the criterion standard should be established. Refinement using the error of measurement could then be done.

This method of criterion standard setting is useful only when standardized cases, projects, or questions are developed in advance and assessed prior to the examination. These cases are often used in an oral examination that involves an interview with the candidate. For this type of examination, the cases may be considered the most standardized and, therefore, controllable facet of the examination, making them appropriate for standard setting. It is noteworthy that the predicated average is very close to the fair average for satisfactory performance, which would be 2.00 if "satisfactory" performance were expected.

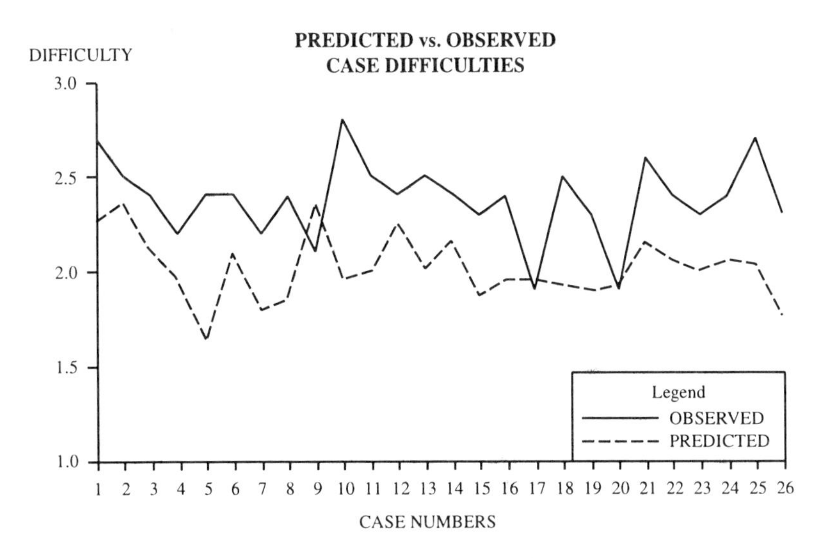

PREDICATED VS OBSERVED PROTOCOL DIFFICULTY

Protocol Number	Obsvd Score	Obsvd Count	Obsvd Average	Predicted Average	
1	808	300	2.7	2.26	
2	456	180	2.5	2.36	
3	404	165	2.4	2.10	
4	67	30	2.2	1.96	
5	374	156	2.4	1.63	
6	759	315	2.4	2.10	
7	67	30	2.2	1.80	
8	362	150	2.4	1.86	
9	63	30	2.1	2.36*	
10	415	150	2.8	1.96	
11	416	165	2.5	2.00	
12	355	150	2.4	2.26	
13	416	165	2.5	2.01	
14	374	156	2.4	2.16	
15	373	165	2.3	1.86	
16	743	306	2.4	1.96	
17	56	30	1.9	1.96*	
18	382	150	2.5	1.93	* predicted average higher than
19	70	30	2.3	1.90	observed average
20	56	30	1.9	1.93	
21	810	315	2.6	2.16	
22	371	156	2.4	2.06	2.02 is the overall predicted
23	361	156	2.3	2.00	average used to set the stand-
24	441	186	2.4	2.06	dard. It must be located on the
25	448	165	2.7	2.03	scale and transferred to a
26	373	165	2.3	1.76	logit value.
Mean	377.7	153.7	2.4	2.02	
S.D.	220.7	85.7	0.2	0.3	

FIGURE 8. Results of pilot study.

CONCLUSION

All of the methods presented require the use of the multifacet Rasch model for examination analysis. Even with norm-referenced standard setting approaches, it is logical to account for the differences in the parallel examination forms before setting a standard. Three of the four methods pilot tested incorporate some expert judgment during the standard setting process. Since a group of items is not available to document the standard for a performance examination, standard setting performances (for example, essays), the synthetic candidate record, protocol or case assessments, or the category definitions (fair average method) are the alternatives for documentation. All of these methods are based on the premise that experts can distinguish among case or task difficulties and/or candidate abilities. These standard setting methods use the multifacet Rasch model to account for the characteristics of the parallel examinations and to construct a scale on which a pass point can be established. Once a criterion standard is established, it can be carried forward using test equating techniques (see Lunz, Wright, Stahl, & Linacre, 1989).

REFERENCES

Angoff, W. H. (1971). Scales, norms and equivalent scores. In R. L. Thorndike (Ed.), *Educational measurement* (pp. 508–600). Washington, DC: American Council on Education.

Cizek, G. J. (1996). Standard setting guidelines. *Educational Measurement Issues and Practice, 15*(1), 13–21.

Fagot, R. (1991). Reliability of rating for multiple judges: Intraclass correlation and metric scales. *Applied Psychological Measurement, 15,* 1–11.

Glass, G. V. (1978). Standards and criteria. *Journal of Educational Measurement, 15,* 237–267.

Grosse, M. E., & Wright, B. D. (1987). Setting, evaluating, and maintaining certification standards with the Rasch model. *Evaluation and the Health Professions, 9*(3), 267–285.

Jaeger, R. M. (1996). Setting standards for complex performances: An iterative, judgmental policy capturing strategy. *Educational Measurement: Issues and Practice, 14*(4), 16–20.

Kortez, D. (1993). New report on Vermont portfolio project documents challenges. *National Council on Measurement in Education, 1*(4), 1–2.

Linacre, J. M. (1989). *Many-facet Rasch measurement.* Chicago: MESA Press.

Linacre, J. M. (1996). *FACETS, a computer program for analysis of examinations with multiple facets.* Chicago: MESA Press.

Lunz, M. E., & Bergstrom, B. A. (1998). *Oral examinations for licensure and certification* (Technical Report No. 100). Chicago: Measurement Research Associates.

Lunz, M. E., Stahl, J. A., & Wright, B. D. (1990). Interjudge reliability and decision reproducibility. *Educational and Psychological Measurement, 54*(4), 913–925.

Lunz, M. E., Wright, B. B., Stahl, J. A., & Linacre, J. M. (1989). *Equating practical examinations.* Paper presented at the annual meeting of the American Educational Research Association, San Francisco.

Nedelsky. L. (1954). Absolute grading standards for objective tests. *Educational and Psychological Measurement, 14,* 3–19.

Norcini, J. J. (1990). Equivalent pass/fail decisions. *Journal of Educational Measurement, 27,* 59–66.

Rasch, G. (1980). *Probabilistic models for some intelligence and attainment tests.* Chicago: University of Chicago Press. (Original work published 1960)

Stone, G. (1995, April). *Objective standard setting.* Paper presented at the annual meeting of the American Educational Research Association, San Francisco.

Tubs, S. L., & Moss, S. (1974). *Human communication: An interpersonal perspective.* New York: Random House.

part III

Measurement Theory

11

A MULTICOMPONENT RASCH MODEL FOR MEASURING COVERT PROCESSES: APPLICATION TO LIFE SPAN ABILITY CHANGES

Susan E. Embretson
Karen M. Schmidt McCollam
University of Kansas

INTRODUCTION

It is widely recognized that solving ability test items depends on successful outcomes to a series of information processing stages. Identifying and measuring the cognitive processes that underlie test performance is a major direction in cognitive ability research. Identifying the component processes, strategies, and knowledge structures involved in item solving explicates the construct representation aspect of construct validity (see Embretson, 1983). Measuring individual differences in processing abilities can explicate the relative importance of different sources of ability differences. For example, life span ability declines may be explained by comparing the relative decline of specific information processes. Explicating the impact of spe-

This research was partially supported by a University of Kansas General Research Fund Grant to to the first author. Susan E. Embretson has also published as Susan E. Whitely.

cific information processes not only provides possible explanations of aging effects, but also may enable improved intervention for modifying these deficits.

The multicomponent latent trait model (MLTM; Whitely, 1980) and the general component latent trait model (GLTM; Embretson, 1984) were developed to measure individual differences in processing components involved in solving ability test items. Both the MLTM and the GLTM are multidimensional Rasch models in which all processing components must be executed successfully to solve the item. Although both the MLTM and the GLTM could estimate several processing abilities simultaneously, subtask data were required to identify the components.

Recent developments in cognitive psychometric modeling permit component abilities to be measured directly from ability test items, without subtask level data. In this chapter, the GLTM is extended to estimate individual differences in fully covert processes. We will show how an improved estimation algorithm may be combined with special identification procedures to link the parameters to cognitive processes.

This chapter has two goals: (1) to present a multicomponent latent trait model for covert processes (GLTM-CP), with estimation and identification procedures that allow direct decomposition of covert processing abilities; and (2) to illustrate how a GLTM-CP can be applied to understand age-related declines in ability. For the first goal, the GLTM-CP not only permits direct decomposition of processes on actual psychometric items, but since GLTM-CP belongs to the Rasch model family, it also has justifiable interval scale properties. For the second goal, the GLTM-CP is applied to spatial ability data (McCollam, 1997) to estimate life span differences in underlying processes. Two covert processing abilities that have been hypothesized to explain age-related ability declines are estimated; control processes (or executive processes) and working memory capacity (a resource-limited process).

GENERAL COMPONENT LATENT TRAIT MODEL

The GLTM (Embretson, 1984) was developed to measure individual differences in underlying processing components on complex aptitude tasks. The GLTM is a conjunctive multidimensional Rasch model for tasks that depend on solving two or more underlying components. Success on each component, in turn, depends on the item's difficulty and the person's ability on the component. Thus, the GLTM specifies both between-component and within-component relationships.

The between-component relationship specifies how outcomes on the underlying components relate to solving the whole task. The GLTM between-component relationship is conjunctive; the task is solved only if each underlying component is correct. Given an exhaustive set of components, a simple GLTM gives the probability of item solving as the product of the component probabilities as follows:

$$P(X_{ijT} = 1 \mid \theta_j, b_i) = \Pi_k P(X_{ijk} = 1 \mid \theta_{jk}, b_{ik}), \tag{1}$$

where θ_{jk} is the ability for person j on component k, b_{ik} is the difficulty of item i on component k, X_{ijk} is the response for person j on the kth component of item i, and X_{ijT} is the response for person j on the total task for item i.

The within-component model belongs to the Rasch family of models. Here, item difficulty depends on a design structure. In the most simple GLTM, each item defines a separate factor. In this case, GLTM reduces to MLTM (Whitely, 1980) in which component outcomes depend on the person's component ability θ_{jk} and item's component difficulty b_{ik} as follows:

$$P(X_{ijT} = 1 \mid \boldsymbol{\theta}_j, \mathbf{b}_i) = \Pi_k \frac{\exp(\theta_{jk} - \mathrm{b}_{ik})}{1 + \exp(\theta_{jk} - \mathrm{b}_{ik})} . \tag{2}$$

A cognitive model of task difficulty, based on the item stimulus features, may be specified in the GLTM. In experimental cognitive studies on complex problem solving, such stimulus features are typically manipulated to operationalize the difficulty of specific processes. Thus, a cognitive model can define constraints for item difficulty, similar to the linear logistic latent trait model (LLTM; Fischer, 1973). Unlike the LLTM, in which only the constraints apply to total item difficulty, in the GLTM constraints apply within the components. Thus, GLTM component item difficulties are replaced by a model of stimulus complexity features, as follows:

$$P(X_{ijT} = 1 \mid \boldsymbol{\theta}_j, \alpha_k, \boldsymbol{\beta}_k) = \Pi_k \frac{\exp(\theta_{jk} - \Sigma_m \beta_{mk} v_{imk} + \alpha_k)}{1 + \exp(\theta_{jk} - \Sigma_m \beta_{mk} v_{imk} + \alpha_k)} , \tag{3}$$

where v_{imk} is the difficulty of item i on stimulus factor m in component k, $\beta_k = (\beta_{1k}, \beta_{2k} \ldots \beta_{Mk})$ are the weights of the M stimulus complexity factors in item difficulty on component k, α_k is the normalization constant for component k, and θ_{jk} is defined as in Equation 2.

The GLTM, as well as the MLTM, requires subtask data to identify the component parameters. That is, subtasks must be constructed from the whole item, according to a theory of task processing, to represent the process outcomes. For example, consider the following analogy:

Event : Memory :: Fire : ?
(1) Matches (2) Ashes* (3) Camera (4) Heat.

A rule-oriented process theory (see Embretson, 1984) would contain two components, Rule Construction and Response Evaluation. In the preceding analogy, subtasks that operationalize the two processing components are: (1) a rule construction subtask, in which the person specifies the analogical rule in the item stem (Event : Memory :: Fire : ?), without the alternatives given and (2) a response evaluation subtask, in which the person selects the correct alternative, given the total task plus the correct rule (for example, "Remainder or Left-over" for this item).

When subtask responses are available to define component process responses, it can be shown that the processing parameters for both the MLTM and the GLTM may be readily estimated from the marginal subtask responses. The estimation equations for the MLTM (Embretson, 1996) and the GLTM (Embretson, 1984) involve only the corresponding subtask responses. Conditional maximum likelihood or marginal maximum likelihood estimation may be employed for either model.

EM Algorithm for MLTM

Maris's (1995) development of an EM procedure for latent response models broadens the type of data to which the MLTM may be applied. The EM algorithm (Dempster, Laird, & Rubin, 1977) is well suited for missing data problems. The EM algorithm involves iterations between two steps: (1) an E-step in which expectations for the missing data are generated from the observed data and the model, and (2) an M-step in which parameters are estimated to maximize the likelihood of the complete (expected and observed) data.

Maris (1995) formulates various models, including the MLTM, as latent response models in which the outcomes of primary interest are missing. In the case of the MLTM, component subtask responses, X_{ijk}, may be missing. However, expectations for the missing data can be given from the model and the observed total task data. Estimation of person and item parameters are obtained by maximizing the likelihood expected data, based on the model and the observed data. Thus, the EM algorithm involves the maximization of a loglikelihood which includes both latent (that is, unobserved) and observed X_{ijk}'s as follows.

$$\text{ln } L(\mathbf{X}\,;\mid \theta j, \boldsymbol{b}) = \Sigma_k\,[\Sigma_i\, b_{ik}\, X_{i.k} + \Sigma_j\, \theta_{jk}\, X_{.jk} - \Sigma_i \Sigma_j\,[\ln(1+\exp(\theta_{jk}-b_{ik})]], \quad (4)$$

where $X_{.jk}$ and $X_{i.k}$ are defined as follows:

$$X_{i.k} = \Sigma_j\, X_{ijk}$$
$$X_{.jk} = \Sigma_i\, X_{ijk}.$$

$X_{i.k}$ and $X_{.jk}$ are sufficient statistics for estimation item component difficulties and component abilities, respectively, in the Rasch model. Thus, Equation 4 is the sum of K loglikelihoods, each of which is an ordinary Rasch model.

In the E-step of the EM algorithm, expectations for the latent responses, X_{ijk}, given the outcome of X_{ijT} and the current estimates of the model parameters, θ_{jk} and b_{ik} are calculated. If the total task was solved, such that $X_{ijT} = 1$, then all components must have been solved. However, if the total task is not solved, then at least one of the components is not solved. In fact, the probability of the least likely component limits the probability of solving the full task. For example, if the probability of solving one component is .80, then the highest possible probability for the full task is .80, according to conjunctive model in Equation 1.

Maris (1995) gives the expectations for the unobserved component outcomes as follows:

$$E\,(X_{ijk} \mid \theta_j, b_i\, X_{ijT}) = 1, \text{ if } X_{ijT} = 1,$$

and:

$$E\,(X_{ijk} \mid \theta_j, b_i\, X_{ijT}) = [P(X_{ijk} = 1) - P(X_{ijT} = 1)] \,/\, [1 - P(X_{ijT} = 1], \text{ if } X_{ijT} = 0. \qquad (5)$$

The expectations from the E-step are then submitted to an M-step to solve for the parameters. The solution iterates between E-steps and M-steps until a convergence criterion is reached.

Maris's (1995) EM algorithm has several advantages for the MLTM. First, if the data set includes responses to both the full task and the corresponding subtasks, the parameter estimates are based on the full set of data. In the original MLTM (Whitely, 1980), the full task responses were used for testing model fit but not for estimating component parameters. Second, mixed data sets in which different items are presented in the full task and subtask conditions are readily accommodated by Maris's (1995) EM formulation of the MLTM. In this case, the item and person marginals in Equation 5 would contain a mixture of observed and expected data. Third, the MLTM with the EM algorithm may be applied to data without subtasks. In this application, components are extracted to improve MLTM fit to the data, analogous to an exploratory factor analysis. Although the number of required components is generally interesting, the identification of the extracted components would be unclear. Analogous to rotational indeterminancy in exploratory factor analysis, the extracted components probably would not be invariant.

GLTM-CP

Subtasks are rarely available to define unique components for complex tasks. In fact, constructing subtasks may be impossible for some processes, such as metacomponent or executive processes. Unfortunately, covert processes for which no subtasks are available may not be measured with any model described earlier. Both the MTLM and the GLTM require subtasks to identify each component. The MLTM with the EM algorithm can be used without subtasks, but the theoretical identity of the components would be questionable, as previously noted.

This section describes a general multicomponent latent trait model for covert processes, GLTM-CP. GLTM-CP is an extension of the GLTM that permits direct decomposition of processes from ability test performance without the use of subtasks or external measurements. Multiple processing abilities may be meaningfully estimated from traditional binary ability test items if two conditions are met:

(1) covert processing outcomes are treated as "missing" data in the estimation algorithm, and (2) processing components can be identified by placing appropriate constraints on the solution. While the MLTM with the EM algorithm, as previously described, is analogous to an *exploratory* factor analysis, GLTM-CP is analogous to a *confirmatory* factor analysis.

Contemporary cognitive theory and research on aptitude leads to expectations about the processes involved on many ability test items. Processes are operationalized in cognitive experiments by manipulating task stimulus features. Then, the contribution of the various features to component process difficulty is estimated by mathematical modeling. Stimulus features in test items can often be scored for process complexity (for example, see Embretson, 1985) because many item types have been studied experimentally. A unidimensional Rasch model, such as the linear logistic latent trait model LLTM (Fischer, 1973), is appropriate if processing complexity combines linearly to influence item difficulty in the same way for all persons. However, if the stimulus features influence independent processes, and if people differ in their pattern of processing abilities, then an MLTM may be appropriate. Further, the item stimulus scores for the different processes may be used to identify separate processing components.

In the GLTM-CP, the covert responses are regarded as missing data. However, component abilities are identified by constraining component difficulties to reflect known sources of processing complexity. That is, component difficulty scores are derived from prior mathematical modeling research on the particular item type. Thus, item difficulty is set within the components by scores, u_{ik}, which reflect the predicted difficulty from the stimulus complexity factors. The GLTM-CP is given as follows:

$$P(X_{ijT} = 1 \mid \theta_j,\alpha_k,\beta_k) = \Pi_k \frac{\exp(\theta_{jk} - \beta_k u_{ik} + \alpha_k)}{1 + \exp(\theta_{jk} - \beta_k u_{ik} + \alpha_k)} \,. \tag{6}$$

For each specified component, only two parameters are estimated for items: an intercept, α_k, and a slope, β_k, for the (weighted) sum of the stimulus complexity factors on process difficulty.

APPLICATION TO LIFE SPAN DIFFERENCES IN SPATIAL ABILITY

It is almost axiomatic in life span psychology that cognitive abilities decline in old age. Various research designs find that nonverbal abilities decline somewhat earlier than verbal abilities. For example, Schaie and Strother's (1968) cross-sequential study found that nonverbal (for example, spatial ability) declines earlier than verbal ability (for example, vocabulary). Longitudinal (Schaie & Hertzog, 1983) and cross-sectional studies (Kaufman, Reynolds, & McLean, 1989) further support the earlier decline of nonverbal reasoning.

Among the various nonverbal abilities, spatial ability declines particularly early. Both longitudinal (see, for example, Schaie & Willis, 1993) and cross-sectional (see, for example, Persaud, 1991) studies of spatial ability show decreases in performance for adults across the life span. On average, the correlation between age and spatial performance in the life span literature is about –.40, and the average 60-year-old performs between 0.5 and 1.5 standard deviations below the average 20-year-old (Salthouse, 1991).

The underlying sources for age-related declines in abilities, such as spatial ability, however, remain controversial. Two general types of explanations have been offered: executive process deficits and nonexecutive process deficits. For executive process deficits, or general control processing deficits, activities such as deciding, predicting, allocating, and monitoring cognitive processes (Brown, 1978; Sternberg, 1980) are studied. For nonexecutive process deficits, resource-based processes, such as working memory capacity, have been studied. Working memory capacity is defined as memory available for temporary storage and manipulation of information (Baddeley, 1986).

Both general control processes and working memory capacity have been shown to affect cognitive performance. For general control processes, rule application (Babcock, 1994), strategy application (Cavanaugh, 1987), and training (Baltes & Kleigl, 1992; Lindenberger, Kleigl, & Baltes, 1992; Willis & Nesselroade, 1990) have been shown to affect older adults' performance. For working memory, older adults' performance has also been shown to be affected by reduced working memory capacity (Salthouse, Mitchell, & Palmon, 1989; Salthouse, Mitchell, Skovronek, & Babcock, 1989; Salthouse & Skovronek, 1992).

However, the available studies are limited in several ways. First, most studies isolate cognitive processes by using simple laboratory tasks (for example, memory tasks). The laboratory tasks do not necessarily measure the processes that are involved in ability tests and most laboratory tasks, in fact, are only weakly correlated with cognitive abilities. Thus, it would be uncertain whether age-related declines in these functions explain age-related declines in ability. Second, although task decomposition of ability items is more relevant to diagnosing the source of ability decline, little research is available. Further, the few available studies use simplified versions of the ability items, rather than actual ability test items. Simplified items probably involve different processes than the more complex psychometric items (Pellegrino & Lyons, 1979; Embretson & Schneider, 1989). Third, comparisons of process impact on age-related decline requires interval scale level measurement. It is well known that effect interpretations are not reliable if interval scale measurement has not been achieved.

Sources of Age-Related Decline in Spatial Ability

Examining the sources of age-related changes is particularly important for spatial ability since it declines particularly early. The GLTM-CP can be applied to decompose processing on spatial ability tasks, given the conditions are met, as described earlier. In particular, identifying processing abilities in the GLTM-CP requires constraining component item difficulties to represent processing complexity. Thus, a spatial ability test for component processing that is well understood is required.

The Spatial Ability Learning Test (SLAT; Embretson, 1989) was designed to reflect well-specified sources of cognitive complexity. The SLAT measures spatial visualization. Two variables in spatial visualization tasks, the degrees of rotation and the number of surfaces carried, have been found to be highly related to item response time and accuracy (Cooper & Shepard, 1973; Shepard & Feng, 1972). SLAT item difficulty is well explained by the amount of spatial processing required to solve the items.

SLAT items are cube folding tasks, which are presented in a multiple-choice format, as shown in Figure 1. An attached folding model was developed to account

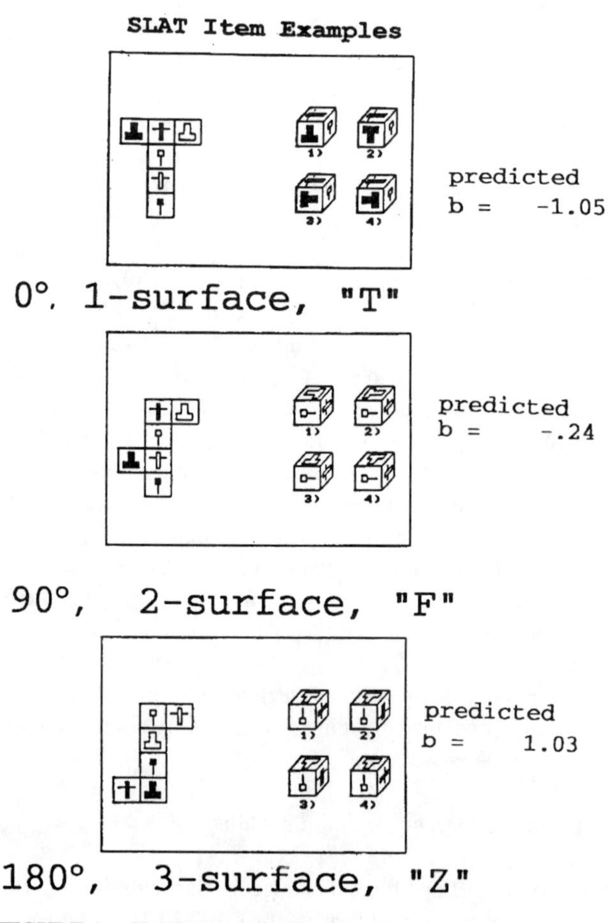

FIGURE 1. SLAT item examples.

for the processes in spatial folding items: (1) encode the item stem and alternatives; (2) select anchoring side and rotate if necessary; (3) fold the stem to meet anchoring side; and (4) confirm mentally folded stem to alternatives (Embretson & Waxman, 1989). SLAT items were designed to operationalize the difficulty of two processes: rotation, which depends on the degree of rotation (0, 90, 180) of the stem to the target; and folding, which depends on the number of surfaces (1, 2, 3) to be carried. Embretson (1994) reported multiple correlations of SLAT design features of $R = .87$ for response time and $R = .88$ for Rasch item difficulty, indicating that spatial processing difficulty can be highly predicted from item stimulus features that reflect rotation and folding.

Component Decomposition of Spatial Ability

Several theories suggest that two global processing items that are involved in complex tasks are (1) nonexecutive processes, which are resource limited, and (2) an executive process, which involves general control processing. For SLAT, the nonexecutive processes would be rotation and folding, while the control process would be successfully applying rotation and folding strategies to the task. A general schematic for these two processes can be given as follows:

$$P(\text{SLAT Task Success}) = P(\text{Apply})P(\text{Success in Rotation and Folding}). \qquad (7)$$

Thus, in terms of GLTM-CP between-components model, the model is the following:

$$P(X_{ijT} = 1 \mid \theta_j, b_i) = P(X_{ij1} = 1)\, P(X_{ij2} = 1) \qquad (8)$$

The component, Success in Rotation and Folding, involves spatial processing to solve the item. This process is resource based, as it involves spatial working memory capacity. The other component, Apply, is a metacomponent that is involved in all items. It includes choosing and maintaining the spatial processing strategy, monitoring processes, and evaluating the results of processing.

Both components must be regarded as latent response variables since they cannot be observed directly. If the response to the SLAT item is incorrect, it could be due either to incorrect Processing or to failure to Apply the Inference Strategy properly. These relationships can be expressed in a simple GLTM-CP, as follows:

$$P(X_{ijT} = 1 \mid \theta_j, \alpha, \beta) = \frac{\exp(\theta_{j1} + \alpha_1)}{1 + \exp(\theta_{j1} + \alpha_1)} \, \frac{\exp(\theta_{j2} - \beta_2 u_{i2} + \alpha_2)}{1 + \exp(\theta_{j2} - \beta_2 u_{i2} + \alpha_2)}, \qquad (9)$$

where:

θ_{j1} is the ability for person j on general control processes,
α_1 is an intercept for general control processing (constant items),
θ_{j2} is the spatial working memory capacity for person j,
u_{i2} is the score for item i on spatial memory load,
β_2 is the weight of spatial memory load in rule inference difficulty, and
α_2 is an intercept for spatial processing difficulty.

Notice that to estimate the model, spatial working memory load values are required for each item. These values are available, as described next, from the SLAT item bank due to its cognitive design.

ANALYSIS

To estimate processing abilities from GLTM-CP, item difficulty on the inference process must be specified by spatial working memory load. The spatial working memory load in SLAT items is defined by the combination of two independent features (surfaces [3] × degrees [3]). Since these two features were manipulated independently in SLAT items, spatial working memory load can be estimated from item bank results on the impact of surfaces and degrees on item difficulty. The LLTM (Fischer, 1973) was applied to SLAT item calibration data on 300 subjects. The independent variables for the LLTM were orthogonal polynomials to define the linear and quadratic impact of both surfaces and degrees, and their interactions. Thus, eight complexity factors were scored for each SLAT item, as follows: degrees linear, degrees quadratic, surfaces linear, surfaces quadratic, degrees linear × surfaces linear, degrees quadratic × surfaces linear, degrees linear × surfaces quadratic, and degrees quadratic × surfaces quadratic. The LLTM weights for item difficulty, b_{i2} in the item bank data (Embretson, 1984) weights are:

b_{i2}' = .771(surfaces linear) + .025(surfaces quadratic) + .185(degrees linear) – .123(degrees quadratic) – .090(degrees linear × surfaces linear) – .041(degrees quadratic × surfaces linear) – .043(degrees linear × surfaces quadratic) – .009(degrees quadratic × surfaces quadratic).

For example, a two-surface, 90-degree item would result in a predicted item difficulty value of –.242 from the following:

$$b_{i2}' = .771(0) + .025(2) + .185(0) - .123(2) - .090(0) - .041(0) - .043(0) - .009(4) = -.235.$$

An example of this item type is shown by the middle item in Figure 1. Note that the complexity of these three items defined by degrees rotation and surfaces carried increases from the top item to the bottom item. The SLAT item bank predicted difficulties are –1.05, .24, and 1.03, respectively, indicating increased item difficulty with increasing SLAT item design complexity.

These predicted item difficulty values from the SLAT item bank were entered as the spatial working memory load value, u_{i2}, in the model defined in Equation 9. Thus, item parameters for the two components were as follows: For general control processes, a constant value, or intercept, was estimated, as it was hypothesized that strategy application would be constant for all items; for working memory capacity, both an intercept and a slope were estimated, as it was postulated that increased item complexity (greater rotation, greater number of surfaces carried) would result in greater spatial memory load for an item. Success in rotation and folding was also postulated to be influenced by spatial working memory capacity. Because the GLTM-CP is a conjunctive model, both spatial processing and strategy application must be successful for the item to be correct.

METHOD

McCollam (1997) administered a test of spatial ability, The SLAT (Embretson, 1989), via microcomputer, to a group of younger ($n = 89$; mean age = 20.47) and older ($n = 89$; mean age = 69.11) adults. Accuracy for the 30 individual item responses was recorded.

ESTIMATION

The COLORA program (Maris, 1993) was used to obtain maximum likelihood estimates for the three item parameters (two intercepts and a slope) and for two processing abilities for each person, one for working memory capacity, and one for general control processes. It was hypothesized that younger adults would have greater working memory capacity and general control processes than older adults, and that these components would explain age differences in spatial ability test performance on the SLAT. Furthermore, it was hypothesized that the components would not have equal impact in explaining age differences.

RESULTS

Mean thetas for working memory capacity and general control processes by age are shown in Figure 2. Younger adults have greater working memory capacity (younger mean = 1.24, $SD = .86$; older mean = .57, $SD = .74$; $t = 5.62$, $p < .001$)

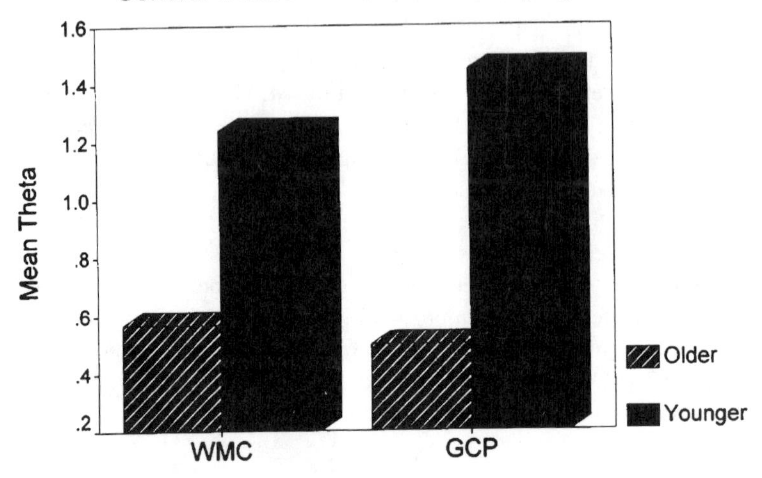

FIGURE 2. Working memory and general control process thetas by age. GCP, general control, processes; WMC, working memory capacity.

TABLE 1.
Correlations between Age, WMC, GCP, and SLAT Ability

	Age	WMC	GCP
WMC	–.406***		
GCP	–.471***	.140	
SLAT	–.579***	.606***	.851***

Note. *** = $p < .001$, two-tailed significance. GCP, general control processes; SLAT, Spacial Learning Ability Test; WMC, working memory capacity.

and general control processes (younger mean = 1.45, $SD = .85$; older mean = .50, $SD = .99$; $t = 6.83$, $p < .001$) than do older adults.

Correlations between general control processes (GCP), working memory capacity (WMC), and SLAT ability are shown in Table 1. Note that GCP and WMC are relatively uncorrelated, but all other variables are strongly related.

Next, results of multiple regressions accounting for SLAT ability by GCP and age are shown in Table 2. Note that after accounting for GCP, the significant, unique age-related variance in SLAT ability was about 4%. Multiple regressions of SLAT ability by WMC and Age are shown in Table 3. Similarly, the unique age-

TABLE 2.
Source Table for Regression of SLAT Ability on General Control Processes and Age

Source	*SS*	*DF*	*MS*	*F*	*R*
GCP	151.23	1	151.23	461.46	.72390
GCP + age	159.76	2	79.88	284.44	.76475

R^2 change due to age = .04085.

TABLE 3.
Source Table for Regression of SLAT Ability on Working Memory Capacity and Age

Source	*SS*	*DF*	*MS*	*F*	*R*
WMC	76.78	1	76.78	102.27	.36751
WMC + age	104.49	2	52.24	87.56	.50018

R^2 change due to age = .13267.

related variance was significant, and accounted for about 13% of the variance in SLAT ability. Thus, the amount of shared variance between age and SLAT ability is 33.5%, but after accounting for WMC and GCP, this shared amount is greatly reduced.

DISCUSSION

The GLTM-CP was applied to estimate covert processing abilities that underlie spatial visualization ability. It was hypothesized that younger adults' components of working memory and general control processes would be greater than those of older adults. Further, it was hypothesized that the combined contribution of these processing resources would greatly attenuate the relationship of age and spatial performance, and this relationship occurred for both working memory capacity and general control processes. Both hypotheses were supported by comparisons of MLTM-CP estimates of working memory capacity and general control strategies in younger versus older adults.

Future suggested directions for cognitive aging research include the incorporation of several processing resources in accounting for performance, and increased use of item response theory models to measure task components. We can then begin to come closer to understanding the nature of the age differences in ability performance across the life span.

CONCLUSION

The GLTM-CP allows decomposition of complex ability items into separate underlying sources. It was shown that covert processes abilities could be decomposed from traditional item response data if two conditions were met: (1) the covert responses are treated as "missing" data, and (2) constraints can be placed on component item difficulties. The item difficulty constraints are necessary to identify the separate processing abilities and must be scored for each item. The item constraints are most justifiable when they operationalize cognitive processing variables in a well supported theory. In the preceding example, the item constraints were LLTM estimates of item difficulty from a prior study. Although other types of estimates can be used, the constraints will be effective in the GLTM-CP only if item difficulty is well predicted. When these conditions are met, an EM algorithm can be applied to estimate the "missing" data; namely, the covert responses to processing components.

The GLTM-CP might be described as a confirmatory factor analysis from a conjunctive model. Like confirmatory factor analysis, constraints must be set to identify the traits. However, unlike confirmatory factor analysis, the GLTM-CP model is conjunctive, such that all underlying components must be correct for the item to be passed. Hence, the GLTM-CP processes combine conjunctively rather than compensatorily, as in factor analysis. Further, GLTM-CP has been formulated expressly for binary data.

Although GLTM-CP has been applied to other research (see, for example, Embretson, 1995), the current example appears to be the first application to cognitive aging. Using item response theory methods in cognitive aging studies will increase our knowledge of the component processes and their modifiability in spatial visualization performance, as well as in other domains.

REFERENCES

Babcock, R. (1994). Analysis of adult differences on the Raven's advanced progressive matrices test. *Psychology and Aging*, *9*, 303–314.

Baddeley, A. (1986). *Working memory* (Vol. 8, pp. 47–90). New York: Oxford University Press.

Baltes, P. B., & Kliegl, R. (1992). Further testing of limits of cognitive plasticity: Negative age differences in a mnemonic skill are robust. *Developmental Psychology, 28*, 121–125.

Brown, A. L. (1978). Knowing when, where, and how to remember: A problem of metacognition. In R. Glaser (Ed.), *Advances in instructional psychology* (Vol. 1, pp. 77–165). Hillsdale, NJ: Erlbaum.

Cavanaugh, J. C. (1987). Age differences in adults' self-reports of memory ability: It depends on how and what you ask. *International Journal of Aging and Human Development, 24*, 271–277.

Cooper, L., & Shepard, R. N. (1973). Chronometric studies of the rotation of mental images. In W. G. Chase (Ed.), *Visual information processing* (pp. 75–176). New York: Academic.

Dempster, A. P., Laird, N. M., & Rubin, D. B. (1977). Maximum likelihood estimating with incomplete data via the EM algorithm. *Journal of the Royal Statistical Society, Series B, 39*, 1–38.

Embretson, S. E. (1983). Construct validity: Construct representation versus nomothetic span. *Psychological Bulletin, 93*, 179–197.

Embretson, S. E. (1984). A general multicomponent latent trait model for response processes. *Psychometrika, 49,* 175–186.

Embretson, S. E. (1989). *Spatial learning ability test (SLAT).* Lawrence, KS: University of Kansas.

Embretson, S. E. (1994). Applications of cognitive design systems to test development. In C. R. Reynolds (Ed.), *Cognitive assessment: A multidisciplinary perspective.* (pp. 107–135). NY: Plenum.

Embretson, S. E. (1996). Multicomponent response models. In W. J. van der Linden & R. K. Hambleton (Eds.), *Handbook of modern item response theory* (pp. 305–322). New York: Springer.

Embretson, S. E., & Schneider, L. M. (1989). Cognitive models of analogical reasoning for psychometric tasks. *Learning and Individual Differences, 1*, 155–178.

Embretson, S. E., & Waxman, M. (1989). *Models for processing and individual differences in spatial folding.* Unpublished manuscript. University of Kansas, Lawrence.

Fischer, G. (1973). Linear logistic test model as an instrument in educational research. *Acta Psychologica, 37*, 359–374.

Kaufman, A. S., Reynolds, C. R., & McClean, J. E. (1989). Age and WAIS-R intelligence in a national sample of adults in the 20-to-74-year age range: A cross-sectional analysis with educational level controlled. *Intelligence, 13*, 235–253.

Lindenberger, U., Kliegl, R., & Baltes, P. B. (1992). Professional expertise does not eliminate age differences in imagery-based memory performance during adulthood. *Psychology and Aging, 7*, 585–593.

Maris, E. M. (1993). *Psychometric models for psychological processes and structures.* Dissertation award lecture at the Psychometric Society meeting, Pompeau Fabre University, Barcelona, Spain.

Maris, E. M. (1995). Psychometric latent response models. *Psychometrika, 60,* 523–547.

McCollam, K. M. (1997). *The modifiability of age differences in spatial visualization.* Unpublished doctoral dissertation, University of Kansas, Lawrence.

Pellegrino, J., & Lyons, D. W. (1979). The components of a componential analysis. *Intelligence, 3*, 169–186.

Persaud, G. (1991). Age and sex differences on four measures of cognitive ability. *Perceptual and Motor Skills, 72*, 1172–1174.

Salthouse, T. A. (1991). *Theoretical perspectives on cognitive aging* (pp. 259–300). Hillsdale, NJ: Erlbaum.

Salthouse, T. A., Mitchell, D. R. D., & Palmon, R. (1989). Memory and age differences in spatial manipulation ability. *Psychology and Aging, 4*, 480–486.

Salthouse, T. A., Mitchell, D. R. D., Skovronek, E., & Babcock, R. L. (1989). Effects of adult age and working memory on reasoning and spatial abilities. *Journal of Experimental Psychology: Learning, Memory and Cognition, 15*, 507–516.

Salthouse, T. A., & Skovronek, E. (1992). Within-context assessment of age differences in working memory. *Journal of Gerontology: Psychological Sciences*, *47*, p110–p120.

Schaie, K. W., & Hertzog, C. (1983). Fourteen-year cohort-sequential analyses of adult intellectual development. *Developmental Psychology, 19*, 531–545.

Schaie, K. W., & Strother, C. R. (1968). A cross-sequential study of age changes in cognitive behavior. *Psychological Bulletin, 70,* 671–680.

Schaie, K. W., & Willis, S. L. (1993). Age difference patterns of psychometric intelligence in adulthood: Generalizability within and across ability domains. *Psychology and Aging*, *8*, 44–55.

Shepard, R. N., & Feng, C. (1972). A chronometric study of mental paper folding. *Cognitive Psychology*, *3*, 228–243.

Sternberg, R. J. (1980). Sketch of a componential subtheory of human intelligence. *Behavioral and Brain Sciences*, *3*, 573–584.

Whitely, S. E. (1980). Multicomponent latent trait models for ability tests. *Psychometrika*, *45*, 479–494.

Willis, S. L., & Nesselroade, C. S. (1990). Long-term effects of fluid ability training in old-old age. *Developmental Psychology*, *26*, 905–910.

12

INTERPRETING THE PARAMETERS OF A MULTIDIMENSIONAL RASCH MODEL

Wen-chung Wang
National Chung Cheng University

Mark Wilson
University of California, Berkeley

Raymond J. Adams
Australian Council for Educational Research

INTRODUCTION

In the last decade or so, there has been considerable work on the development of multidimensional item response theory (IRT) models and on the consequences of applying unidimensional models to simulated or real multidimensional data (Ackerman, 1992; Andersen, 1985; Camilli, 1992; Embretson, 1991; Glas, 1992; Kelderman, 1996; Kelderman & Rijkes, 1994; Luecht & Miller, 1992; Oshima & Miller, 1992; Reckase, 1985; Reckase & McKinley, 1991; Way, Ansley, & Forsyth, 1988; to name a few). Most of the work arises from perspectives such as the following:

- Many standardized tests are based on an assumption that more than one ability is involved (Traub, 1983).

- If the unidimensionality assumption is violated and examinees take different item subsets (such as in computerized adaptive testing) ability estimates may reflect different composites of abilities (Way et al., 1988).
- If two groups of examinees have different underlying multidimensional ability distributions and the test items can discriminate among levels of abilities on these dimensions, any unidimensional scoring scheme has the potential to produce item bias (Ackerman, 1992).
- If our assessments are to be useful in diagnosis and for other forms of feedback, we need to employ multiple types of assessment (for example, selected-response items, constructed-response items, and performance assessments) in order to capture a rich report about the student's understanding (Romberg, 1992; Wilson, 1991).
- In test construction it may be desirable not to be limited to a single trait. Rather, we may need to go further and specify theories that explain what the items are intended to measure. Recent developments in cognitive psychology have urged us to focus on the cognitive process rather than the products measured by the test (Embretson, 1985; Frederiksen, Mislevy, & Bejar, 1993; Sternberg, 1982). Our measurement models should allow us to model not only students' overall proficiency but also their strategies, which are clearly different dimensions. Moreover, by specifying a model for each competing theory and comparing the fit of these models to the data, we can gain substantial knowledge about what the test measures.

Even though multidimensional models are preferable on theoretical grounds, their applications have been limited because (1) interpretation of parameters of the existing multidimensional IRT models have been challenging to many test users, and (2) to date the computer programs available for multidimensional models have also challenged many potential users.

Recently, Adams, Wilson, and Wang (1997) have proposed a multidimensional model, the multidimensional random coefficient multinomial logit model (MRCML), which can be applied to multidimensional polytomous test items. In their paper, they described two basic submodels: the multidimensional between-item models and the multidimensional within-item models. They also described a marginal maximum likelihood estimation algorithm (Bock & Aitkin, 1981) for the model. The multidimensional between-item models are suitable for tests with several unidimensional subtests, such as the Verbal and the Quantitative subtests in the Scholastic Aptitude Test. The multidimensional within-item models are suitable for tests with items that are themselves multidimensional, such as constructed-response items scored on more than one dimension.

What have not been thoroughly discussed in previous works are (1) applications and implications of the MRCML in complex testing situations, such as where strategies of problem solving are hypothesized; (2) clear guidelines for parameter interpretation; and (3) guidelines on choosing scoring rubrics and models. In this chapter, we begin with a brief introduction to the MRCML. Two examples of real data analyses are followed to demonstrate the applications of the MRCML. Along with the two examples, we provide a detailed discussion of interpretation of item parameters, in comparison to those derived from their corresponding unidimensional Rasch measurement models. Finally, we provide some guidelines for constructing scoring matrices and the corresponding models to aid parameter interpretation.

THE MRCML

The MRCML is a multidimensional extension of the random coefficients multinomial logit model (Adams & Wilson, 1996). Assume that a set of D traits underlie the individuals' responses and the individuals' positions are represented by the vector $\boldsymbol{\theta} = (\theta_1,\theta_2,\ldots,\theta_D)'$. Suppose we have I items indexed $i = 1, \ldots, I$ with each item admitting $K_i + 1$ response alternatives indexed $k = 0, 1, \ldots, K_i$. We then use the vector-valued random variable:

$$\mathbf{X}_i = (X_{i1}, X_{i2},\ldots,X_{iK_i})',$$

where:

$$X_{ik} = \begin{cases} 1 \text{ if response to item } i \text{ is in category } k, \\ 0 \text{ otherwise,} \end{cases}$$

to indicate the $K_i + 1$ possible responses to item i.

By convention, a response in the zero category is denoted by a vector of zeroes. This effectively makes the zero category a reference category and is necessary for model identification. The choice of this as the reference category is arbitrary and does not affect the generality of the model. In IRT, the incorrect or least correct response is usually treated as the reference category.

The items are described through a prime vector $\boldsymbol{\xi} = (\xi_1,\xi_2,\ldots,\xi_p)$, of p parameters. Linear combinations of these are used in the response probability model to describe the empirical characteristics of the response categories of each item. These linear combinations are defined by design vectors $\mathbf{a}_{ik} = (a_{i1}, a_{i2},\ldots,a_{iK_i})'$ each of length p, which can be collected to form a design matrix, $\mathbf{A} = (\mathbf{a}_{11}, \mathbf{a}_{12},\ldots,\mathbf{a}_{1K_i},\mathbf{a}_{21},\ldots,\mathbf{a}_{2K_2},\ldots,\mathbf{a}_{1K_1})'$. For the Rasch model, every K_i is 1 and a_{i1} is a column vector with 1 in the ith place and 0 otherwise. The design matrix $\mathbf{A}$ becomes a diagonal matrix.

An additional feature of the MRCML is the introduction of a scoring function that allows the specification of the score or "performance level" that is assigned to each possible response to each item. A response in category k in dimension d of item i is scored b_{ikd}. The scores across D dimensions can be collected into a column vector $\mathbf{b}_{ik} = (b_{ik1},b_{ik2},\ldots,b_{ikD})'$—(by definition, the score for a response in the zero category is zero, but other responses may also be scored zero)—and can again be collected into a scoring submatrix for item i, $\mathbf{B}_i = (\mathbf{b}_{i1}, \mathbf{b}_{i2}, \ldots, \mathbf{b}_{iD})'$, and then collected into a scoring matrix $\mathbf{B} = (\mathbf{B}'_1, \mathbf{B}'_2, \ldots, \mathbf{B}'_I)'$ for the whole test. For the Rasch model where k and D are equal to 1, $\mathbf{b}_{ik}$ as well as $\mathbf{B}_i$ is a scalar of 1. $\mathbf{B}$ is thus a column vector of 1.

In the MRCML the probability of a response in category k of item i depending on both the two matrices, $\mathbf{A}$ and $\mathbf{B}$, and item parameters ξ, and conditional on person abilities, θ, is modeled as follows:

$$f(X_{ik} = 1; \mathbf{A}, \mathbf{B}, \boldsymbol{\xi} | \boldsymbol{\theta}) = \frac{\exp(\mathbf{b}'_{ik}\boldsymbol{\theta} + \mathbf{a}'_{ik}\boldsymbol{\xi})}{\sum_{j=0}^{K_1} \exp(\mathbf{b}'_{ij}\boldsymbol{\theta} + \mathbf{a}'_{ij}\boldsymbol{\xi})}.$$

Note that the item score vector $\mathbf{b}_{ik}$ in the MRCML is not a free parameter, but is known a priori. Also, the notation ";" and "|" are used to denote dependence and conditionality respectively, where we are using conditionality to mean dependence on a nuisance parameter that will later be integrated out.

MARGINAL MAXIMUM LIKELIHOOD ESTIMATION

A marginal likelihood estimation with EM algorithm was implemented. This has been described by Adams and colleagues (1997), but will be summarized here for the readers' convenience. We assume that the latent traits vector $\boldsymbol{\theta}$ is sampled from a population in which the distribution of $\boldsymbol{\theta}$ is given by a multivariate density function $g(\boldsymbol{\theta}; \boldsymbol{\alpha})$, and a corresponding cumulative distribution $G(\theta; \alpha)$, where α is a set of parameters characterizing the distribution.

Based on the assumption of conditional independence among items and persons, the probability of a response vector conditioned on the random quantities, $\boldsymbol{\theta}$, is:

$$f(\mathbf{X} = \mathbf{x}; \mathbf{A}, \mathbf{B}, \boldsymbol{\xi} | \boldsymbol{\theta}) = \frac{\exp(\mathbf{x}\,(\mathbf{B}\boldsymbol{\theta} + \mathbf{A}\boldsymbol{\xi}))}{\Psi(\boldsymbol{\theta}, \boldsymbol{\xi})},$$

with:

$$\Psi(\boldsymbol{\theta}, \boldsymbol{\xi}) = \sum_{\mathbf{z} \in \Omega} \exp(\mathbf{z}\,(\mathbf{B}\boldsymbol{\theta} + \mathbf{A}\boldsymbol{\xi})),$$

where Ω is the set of all possible response vectors (hereafter $\mathbf{A}$ and $\mathbf{B}$ are omitted for brevity if not necessary).

The marginal density of the response $\mathbf{x}_n$ of person n is:

$$f(\mathbf{X} = \mathbf{x}_n) = \int_{\boldsymbol{\theta}} \frac{\exp(\mathbf{x}'_n\,(\mathbf{B}\boldsymbol{\theta} + \mathbf{A}\boldsymbol{\xi}))}{\Psi(\boldsymbol{\theta}, \boldsymbol{\xi})}\, dG(\boldsymbol{\theta}; \boldsymbol{\alpha}),$$

and the likelihood for a set of N response vector is:

$$\Lambda(\boldsymbol{\xi},\boldsymbol{\alpha} \mid \mathbf{X}) = \prod_{n=1}^{N} \int_{\boldsymbol{\theta}} \frac{\exp(\mathbf{x}'_n (\mathbf{B}\boldsymbol{\theta} + \mathbf{A}\boldsymbol{\xi}))}{\Psi(\boldsymbol{\theta},\boldsymbol{\xi})} \, dG(\boldsymbol{\theta};\boldsymbol{\alpha}).$$

The likelihood equations for the item parameters are:

$$\frac{\partial \log \Lambda(\boldsymbol{\xi},\boldsymbol{\alpha} \mid \mathbf{X})}{\partial \boldsymbol{\xi}} = \sum_{n=1}^{N} \int_{\boldsymbol{\theta}} \frac{\partial \log f(\mathbf{x}_n;\boldsymbol{\xi},\boldsymbol{\alpha})}{\partial \boldsymbol{\xi}} \, dH(\boldsymbol{\theta};\boldsymbol{\xi},\boldsymbol{\alpha} \mid \mathbf{x}_n) = \mathbf{0}.$$

where $H(\boldsymbol{\theta};\boldsymbol{\xi},\boldsymbol{\alpha} \mid \mathbf{x}_n)$ is the distribution function of the marginal density of $\boldsymbol{\theta}$ given $\mathbf{x}_n$, with a density function:

$$h(\boldsymbol{\theta};\boldsymbol{\xi},\boldsymbol{\alpha} \mid \mathbf{x}_n) = \frac{f(\mathbf{x}_n;\boldsymbol{\xi} \mid \boldsymbol{\theta}) \; g(\boldsymbol{\theta} ; \boldsymbol{\alpha})}{f(\mathbf{x}_n;\boldsymbol{\xi})} .$$

Assuming the population distribution is multivariate normal so that $\boldsymbol{\alpha} = (\boldsymbol{\mu}, \boldsymbol{\Sigma})$, the likelihood equations are:

$$\frac{\partial \log \Lambda(\boldsymbol{\xi},\boldsymbol{\mu},\boldsymbol{\Sigma} \mid \mathbf{X})}{\partial \boldsymbol{\mu}} = \sum_{n=1}^{N} \int_{\boldsymbol{\theta}} \frac{\partial \log g(\boldsymbol{\theta};\boldsymbol{\mu},\boldsymbol{\Sigma})}{\partial \boldsymbol{\mu}} \, dH(\boldsymbol{\theta};\boldsymbol{\xi},\boldsymbol{\mu},\boldsymbol{\Sigma} \mid \mathbf{x}_n) = \mathbf{0},$$

and:

$$\frac{\partial \log \Lambda(\boldsymbol{\xi},\boldsymbol{\mu},\boldsymbol{\Sigma} \mid \mathbf{X})}{\partial \boldsymbol{\Sigma}} = \sum_{n=1}^{N} \int_{\boldsymbol{\theta}} \frac{\partial \log g(\boldsymbol{\theta};\boldsymbol{\mu},\boldsymbol{\Sigma})}{\partial \boldsymbol{\Sigma}} \, dH(\boldsymbol{\theta};\boldsymbol{\xi},\boldsymbol{\mu},\boldsymbol{\Sigma} \mid \mathbf{x}_n) = \mathbf{0},$$

If it is not reasonable to assume the population distribution is normal, a step distribution defined on a prespecified set of nodes can be used. For both the normal case and the step distribution case, Bock and Aitkin's (1981) formulation of EM algorithm (Dempster, Laird, & Rubin, 1977) is implemented to estimate the model parameters. The marginal posterior density of $\boldsymbol{\theta}$ given $\mathbf{x}_n$ at iteration t is calculated. Then, a Newton-Raphson method is used to solve the likelihood equations for the item parameters. Regarding the population parameters, in the normal case direct estimates of $\hat{\boldsymbol{\mu}}^{(t+1)}$ and $\hat{\boldsymbol{\Sigma}}^{(t+1)}$ or in the step distribution case estimates of weights of the set of nodes at iteration $t + 1$ are produced. Moreover, in the normal case, the distribution can also be estimated by using the Monte Carlo method whereby the nodes are readjusted according to the recent estimates for better integration. Interested readers are referred to the manual for ConQuest (Wu, Adams, & Wilson, 1997), a computer program accompanying the MRCML, and to the paper by Adams and colleagues (1997) for details.

SOME SPECIFICATIONS OF THE SCORING MATRIX AND THE DESIGN MATRIX

The way we design scoring rubrics or, equivalently, scoring matrices is based on our knowledge about the test constructs. Changing the scoring rubrics may substantially change the construct definition. For items measuring several dimensions, the choice between scoring rubrics may be highly controversial because of ambiguity of test constructs. If users would like to compare several possible models so that the underlying cognitive processes could be validated, they should carefully examine scoring rubrics and prepare to justify the rationales. Using the MRCML, users are enabled to compare model fit to gain knowledge about the way theories are reflected in data.

For dichotomous or polytomous items, the scores given to item responses are ordinal per se (for example, 0, 1, 2, …). In the framework of the unidimensional Rasch models, this ordering is embedded with respect to the underlying dimension. For multidimensional items, the ordering becomes more complex. It can be embedded within dimensions only (referred to as the *internal ordering*) or both within and between dimensions (referred to as the *cross ordering*). The ordering for unidimensional models is internal because only one dimension is involved. These two kinds of ordering together with their corresponding design matrices are described in the next sections.

Internal Ordering

Here are some examples of what we would call "internal ordering." First, consider the following constructed-response item (Kelderman & Rijkes, 1994):

$$2X^2 + X - 1 = 0, X = ?$$

It can be solved by either of the following approaches:

- "Complete the product," $(2X - 1)(X + 1) = 0$; hence $X = 1/2$, or -1;
- "Quadratic formula," $X = \dfrac{-b \pm \sqrt{b^2 - 4ac}}{2a}$, where $a = 2$, $b = 1$, and $c = -1$; hence $X = 1/2$, or -1.

These two strategies are treated as dimensions. Examinees are assumed to have some proficiency on both dimensions.

Second, Pellegrino, Mumaw, and Shute (1985) have developed a theory of spatial aptitude based on three forms of representation—isometric drawing, orthographic drawing, and verbal description—and have combined them to produce three problem types. In one of the problem types, for example, examinees were first shown a slide containing three orthographic projections with one view replaced by an empty frame. The second slide showed a possible third view for that orthographic drawing. The examinee was asked to determine whether the third view given was compatible with the first two views.

After the testing, each examinee was asked to describe the strategy that had been used on each item. One major strategy found was to construct the representation mentally to mediate problem solving. The alternative solution was an analytical feature-matching strategy, which requires identifying and comparing local features of the representations. In this example, both the strategies and the performances (correctness) were observed.

Third, consider the conservation tasks of Piagetian scales. The test administrator begins with two standard glasses containing equal amounts of water or grains of corn and pours the contents into several small glasses or a flat dish. The child judges their similarity or difference. After responding, the child is asked to explain how the response is reached. From a Piagetian point of view, conservation tasks are used to measure the child's transition from the preoperational to the concrete operational stage of thinking. These two stages can be treated as two strategies or dimensions. Each child has some proficiency on each dimension.

In the preceding three examples, suppose the two strategies are exhaustive and mutually exclusive. For illustration, assume the performances are scored on a three-point scale: incorrect, partially correct, and correct. When the first strategy is applied, it may come out with an incorrect response, a partially correct response, and a correct response. Likewise, when the second strategy is applied, it may also come out with the three kinds of responses. The examinees are free to apply either strategy and the two strategies are given equal scores.

In such cases, five response categories are observed. They are:

Category zero: an incorrect response either by applying the first strategy or the second strategy and scored 0 on both dimensions

Category 1: a partially correct response when the first strategy was applied and scored 1 and 0 on the two dimensions, respectively

Category 2: a correct response when the first strategy was applied and scored 2 and 0 on the two dimensions, respectively

Category 3: a partially correct response when the second strategy was applied and scored 0 and 1 on the two dimensions, respectively

Category 4: a correct response when the second strategy was applied and scored 0 and 2 on the two dimensions, respectively.

Let the log-odds be modeled as

$$\log (p_1 / p_0) = \theta_1 - \delta_1,$$
$$\log (p_2 / p_1) = \theta_1 - \delta_2,$$
$$\log (p_3 / p_0) = \theta_2 - \delta_3,$$
$$\log (p_4 / p_3) = \theta_2 - \delta_4,$$

where:

p_0, p_1, p_2, p_3, and p_4 denote the probabilities of being in categories zero, 1, 2, 3, and 4, respectively;

θ_1 and θ_2 denote the abilities on the first and the second dimensions, respectively;

δ_1 and δ_2 denote the first step and the second step difficulties with respect to the first dimension, respectively; and,

δ_3 and δ_4 denote the first step and the second step difficulties with respect to the second dimension, respectively.

These difficulty parameters can be interpreted within the framework of the unidimensional partial credit model. Setting category zero for reference, the two submatrices for item i are as follows:

$$\mathbf{A}_i = \begin{bmatrix} -1 & 0 & 0 & 0 \\ -1 & -1 & 0 & 0 \\ 0 & 0 & -1 & 0 \\ 0 & 0 & -1 & -1 \end{bmatrix} \text{ and } \mathbf{B}_i = \begin{bmatrix} 1 & 0 \\ 2 & 0 \\ 0 & 1 \\ 0 & 2 \end{bmatrix}$$

With respect to the first dimension, the ordering occurs in categories zero, 1, and 2. Regarding the second dimension, the ordering occurs in categories zero, 3, and 4. Since the ordering is embedded within each dimension, the model is referred to as the internal ordering model.

Suppose incorrect responses can be further partitioned into two categories: Incorrect by using the first strategy, and incorrect by using the second strategy. These three categories are all scored 0 on both dimensions because no credits are given. Let the three response categories be indexed as zero, and 1, respectively and the other response categories be indexed from 2 in sequence. Altogether six response categories are formed. In such cases, at most five parameters can be estimated.

Let the log-odds be modeled as:

$$\begin{aligned} \log(p_2/p_0) &= \theta'_1 - \delta'_1, \\ \log(p_3/p_2) &= \theta'_1 - \delta'_2, \\ \log(p_4/p_1) &= \theta'_2 - \delta'_3, \\ \log(p_5/p_4) &= \theta'_2 - \delta'_4, \\ \log(p_0/p_1) &= \delta'_5, \end{aligned}$$

Using the MRCML, it follows that the two submatrices are:

$$\mathbf{A}_i = \begin{bmatrix} 0 & 0 & 0 & 0 & -1 \\ -1 & 0 & 0 & 0 & 0 \\ -1 & -1 & 0 & 0 & 0 \\ 0 & 0 & -1 & 0 & 0 \\ 0 & 0 & -1 & -1 & -1 \end{bmatrix} \text{ and } \mathbf{B}_i = \begin{bmatrix} 0 & 0 \\ 1 & 0 \\ 2 & 0 \\ 0 & 1 \\ 0 & 2 \end{bmatrix}.$$

This model is a supermodel of the previous one. By comparing these two models, one could detect the influence of different kinds of incorrectness.

For instance, consider the following situation. Suppose there are 500 incorrect responses, 200 partially correct responses when the first strategy was applied, and 100 partially correct responses when the second strategy was applied. Based on the first model, rough estimates of the first step difficulty parameters for the two strategies are .9 (= log (500 / 200) and 1.6 (= log (500 / 100) logits, respectively. With these estimates, one might assume that the first strategy is "better" than the second strategy, given that examinees master both strategies equally. But this may not be the case.

Suppose, among the 500 incorrect responses, that 400 responses come from the first strategy and the other 100 come from the second strategy. Hence, rough estimates of the first step difficulty parameters for the two strategies become .7 (= log (400 / 200) and 0.0 (= log (100 / 100) logits, respectively. It turns out that the second strategy is better, given that examinees master both strategies equally. Therefore, when the incorrect responses can be further categorized, the relative frequency parameters should be added to clarify the item difficulty parameters.

If only the unidimensional models are available, one might analyze each dimension (strategy) separately. For example, the three response categories (incorrect, partially correct, and correct responses) in the first dimension can be analyzed by applying the partial credit model. The other three response categories in the second dimension can be analyzed similarly. In so doing, however, two major problems arise. First, the item in fact has six response categories rather than two sets of three response categories. The probabilities of the six response categories should sum to one rather than two, as when the unidimensional models are applied. Second, it is the differences between ability levels and item difficulties within dimensions that determine which strategy to be applied. If the difference of the first dimension is larger than that of the second dimension, the probability of getting a correct response using the first strategy is increased, and vices versa. Using the unidimensional models, we are not able to determine these probabilities. Without this information, the diagnostic function of the items will be limited.

Cross Ordering

Sometimes, problem-solving strategies or cognitive components are not directly observed but can only be inferred in theory. For example, consider the following item:

Define the concepts of confidence intervals and hypothesis testing and then the relationship between them.

Suppose to solve the item, two components are needed: (1) low-level thinking (knowledge of definition of confidence interval and hypothesis testing), and (2) high-level thinking (Understanding the relationship). Note that these two components are ordered, because in theory one cannot understand the relationship without comprehending the individual concepts. Moreover, examinees who comprehend the concepts but do not understand the relationship can only give partially correct responses (such as stating the individual concepts only). Only those who not only comprehend the concepts but also understand the relationship can give completely correct responses. These two components can be viewed as dimensions. Each examinee has some proficiency on each dimension. The ordering occurs not only within dimensions but also between dimensions, which is referred to as *cross ordering*.

Now, suppose the responses are judged to these categories as described earlier. For example, responses involving only descriptions of the individual concepts are judged into three categories: 0, 1, and 2. Likewise, statements involving the relationship are also judged on a three-point scale. In such cases, five response categories could be observed:

Category zero: The individual concepts are not even correctly mentioned, and thus are scored 0 on both dimensions.

Category 1: The individual concepts are partially correctly stated and thus scored 1 and 0 on both dimensions, respectively. (The relationship cannot be correctly stated because in theory, only when the concepts are completely comprehended can the relationship be stated.)

Category 2: The individual concepts are correctly stated but the relationships are incorrect, and thus scored 2 and 0 on both dimensions, respectively.

Category 3: The individual concepts are correctly stated but the relationship is partially correctly stated, and thus scored 2 and 1 on the two dimensions, respectively.

Category 4: Both the concepts and the relationship are correctly stated and thus scored 2 on both dimensions.

Let the log-odds be modeled as follows:

$$\log (p_1 / p_0) = \theta_1'' - \delta_1'',$$
$$\log (p_2 / p_1) = \theta_1'' - \delta_2'',$$
$$\log (p_3 / p_2) = \theta_2'' - \delta_3'',$$
$$\log (p_4 / p_3) = \theta_2'' - \delta_4''.$$

Setting category zero for reference, we establish the two submatrices for item i as follows:

$$\mathbf{A}_i = \begin{bmatrix} -1 & 0 & 0 & 0 \\ -1 & -1 & 0 & 0 \\ -1 & -1 & -1 & 0 \\ -1 & -1 & -1 & -1 \end{bmatrix} \text{ and } \mathbf{B}_i = \begin{bmatrix} 1 & 0 \\ 2 & 0 \\ 2 & 1 \\ 2 & 2 \end{bmatrix}$$

This cross ordering as well as the preceding internal ordering for five response categories is depicted in Figure 1.

Note that although each dimension is judged on a three-point scale, only five response categories are possible. This is because some responses are excluded in theory, such as incorrect or partially correct in the first dimension while partially correct or correct in the second dimension. If the two 3-point scales are separately treated and modeled as two unidimensional subtexts, it may come out with some probabilities for responses such as incorrect in the first dimension and correct in the second, which is impossible in theory. The multidimensional model proposed earlier in this chapter allows a resolution of this problem.

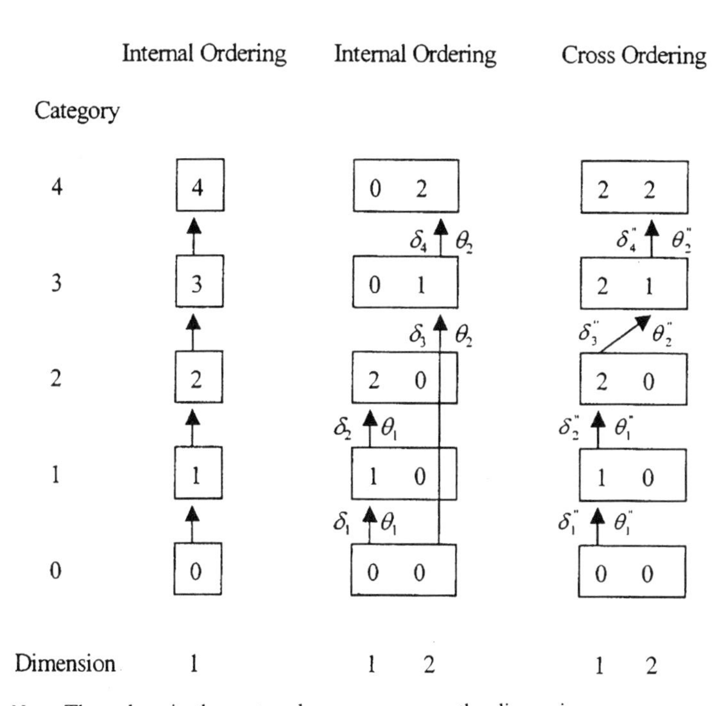

Note. The values in the rectangles are scores on the dimensions

FIGURE 1. Example of internal ordering and cross ordering.

EXAMPLE 1: SIZE CONCEPTS

Van Kuyk (1988), in a classic data set already used as an example in another latent trait model (Kelderman, 1989, 1996), administrated a 15-item test to investigate child development of size concepts, such as "long-short," "high-low," and "thick-thin." His theory is that the general size concept such as "big-small" is developed in advance of the specific size concept such as "long-short," "high-low," and "thick-thin." In each item, a test administrator showed the child several figures in which a feature changed, such as the length of girls' skirts. Then the test administrator asked the child: "Here you see some skirts, they gradually become a bit…?" If the child said "shorter," the response was rated correct since the right size concept (that is, long-short) was given and was correctly applied (that is, shorter rather than longer). Some younger children were not be able to produce the specific size concept but instead used the general size concept "big-small," and gave the response "smaller." Their response was rated partially correct. All other responses were rated incorrect.

Thus, there are three response categories for each item, resulting in at most two item parameters for each item (because one category is treated as reference for model identification). Suppose there are three competing theories. First, one could assume that a single *spatial* ability underlies children's responses. In that case, the unidimensional partial credit model (Masters, 1982) could be applied, where the incorrect response is scored 0, the partially correct response is scored 1, and the correct response is scored 2.

Under the first approach, the partial credit model can be applied. In the partial credit model, it follows that

$$\log (p_1 / p_0) = \theta - \delta_1, \quad (1)$$
$$\log (p_2 / p_1) = \theta - \delta_2, \quad (2)$$

where:

$p_{0,} p_1$, and p_2 denote the probabilities of being the incorrect, the partially correct, and the correct responses, respectively;
θ denotes the unidimensional ability; and
δ_1 and δ_2 denote the first and the second step difficulties, respectively.

Not that the $\mathbf{B}_i$ is a vector rather than a matrix because only a single dimension is assumed.

Using the MRCML, the partial credit model could be formed by specifying the scoring submatrix and the design sub-matrix for item i as follows:

$$\mathbf{A}_i = \begin{bmatrix} -1 & 0 \\ -1 & -1 \end{bmatrix} \text{ and } \mathbf{B}_i = \begin{bmatrix} 1 \\ 2 \end{bmatrix}.$$

Second, two abilities underlay the children's responses: a *general extensiveness* ability (for example, small versus large) and a *specific extensiveness* ability (for example, short versus long). If the children applied the general extensiveness ability, they would have come up with either incorrect or partially correct responses. If they applied the specific extensiveness ability, they would have come up with either incorrect or correct responses. In such cases, the incorrect response is scored 0 on both dimensions, the partially correct response is scored 1 and 0 on the general and specific extensiveness abilities, respectively, and the correct response is scored 0 and 1 on the two dimensions, respectively.

With these assumptions, the model is as follows:

$$\log (p_1 / p_0) = \theta_G - \delta_G, \quad (3)$$
$$\log (p_2 / p_0) = \theta_S - \delta_S, \quad (4)$$

where:

θ_G and θ_S denote the abilities on the general extensiveness and the specific extensiveness, respectively; and

δ_G and δ_S denote the difficulties on these two dimensions, respectively.

This model is multidimensional because two dimensions are involved. Using the MRCML, the scoring submatrix and the design submatrix for item i as follows:

$$\mathbf{A}_i = \begin{bmatrix} -1 & 0 \\ 0 & -1 \end{bmatrix} \text{ and } \mathbf{B}_i = \begin{bmatrix} 1 & 0 \\ 0 & 1 \end{bmatrix}.$$

$\mathbf{B}_i$ becomes a matrix because two dimensions are assumed although Equations 3 and 4 each involve only a single dimension. Comparing Equations 3 and 4 with Equations 1 and 2, two major differences can be found. First, although each equation involves only one dimension, the two dimensions in Equations 3 and 4 are different, whereas those in Equations 1 and 2 are identical. Second, in Equations 3 and 4 the incorrect response is always the category in the denominators, whereas in Equations 1 and 2, the previous response categories are in the denominators.

Keeping these two differences in mind, we are able to interpret the item difficulties, δ_G and δ_S, and the abilities, θ_G and θ_S, by relating them back to the unidimensional partial credit model. For instance, let δ_G and δ_S of item i be 1.0 and 2.0, respectively. For an examinee with 1.0 and 2.0 on θ_G and θ_S, respectively, the log-odds of the partially correct response to the incorrect response on item i are 0 (= 1.0 – 1.0); the log-odds of the correct response to the incorrect response on item i are also 0 (= 2.0 – 2.0). Note that since θ_G and θ_S are not constrained to be orthogonal, these two dimensions can be correlated.

Third, assume that two abilities—the general extensiveness ability and a *specific concept* ability—underlie children's responses. For the partially correct response, the child should have operated successfully on the general extensiveness but unsuccessfully on the specific concept. For the correct response, the child should not only have operated successfully on the general extensiveness but also have extra effort on the specific concept. For the incorrect response, the child should have operated unsuccessfully on both the general extensiveness and the specific concept dimensions, because only if the first dimension is successfully operated could the second dimension be operated. Under this circumstance, the incorrect response is scored 0 on both dimensions, the partially correct response is scored 1 and 0, respectively, and the correct response is scored 1 on both dimensions.

With these assumptions, the model is as follows

$$\log (p_1/ p_0) = \theta_G - \delta_G, \tag{5}$$
$$\log (p_2/ p_1) = \theta_{SP} - \delta_{SP}, \tag{6}$$

where:

θ_G and θ_{SP} denote the abilities on the general extensiveness and the specific concept, respectively; and
δ_G and δ_{SP} denote the difficulties on these two dimensions, respectively.

Note that on the left-hand side of Equations 5 and 6, the comparison is with respect to successive response categories, rather than the first response, as it was in Equations 3 and 4. The log-odds of the partially correct response to the incorrect response depend on the difference between θ_G and δ_G, and that of the correct response to the partially correct response depend on the difference between θ_{SP} and δ_{SP}. Therefore, both the abilities and item difficulties can be interpreted similarly to those in the unidimensional partial credit model as well. This model is also multidimensional because two dimensions are involved. Using the MRCML, the scoring submatrix and the design submatrix for item *i* are as follows:

$$\mathbf{A}_i = \begin{bmatrix} -1 & 0 \\ -1 & -1 \end{bmatrix} \text{ and } \mathbf{B}_i = \begin{bmatrix} 1 & 0 \\ 1 & 1 \end{bmatrix}.$$

Figure 2 show the scoring functions and the structures of the three models, respectively. The second and the third models are internal ordering and cross ordering, respectively.

These two models are conceptually different since somewhat different structures of dimensionality are assumed. However, they are mathematically equivalent because there are identical numbers of dimensions and item difficulties are involved in the two models. More specifically, Equation 5 plus Equation 6 is:

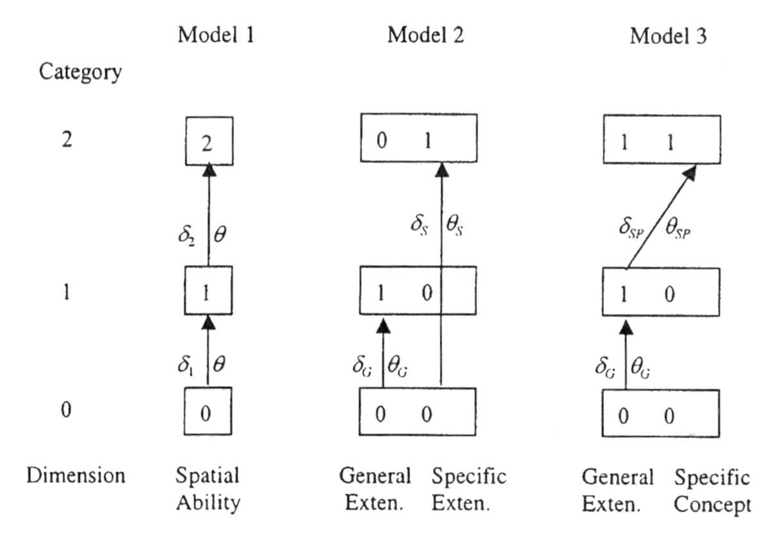

Note. The values in the rectangles are scores on the dimensions.

FIGURE 2. Scoring functions and structures of the three models in Example 1.

$$\log (p_2/p_0) = (\theta_G + \theta_{SP}) - (\delta_G + \delta_{SP})$$
$$= \theta_S - \delta_S,$$

which is identical to Equation 4. That is, the specific extensiveness of the second model, θ_S, is decomposed into two parts: the general extensiveness, θ_G, and the specific concept, θ_{SP}, of the third model. Likewise, the item parameter δ_S in the second model is decomposed into δ_G and δ_{SP}. Therefore, θ_{SP} can be treated as an extra effort needed to reach the correct response from the partially correct response. The parameters in one model can be derived from the other through reparameterization. For instance, once δ_G and δ_S of the second model are estimated, δ_{SP} of the third model can be calculated by simply subtracting δ_G from δ_S.

In van Kuyk's data the sample size is 263, with 66 children aged 4 to 5 years, 132 children aged 5 to 5½ years, and 65 children aged 5½ to 6 years. For illustrative purpose, we analyzed this data set using all the three models. As shown in Table 1, the deviances (G^2, Bishop, Fienberg, & Holland, 1975) of these three models are 5,987.45, 5,915.67, and 5,915.30, with 31, 33, and 33 parameters estimated, respectively. The two multidimensional models are very close because as just described, they are mathematically equivalent. Akaike's information criterion (*AIC*; Akaike, 1974) is used for model comparison. The *AIC*s for these three models are 6,049.45, 5,981.67, and 5,981.30, respectively. Therefore, these multidimensional models are better fitting than the unidimensional model. We interpret this to mean that the assumption of two abilities underlying the responses is confirmed by the data.

Parameter estimates of the three models are also shown in Table 1. Regarding item parameters in the two multidimensional models, as expected, the first step difficulties in the two models are basically the same. The second step difficulties in the second model (for example, –3.65 for item 1) are close to the sums of the first and the second step difficulties in the third model (for example, –3.65 = –1.40 – 2.25, for item 1), as shown in the last two columns of Table 1.

TABLE 1.
Summarized Results of the Three Models in Example 1.

	Model 1: Step Difficulty		Model 2: Step Difficulty		Model 3: Step Difficulty		Difference Between Models 2 and 3: Step Difficulty	
Item	1st	2nd	1st	2nd	1st	2nd	1st	2nd
1	–1.55	–2.10	–1.41	–3.65	–1.40	–2.25	.01	.00
	(.34)	(.21)	(.34)	(.32)	(.35)	(.22)		
2	–.24	1.60	–.23	1.34	–.22	1.57	.00	.01
	(.15)	(.20)	(.16)	(.24)	(.16)	(.21)		
3	–2.85	–.99	–2.81	–3.88	–2.81	–1.06	.00	.00
	(.37)	(.16)	(.38)	(.38)	(.38)	(.17)		
4	–.50	–2.10	–.34	–2.60	–.33	–2.26	.01	.01
	(.28)	(.22)	(.29)	(.25)	(.29)	(.23)		
5	–1.60	–1.69	–1.49	–3.30	–1.48	–1.81	.01	.00
	(.30)	(.19)	(.31)	(.30)	(.31)	(.19)		
6	–.82	.56	–.80	–.28	–.80	.53	.01	.01
	(.17)	(.16)	(.19)	(.22)	(.18)	(.17)		
7	–.23	.55	–.18	.30	–.18	.49	.01	.01
	(.16)	(.17)	(.18)	(.21)	(.18)	(.18)		
8	1.94	–2.44	2.13	–.50	2.15	–2.64	.01	.01
	(.33)	(.32)	(.33)	(.19)	(.33)	(.33)		
9	–.69	–1.03	–.58	–1.73	–.58	–1.15	.01	.01
	(.22)	(.17)	(.23)	(.22)	(.23)	(.18)		
10	–1.67	1.40	–1.74	–.31	–1.74	1.44	.01	.01
	(.19)	(.17)	(.20)	(.25)	(.20)	(.18)		
11	–1.64	–.71	–1.58	–2.37	–1.58	–.79	.01	.01
	(.24)	(.16)	(.25)	(.26)	(.25)	(.17)		
12	–1.75	–1.33	–1.66	–3.11	–1.66	–1.44	.01	.01
	(.29)	(.17)	(.30)	(.29)	(.30)	(.18)		
13	–2.16	1.20	–2.24	–1.00	–2.24	1.25	.01	.01
	(.22)	(.16)	(.23)	(.27)	(.23)	(.17)		
14	–.83	–1.72	–.69	–2.55	–.69	–1.86	.01	.01
	(.26)	(.19)	(.27)	(.25)	(.27)	(.20)		
15	–3.07	2.35	–3.24	–.78	–3.23	2.46	.01	.01
	(.28)	(.21)	(.29)	(.36)	(.29)	(.22)		
Deviance	5987.45		5915.67		5915.30			
Variance	.73		1.28		1.29			
Variance			2.88		1.12			
Covariance			1.53		.26			

Note. Values in parentheses are standard errors.

Consider the population distribution. The means of the two dimensions are set to be zero because of model identification. The variances of the general extensiveness of the two models, 1.28 and 1.29, respectively, are basically identical. The variance of the specific extensiveness of the second model, 2.88, is about equal to the sum of the variances of the two dimensions of the third model plus two times of their covariance (= 1.29 + 1.12 + 2 × .26 = 2.93). This is expected because as just described:

$$\theta_S = \theta_G + \theta_{SP},$$

hence:

$$\mathrm{Var}(\theta_S) = \mathrm{Var}(\theta_G) + \mathrm{Var}(\theta_{SP}) + 2\mathrm{Cov}(\theta_G, \theta_{SP}).$$

To illustrate the interpretation of the multidimensional parameters, consider the third model (readers can easily generalize the interpretation to the second model). The log-odds of the partially correct response to the incorrect response depend on the general extensiveness and the first step difficulty (for example, –1.40 logits for the first item). The log-odds of the correct response to the partially correct response depend on the specific concept and the second step difficulty (for example, –2.25 logits for the first item). Therefore, for the first item, a child with –1.40 logits on the general extensiveness ability will have a conditional probability of .5 of giving a partially correct response rather an incorrect response on item 1. However, he or she only needs –2.25 logits of the specific concept ability to have a conditional probability of .5 of giving a correct response rather than a partially correct response.

Hence, for item 1 the first step (reaching the partially correct response level from scratch) is relatively more difficult than the second step (reaching the correct response level from the partially correct response level), given that the proficiency levels of the two abilities are equal, although this will not happen very often as their correlation is only .21. Other items can be interpreted in the same way. It should be noted that these two difficulties are compared on two different dimensions, rather than on one dimension, as in the unidimensional partial credit models.

Regarding the relationship between the two dimensions, the covariance and the correlation coefficient of them are .26 and .21, respectively. Treating these two dimensions as two vectors, we derive the angle between them of 78 degrees, which means that these two dimensions are near to but not quite orthogonal.

EXAMPLE 2: ADDITION PROBLEMS

The data set for this example is from Siegler's simple addition problems. Siegler (1987) has described five strategies which elementary school students apply to solve simple addition problems (for example, 3 + 7 = ?).

1. Guessing: The child says she guessed or did not know the answer.
2. Counting All: The child counts from one the number of times indicated by the sum.
3. Min Strategy: The child counts up from the larger addend the number of times indicated by the small addend.
4. Decomposition: The child transforms the problem into two or more simpler problems.
5. Retrieval: The child retrieves the answer from memory.

Typically, one and only one strategy is applied for each item. Children were given problems one at a time. Once they gave responses, they were asked how they solved the problems. Their responses to the items were then judged individually on mastery (correctness) and strategy used. Mastery was scored dichotomously, either right or wrong.

We picked six items for illustration. There are 67 students from kindergarten, grade 1, and grade 2 with complete data for these six items. Because of the small sample size and also for ease of illustration, let the Min Strategy, Decomposition, and the Retrieval be combined into a category called *Advanced Strategies*, compared to *Guessing* and *Counting All.* In such a case, six response categories are observed for each item: (zero) guessing-wrong (referred to as *GW*), (1) guessing-right (referred to as *GR*), (2) counting-wrong (referred to as *CW*), (3) counting-right (referred to as *CR*), (4) advanced-wrong (referred to as *AW*), and (5) advanced-right (referred to as *AR*).

These three strategies are treated as dimensions and denoted by *G*, *C*, and *A*, respectively. The response categories depend on both the proficiency levels and the item difficulties of these three dimensions. As a starting point, we propose a model where each log-odds depends on a single dimension. The model has the following log-odds:

$$\log(p_{GR} / p_{GW}) = \theta_G - \delta_1, \tag{7}$$
$$\log(p_{CR} / p_{CW}) = \theta_C - \delta_2, \tag{8}$$
$$\log(p_{AR} / p_{AW}) = \theta_A - \delta_3, \tag{9}$$
$$\log(p_{GW} / p_{CW}) = \delta_4, \tag{10}$$
$$\log(p_{GW} / p_{AW}) = \delta_5, \tag{11}$$

where p_{GW} denotes the probability of getting a *GW* response, p_{GR} denotes the probability of getting a *GR* response, and so on. θ_G, θ_C, and θ_A are abilities of the three dimensions, respectively, and where δ_i denotes an item parameter. For model identification, the population means of the three dimensions are set to be zero. In the above matrices, *GW* is the zero category and treated as reference which makes the corresponding row vector in both the $\mathbf{A}_i$ and the $\mathbf{B}_i$ zero.

Using the MRCML, the design submatrix $\mathbf{A}_i$ and the scoring submatrix $\mathbf{B}_i$ for the six response categories to item *i* are:

$$
\begin{array}{l}
\textit{Category} \\
GW \\ GR \\ CW \\ CR \\ AW \\ AR
\end{array}
\quad
\mathbf{A}_i = \begin{array}{c} \begin{array}{ccccc} 1 & 2 & 3 & 4 & 5 \end{array} \\ \begin{bmatrix} 0 & 0 & 0 & 0 & 0 \\ -1 & 0 & 0 & 0 & 0 \\ 0 & 0 & 0 & -1 & 0 \\ 0 & -1 & 0 & -1 & 0 \\ 0 & 0 & 0 & 0 & -1 \\ 0 & 0 & -1 & 0 & -1 \end{bmatrix} \end{array}
\quad \text{and } \mathbf{B}_i = \begin{array}{c} \begin{array}{ccc} G & C & A \end{array} \\ \begin{bmatrix} 0 & 0 & 0 \\ 1 & 0 & 0 \\ 0 & 0 & 0 \\ 0 & 1 & 0 \\ 0 & 0 & 0 \\ 0 & 0 & 1 \end{bmatrix} \end{array}
$$

These five parameters of item i can be easily interpreted within the unidimensional framework. For instance, δ_1, δ_2, and δ_3 are the item difficulties with respect to dimensions G, C, and A, respectively; δ_4 and δ_5 are *relative frequency* parameters of GW, compared to CW and AW, respectively. These two parameters are the log-odds of an incorrect response using *Guessing* to *Counting All* and *Guessing* to *Advanced Strategies*, respectively. If δ_4 is positive, the probability of using *Guessing* is larger than that of using *Counting All*, given the responses are incorrect. δ_5 can be interpreted similarly.

These two parameters are used to clarify the three difficulty parameters. Without these two adjustment parameters, the three difficulty parameters cannot be compared directly if the frequencies of GW, CW, and AW are very different. Only when the frequencies are quite close can the relative frequency parameters be ignored.

The relative frequency parameter of *Counting All* to *Advanced Strategies* can be found by simple algebraic calculation. From Equations 10 and 11, we have

$$\log (p_{CW} / p_{AW}) = \delta_5 - \delta_4.$$

Thus, $\delta_5 - \delta_4$ is the relative frequency parameter of *Counting All* to *Advanced Strategies*.

Returning to Siegler's data, the full model has a deviance of 897.58 with 36 parameters. The item parameters are shown in Table 2. Consider the three diffi-

TABLE 2.
Item Estimates and Standard Errors for the Multidimensional Full Model in Example 2.

Item	Diff. G	Diff. C	Diff. A	Rel. G/C	Rel. G/A
1	.83 (.69)	2.67 (.87)	−1.15 (.57)	−.12 (.49)	−.12 (.49)
2	.38 (.70)	.97 (.95)	−3.41 (.74)	1.10 (.67)	1.10 (.67)
3	.87 (.60)	2.06 (.92)	−1.20 (.60)	1.03 (.52)	.69 (.46)
4	−.03 (.62)	2.85 (1.02)	−1.40 (.59)	.41 (.53)	.12 (.49)
5	1.00 (.75)	1.80 (.97)	−2.76 (.64)	.81 (.60)	.59 (.56)
6	.18 (.58)	2.84 (1.01)	−.34 (.56)	.61 (.51)	−.00 (.43)

Note. The values in the parentheses are standard errors.
Diff, difficulty; *Rel*, relative frequency; *G*, *guessing*; *C*, *Counting All*; *A*, *Advanced Strategies*.

culty parameters for each item. Those difficulty parameters of the *Advanced Strategies* are the smallest across items, and those of *Counting All* are the largest across items, even larger than those of *Guessing*. Apparently, for these particular items, using *Counting All* is less efficient than *Guessing*. The *Advanced Strategies* is the most efficient among the three, given that the proficiency levels of the three dimensions are identical.

More specifically, consider the three difficulty parameters of the first item. They are .83, 2.66, and –1.15 logits, respectively. For an examinee with equal proficiency level across the three dimensions, say all equal to 0.0 logits, if she applies *Guessing*, the odds of being correct to incorrect are .44. If she applies *Counting All*, the odds become .07. If she applies *Advanced Strategies*, they are as high as 3.16. Obviously, *Advanced Strategies* is the best among the three, given the proficiency levels of the three dimensions are equal. For examinees' with different levels on the three dimensions, the best strategy to be applied can be found by checking the values of $\theta_G - \delta_1$, $\theta_C - \delta_2$, and $\theta_A - \delta_3$. The larger is the value, the better the strategy.

Consider the relative frequency parameters: Except for item 1 where the two parameters are around zero, they are all positive. In addition, those of *G* to *C* are somewhat larger than the corresponding parameters of *G* to *A*, meaning that given the responses are incorrect, we expect the probability of being *Guessing* is larger than that of *Counting All* or *Advanced Strategies*.

Suppose the frequencies of *GW*, *CW*, and *AW* are quite close so that δ_4 and δ_5 can be constrained to be zero, resulting in only three difficulty parameters being estimated. In that case, the model has the following log-odds:

$$\begin{aligned}\log (p_{GR} / p_{GW}) &= \theta'_G - \delta'_1,\\ \log (p_{CR} / p_{CW}) &= \theta'_C - \delta'_2,\\ \log (p_{AR} / p_{AW}) &= \theta'_A - \delta'_3.\end{aligned}$$

The design submatrix and the scoring submatrix for item *i* become:

$$\begin{array}{l} \textit{Category} \\ GW \\ GR \\ CW \\ CR \\ AW \\ AR \end{array} \quad \mathbf{A}_i = \begin{array}{c} \begin{matrix} 1 & 2 & 3 \end{matrix} \\ \begin{bmatrix} 0 & 0 & 0 \\ -1 & 0 & 0 \\ 0 & 0 & 0 \\ 0 & -1 & 0 \\ 0 & 0 & 0 \\ 0 & 0 & -1 \end{bmatrix} \end{array} \text{ and } \mathbf{B}_i = \begin{array}{c} \begin{matrix} G & C & A \end{matrix} \\ \begin{bmatrix} 0 & 0 & 0 \\ 1 & 0 & 0 \\ 0 & 0 & 0 \\ 0 & 1 & 0 \\ 0 & 0 & 0 \\ 0 & 0 & 1 \end{bmatrix} \end{array}$$

Note that there are no relative frequency parameters in the model. Since the reduced model is a submodel of the previous full model, the usual likelihood ratio test can be applied to compare models.

In the first model, the relative frequency parameters are not significantly different from zero; hence, the reduced model was also formed where all the relative frequency parameters were constrained to be zero. This reduced model has a deviance of 908.24 with 24 parameters estimated. It is not significantly different from the preceding full model, based on the usual likelihood ratio test ($\Delta G^2 = 10.66$, $df = 12$, $p > .05$). Thus, by Occam's razor, the reduced model is preferred.

In summary, using the MRCML, we are able to treat the strategies as dimensions. Each examinee derives three ability estimates. It is interesting and somewhat surprising that *Counting All* is worse than *Guessing*. This may be because the examinees are too young to master *Counting All* very well. Therefore, using this strategy leads to a higher error rate. Or, it may be a step up to *Advanced Strategies*.

Wilson (1992) applied the ordered partition model, an extension of the partial credit model, to analyze parts of the same data set. His approach was unidimensional. All the incorrect responses were combined into one category and the strategies used were treated as steps. We also conducted a unidimensional model for this kind of scoring. For the six response categories, we assign the three incorrect response categories, *GW*, *CW*, and *AW*, and the correct response using *Guessing* (*GR*) a score of 0; the correct responses using *Counting All* (*CR*) a score of 1; and correct responses using *Advanced Strategies* (*AR*) a score of 2.

In that case, it follows that:

$$\begin{aligned}
&\log (p_{GW} / p_{GR}) = \delta''_1,\\
&\log (p_{CR} / p_{CW}) = \theta - \delta''_2,\\
&\log (p_{AR} / p_{AW}) = \theta - \delta''_3,\\
&\log (p_{GW} / p_{CW}) = \delta''_4,\\
&\log (p_{GW} / p_{AW}) = \delta''_5.
\end{aligned}$$

Using the MRCML, the design submatrices for item i are as follows:

$$\begin{array}{l} \textit{Category} \\ GW \\ GR \\ CW \\ CR \\ AW \\ AR \end{array}
\quad \mathbf{A}_i = \begin{array}{c} \begin{array}{ccccc} 1 & 2 & 3 & 4 & 5 \end{array} \\ \begin{bmatrix} 0 & 0 & 0 & 0 & 0 \\ -1 & 0 & 0 & 0 & 0 \\ 0 & 0 & 0 & -1 & 0 \\ 0 & -1 & 0 & -1 & 0 \\ 0 & 0 & 0 & 0 & -1 \\ 0 & 0 & -1 & 0 & -1 \end{bmatrix} \end{array}
\quad \text{and } \mathbf{B}_i = \begin{bmatrix} 0 \\ 0 \\ 0 \\ 1 \\ 0 \\ 1 \end{bmatrix}$$

This unidimensional model yields a deviance of 1,011.36 with 31 parameters. This unidimensional model is not a submodel of the previous two multidimensional models. Based on the *AIC*s, the multidimensional models are better than the unidimensional model. Not only do the multidimensional models provide better fit, but they also give information about children's ability levels on all the three individual dimensions, rather than a single composite dimension.

CONCLUSIONS

We have briefly introduced the MRCML and discussed the interpretation of item parameters and scoring schemes and have shown how they can be interpreted within the framework of unidimensional models.

The scoring matrix **B** is one element of the MRCML. It is especially important in gaining a richer and deeper understanding about examinees' performances. For instance, many performance-based items require examinees to spend much testing time yet are judged on a relatively rough scale, say 0 to 5. Suppose we can develop several thorough and applicable scoring scales for each item; then, these items become multidimensional and may provide much richer information (for example, a profile) about the examinees. Thus, instead of judging the overall performance on a composition, we may judge it in terms of content, grammar, rhetoric, and so on.

The design matrix **A** is another element. In standard Rasch models, the matrix **A** is implicitly determined. Because test users cannot manipulate these models, flexibility is limited. We have found that quite frequently test users are forced to sacrifice or ignore the complexity of their real test data in order to apply standard Rasch models. Using the MRCML, test users are offered a great deal of flexibility and responsibility to establish their own interpretable models.

Both matrices **B** and **A** are manipulated and defined by test users. They are what make the MRCML flexible and powerful. To interpret the parameters derived from the MRCML, one should focus on the conditional probabilities with respect to the proper response categories, rather than on the unconditional probabilities. In doing so, the interpretation of parameters is similar to that of the unidimensional Rasch models.

In Example 1, we demonstrate how the MRCML can be applied to analyze three response category items, by assuming the item responses are dominated by (1) a unidimensional ability, (2) two internal ordering abilities, or (3) two cross ordering abilities. Comparing the model fits as well as model interpretability, one can select the best model to account for the complexities of the underlying cognitive processes revealed in the data set. In Example 2, we adopt a three-dimensional model to account for the three strategies when solving simple addition problems. When the relative frequency parameters are modeled, the remaining difficulty parameters become more interpretable and can be easily used to depict strategy utility, even though, in fact, these relative frequency parameters are not statistically significant for this particular data set.

These models and examples are merely a few of the many which can be addressed with the MRCML. We have found that analyzing measurement situations with this model and its accompanying software (ConQuest; Wu, Adams, & Wilson, 1998), is complex, yet at the same time, it addresses many of the concerns we have had in applying standard item response models to testing data, such as dependence across items and rater severities in constructed-response items.

REFERENCES

Ackerman, T. A. (1992). A didactic explanation of item bias, item impact, and item validity from a multidimensional perspective. *Journal of Educational Measurement, 29,* 67–91.

Adams, R. J., & Wilson, M. R. (1996). Formulating the Rasch model as a mixed coefficients multinomial logit. In G. Engelhard & M. Wilson (Eds.), *Objective measurement: Theory into practice,* (Vol. 3, pp. 143–166*).* Norwood, NJ: Ablex.

Adams, R. J., & Wilson, M. R., & Wang, W. (1997). The multidimensional random coefficients multinomial logit model. *Applied Psychological Measurement, 21*, 1–23.

Akaike, H. (1974). A new look at the statistical identification model. *IEEE Transactions on Automatic Control, 19,* 716–723.

Andersen, E. B. (1985). Estimating latent correlations between repeated testings. *Psychometrika, 50*, 3–16.

Bishop, Y. M. M., Fienberg, S. E., & Holland, P. W. (1975). *Discrete multivariate analysis.* Cambridge, MA: MIT Press.

Bock, R. D., & Aitkin, M. (1981). Marginal maximum likelihood estimation of item parameters: An application of the EM algorithm. *Psychometrika*, *46*, 443–459.

Camilli, G. (1992). A conceptual analysis of differential item functioning in terms of a multidimensional item response model. *Applied Psychological Measurement, 16,* 129–147.

Dempster, A. P., Laird, N. M., & Rubin, D. B. (1977). Maximum likelihood from incomplete data via the EM algorithm. *Journal of the Royal Statistical Society, Series B, 39*, 1–38.

Embretson, S. E. (Ed.). (1985). *Test design: Developments in psychology and psychometrics*. Orlando, FL: Academic.

Embretson, S. E. (1991). A multidimensional latent trait model for measuring learning and change. *Psychometrika*, *56*, 495–515.

Frederiksen, N., Mislevy, R. J., & Bejar, I. I. (Eds.). (1993). *Test theory for a new generation of tests*. Hillsdale, NJ: Erlbaum.

Glas, C. A. W. (1992). A Rasch model with a multivariate distribution of ability. In M. Wilson (Ed.), *Objective measurement: Theory into practice* (Vol. I, pp. 236–258). Norwood, NJ: Ablex.

Kelderman, H. (1989). *Loglinear multidimensional IRT models for polytomously scored items* (ERIC Document Reproduction Service No. ED 308238). Paper presented at the Fifth International Objective Measurement Workshop, University of California, Berkeley.

Kelderman, H. (1996). Multidimensional Rasch models for partial-credit scoring. *Applied Psychological Measurement, 20*, 155–168.

Kelderman, H., & Rijkes, C. P. M. (1994). Loglinear multidimensional IRT models for polytomously scored items. *Psychometrika, 59,* 149–176.

Luecht, R. M., & Miller, R. (1992). Unidimensional calibrations and interpretations of composite traits for multidimensional tests. *Applied Psychological Measurement, 16*(3), 279–293.

Masters, G. N. (1982). A Rasch model for partial credit scoring. *Psychometrika*, *47*, 149–174.

Oshima, T. C., & Miller, M. D. (1992). Multidimensionality and item bias in item response theory. *Applied Psychological Measurement, 16*(3), 237–248.

Pellegrino, J. W., Mumaw, R. J., & Shute, V. J. (1985). Analyses of spatial aptitude and expertise. In S. E. Embretson (Ed.), *Test design: Developments in psychology and psychometrics*. San Diego, CA: Academic Press.

Reckase, M. D. (1985). The difficulty of test items that measure more than one ability. *Applied Psychological Measurement, 9,* 401–412.

Reckase, M. D., & McKinley, R. L. (1991). The discriminating power of items that measure more than one dimension. *Applied Psychological Measurement, 15,* 361–373.

Romberg, T. (1992). Evaluation: A coat of many colors. In T. Romberg (Ed.), *Mathematics assessment and evaluation: Imperatives for mathematics education.* Albany, NY: SUNY Press.

Siegler, R. S. (1987). The perils of averaging data over strategies: An example from children's addition. *Journal of Experimental Psychology: General, 116*, 250–264.

Sternberg, R. J. (Ed.). (1982). *Advances in the psychology of human intelligence* (Vol. I). Hillsdale, NJ: Erlbaum.

Traub, R. E. (1983). A prior consideration in chossing an item response model. In R. K. Hambleton (Ed.), *Applications of item response theory*. Vancouver, BC: Educational Research Institute of British Columbia.

van Kuyk, J. J. (1988). Verwerven van grootte-begrippen [Acquiring size concepts.] *Pedagogische Studi*, *65,* 1–10.

Way, W. D., Ansley, T. N., & Forsyth, R. A. (1988). The comparative effects of compensatory and noncompensatory two-dimensional data on unidimensional IRT estimation. *Applied Psychological Measurement, 12,* 239–252.

Wilson, M. R. (1991). *Measurement models for new forms of student assessment in mathematics education*. Keynote address at the First National Conference on Assessment in the Mathematical Sciences, Australian Council for Educational Research, Geelong, Australia.

Wilson, M. R. (1992). The partial order model: An extension of the partial credit model. *Applied Psychological Measurement, 16,* 309–325.

Wu, M., Adams, R., & Wilson, M. (1998). *ConQuest* [Computer software]. Camberwell, Victoria: Australian Council for Educational Research.

13

THE IMPLICATIONS OF HALO EFFECTS AND ITEM DEPENDENCIES FOR OBJECTIVE MEASUREMENT

T. F. McNamara
University of Melbourne

Raymond J. Adams
Australian Council for Educational Research

INTRODUCTION

The analysis of data from performance-based language tests presents the problem of dealing appropriately with bundles of items that are linked to particular tasks. For example, clusters of items in reading tests are often linked to a common pas-

This is a revised version of a paper presented in the symposium "Empirical Examinations of Some Threats to Objectivity in Assessment that Relies on Professional Judgment" at the Annual Meeting of the American Educational Research Association, Chicago, Illinois, March 24–28, 1997. An earlier version of the first part of this paper was presented at the Language Testing Research Colloquium held in Tampere, Finland in August 1996 and has appeared in the proceedings under the title "New Approaches to the Analysis of Task- and Rater-related Dependencies in Performance Assessments." We are grateful to the Test Development Committee of the **access**: project for permission to use data from the test in this research.

sage, and in writing and speaking tests, multiple aspects of performance on a single task may be rated. It has long been recognized that such items almost certainly violate the local independence assumptions implicit in most analytical techniques. The reliance of the items on a common stimulus or performance is likely to introduce a dependence between the item responses. Similar dependencies are likely to occur between the ratings made by a single rater when considering the qualities of a single performance or task with respect to multiple criteria. Typically in both the analysis and reporting of subjects' achievements the effect of these intratask and intrajudge dependencies are overlooked, although where design considerations permit, approaches using Generalizability Theory may be used to analyze some aspects of task and judge effects (for example, Lynch & McNamara, 1998; McNamara & Lynch, 1997).

Put simply, the problem of lack of independence is that in the case where items or ratings are dependent, the standard analysis inflates the information value of each item score by assuming that each such score provides entirely independent further evidence of the candidate's ability. The claim to precision of measurement is exaggerated as a result; acknowledgment of the dependency typically results in less precision, reflected through an increase in the standard error of candidate ability estimates.

Recent developments in Rasch modeling permit a new approach to this classic problem. They allow the analysis of bundles of items (or sets of ratings) where the items or ratings lack the standard independence assumptions of generally applied analysis methods. In this chapter, we use these methods to examine the extent of dependence between items and ratings, and we explore the implications of those dependencies on the assessments. Data from two performance assessments are used in two studies of this question:

Study 1: data from the Writing module of the **access:** test, a test of English as a Second Language (ESL) for certain categories of intending adult immigrants to Australia

Study 2: data from a test of writing in English as a mother tongue by Australian students in Year 6 (average age 11)

For Study 1, two analyses are reported: one in which the task and rater dependencies are ignored, and one in which the dependencies are recognized. The conclusions about the effect of rater dependencies reached in Study 1 are supported by the results of Study 2.

ADVANCES IN RASCH MODELING

The analyses reported in this chapter are made possible by recent advances in Rasch modeling, specifically the development of a family of generalized Rasch item response models as described in Wilson and Adams (1995); Adams and Wilson (1996); Adams, Wilson, and Wu (1997) and Adams, Wilson, and Wang (1997) and their implementation through new computer software. These generalized models represent a way of writing most existing Rasch models so that they become particular realizations of the general model, in much the same way that KR-20 is a simpler and more specific version of the more general and powerful Cronbach's α, or the t-test is a simpler and more specific version of the more powerful and general ANOVA. The generalized model (the random coefficients multinomial logit, or RCML; Wilson & Adams, 1995) that is implemented in the software used in this paper (ConQuest: Wu, Adams, & Wilson, 1997) integrates and extends many existing Rasch-type models, including the simple logistic model (Rasch, 1960), the linear logistic latent trait model (Fischer, 1973), the rating scale model (Andrich, 1978), the partial credit model (Masters, 1982), the ordered partition model (Wilson & Adams, 1993), Linacre's (1989) FACETS model, and Glas and Verhelst's extension to the partial credit model (Glas, 1989; Glas & Verhelst, 1989), among others. In addition, the model provides a great deal of flexibility in allowing the design of customized models for particular test situations. The software is also capable of fitting multidimensional forms of each of the preceding models.

In the context of this chapter we are using the model to fit partial credit, rating scale, and FACETS models along with the ordered partition model (Wilson & Adams, 1993). This latter model relaxes assumptions about the ordering of score categories, so that not all score categories are ordered in relation to each other, but are ordered in relation to at least some other scores categories above or below them. Thus, a number of separate response categories to an item may be given the same score level. The potential of the partial order model is exploited in this chapter in our examination of task- and rater-related dependencies.

STUDY 1: *ACCESS:* TEST-RATINGS OF PERFORMANCES ON ESL WRITING TASKS BY ADULT IMMIGRANTS TO AUSTRALIA

Test Background—Study 1

The Australian government in 1992 introduced a formal test of communicative skills in English for certain classes of intending immigrants, to replace the informal assessments that were the basis of the allocation of points for English language skills in the highly competitive and selective process that results in approval being granted for a person to enter the country as an immigrant. The test, the Australian Assessment of Communicative English Skills (**access:**), is administered at a large number of test centers overseas. It was developed under the overall direction of the National Centre for English Language Teaching and Research at

Macquarie University, Sydney, by a team originally involving experts from Macquarie University and from the Language Australia Language Testing Research Centre at the University of Melbourne and its equivalent at Griffith University, Brisbane.[1] There are four test modules, one for each macroskill. A group of researchers at Macquarie University was responsible for the development of the Writing module, data from trials of which are analyzed in this study. The University of Melbourne team was responsible for the development of the Speaking skills module, and for the analysis of data from the trialling of the Speaking and Writing modules.

Data—Study 1

Subjects

The data in this study are from scripts used as part of the training material for raters of the Writing module; these scripts are from actual test candidates taking Version B of the test at a live administration in 1993. The data are from 49 candidates.

Test Tasks

The **access:** test Writing module contains three tasks, as follows:

Task 1 (8 to10 minutes): Establishing and maintaining interpersonal contacts
Example task: Replying to an invitation
Task 2 (15 to 20 minutes): Giving or requesting information or explanation
Example task: Writing a memo
Task 3 (20 to 30 minutes): Arguing or discussing an issue
Example task: Letter to the editor

Test Scoring

The performances of the 49 candidates were each marked by 19 raters. For each task, separate ratings on a six-point scale were given for each of the following aspects of performance (four per task): *Task fulfilment and appropriacy*, *Conventions of presentation*, *Cohesion and organization*, and *Grammatical control.*

Raters

The 19 raters in this study were all native speakers of English and trained teachers of English as a Second Language, each with extensive experience of teaching at advanced levels. Each rater had attended a 1-day training session at which independent ratings of sample scripts were made and discussed, and had subsequently completed the rating of the set of 49 scripts. Six weeks later, as part of a study on feedback to raters using the bias analysis feedback in multifaceted Rasch measurement (Lunt, Morton, & Wigglesworth,1994), the raters carried out ratings of the 49 scripts again. These latter data were used in the analysis.

Data Analysis—Study 1

All analysis was done using the ConQuest program (Wu, Adams, & Wilson, 1997). Details of models used in data analysis in this and subsequent study are found in the Appendix.

Stage 1: Investigating Task Dependencies

The first stage of analysis investigated the impact of *task* dependencies. This corresponds to the well-known halo effect, whereby ratings of different aspects of a single performance may be similar, that is, lack independence. A simple comparison was made of ability estimates obtained under two conditions (Analyses 1a and 1b):

Analysis 1a (standard approach—task dependencies left unmodeled). This involved standard procedures. Ratings per task were treated as independent, that is, the possible effect of task dependencies was ignored. A standard Rasch Rating Scale analysis of the data was carried out.

Analysis 1b:(alternative approach—task dependencies modeled). This involved exploiting the capabilities of the program so that task dependency could be modeled. Each task was treated as a single "meta-item," with a large number of possible score patterns within it, each possible score pattern constituting a possible response category for the meta-item. Each response category thus represented a string of subscores (the scores on the old, more narrowly defined "items"). Different strings of scores could make up the same total score; that is, they were not ordered in relation to each other, but were ordered in relation to strings making up different total scores (higher or lower). An ordered partition model analysis (Wilson & Adams, 1993) was carried out. This analysis involved a potentially huge number of possible response patterns; in fact, of course, not all of these were observed. Only those score patterns which actually occurred were treated as possible. Thirty response patterns were so identified. The second analysis thus involved three meta-items with 30 categories of response.

The *results* from the two analyses (1a and 1b) were identical. It had been anticipated that the standard error of measurement from the second analysis would be greater than in the first analysis, but this was not found to be so; any differences were negligible. Whatever impact there might be of task dependency was being masked by other factors in these data.

Stage 2: Investigating Rater Dependencies

Given the preceding finding, it was realized that the dependency associated with particular raters might be a more significant issue than the dependency associated with task. That is, the greatest unmodeled dependency in these ratings might be that associated with a rater; the well-known "halo" effect might be rater-specific. In particular, raters may have been judging performances on different tasks as similar when they were asked to make multiple ratings of the same aspect across the three tasks. In order to investigate this possibility, ratings given for the first aspect (*Task fulfilment and appropriacy*) for each of the three tasks for each of the 19 raters were considered; there were thus 57 observations for each of the 49 candidates. A number of analyses were then done, as follows, again with the intention of comparing ability estimates when rater dependencies were and were not modeled.

Analysis 2a (Standard Approach—Rater Dependencies Left Unmodeled). In this analysis, rater effects were estimated as in conventional analyses using FACETS (Linacre, 1989). A model was fitted that treated each rater/task combination as an independent item, as in conventional (rating scale) analysis in a program such as Quest (Adams & Khoo, 1993). Ability estimates were derived. This was followed by the fitting of a sequence of models of rater and task effects, and incorporating possible rater by task interactions, as in FACETS bias analysis. No significant bias was found, that is, the rater × task interaction was not significant. While the FACETS model estimates rater effects or rater by task interactions, it continues to assume a *conditional* independence between ratings. In this context, the term conditional is used to indicate that independence is assumed after accounting for dependencies that result directly from estimated rater characteristics (in this case the rater harshness), task characteristics (in this case the task difficulty and step structure), and the candidate abilities. That is, after taking into consideration the estimated harshness of each of the raters, the estimated difficulties of each of the tasks, and the proficiency level of the candidates, it is assumed that there are no other factors that make the judgments of a particular rater more or less similar within a candidate than between candidates.

Thus, in this analysis, the "rater effect" was modeled as in a FACETS analysis, but there remained a possible dependency between judgments per rater which had not been modeled. This was addressed in the subsequent analyses.

Analysis 2b (Alternative Approach—Rater Dependencies Modeled). To what extent was this possible dependency actually evidenced in the data? That is, how important was it at face value to model this dependency? Before starting out on an analysis in which rater dependency was modeled, we decided to get a feel for its extent. The following preliminary procedure was thus carried out: Assuming independence of observations per rater for a given aspect being rated across tasks, it was possible to calculate the probability of any response string of three ratings (the separate ratings for that aspect for each task) for any rater, given the data of 49 judgments per rater for the aspect considered. We were then in a position to compare the frequency of observed response patterns with the frequency as predicted under the assumption of independence. (This is the same logic as in bias analysis in FACETS runs). The expected versus actual frequency of response patterns was calculated for 5 of the raters, and is presented in Figures 1 through 5.[2]

The dark shading in Figures 1 through 5 shows the actual observations; the light shading shows the distribution as predicted from the assumption of independence. It is clear that there is considerable unmodeled dependency in this data. For exam-

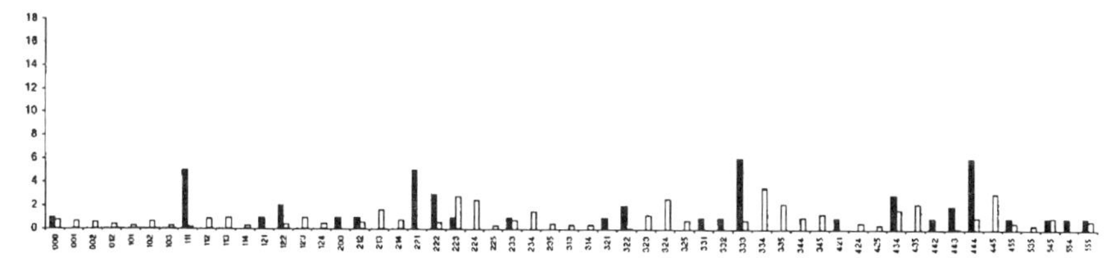

FIGURE 1. Comparison of the actual and model predicted number of occurrences of each response pattern for Rater 1.

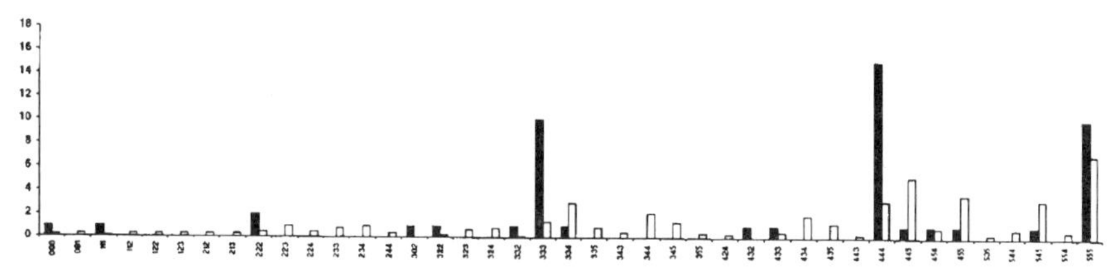

FIGURE 2. Comparison of the actual and model predicted number of occurrences of each response pattern for Rater 2.

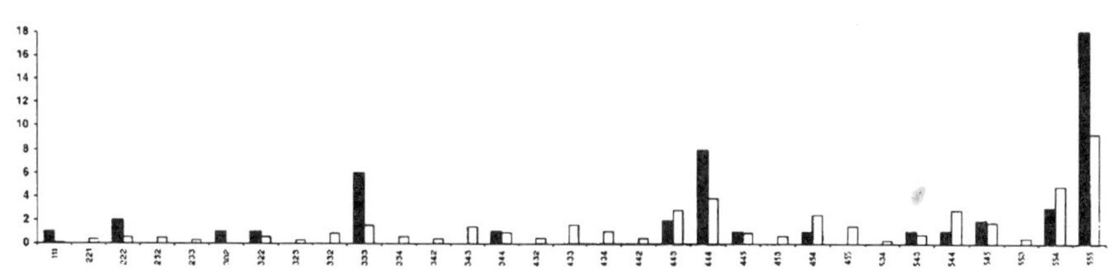

FIGURE 3. Comparison of the actual and model predicted number of occurrences of each response pattern for Rater 3.

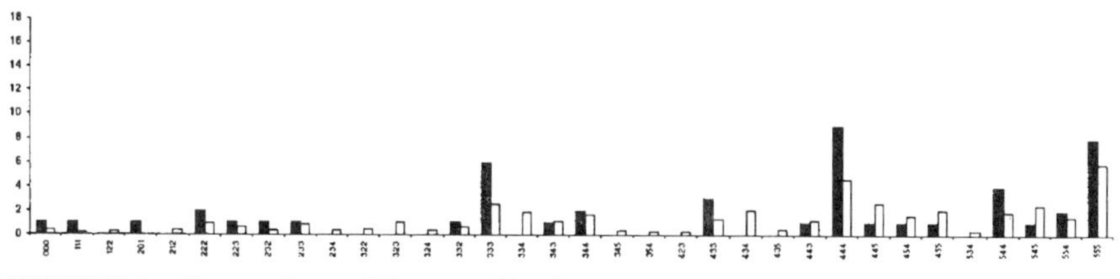

FIGURE 4. Comparison of the actual and model predicted number of occurrences of each response pattern for Rater 4.

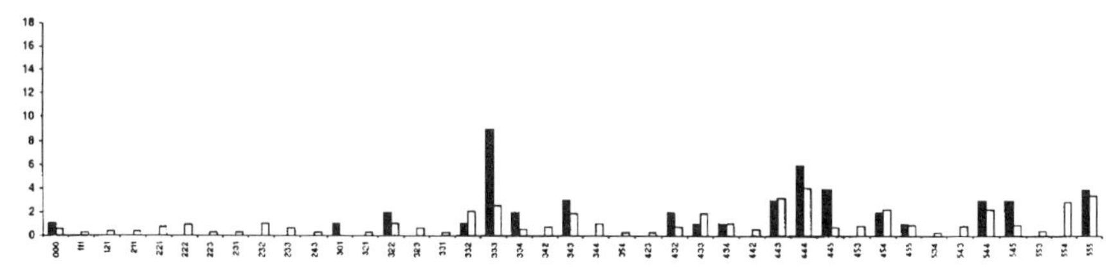

FIGURE 5. Comparison of the actual and model predicted number of occurrences of each response pattern for Rater 5.

ple, the greater than expected incidence of the string 4,4,4 suggests that the rater has decided "this candidate is a 4 on this aspect" and uses this score over and over again, regardless of task.

An analysis was now conducted that would model this dependency and investigate its impact on candidate ability measures. For each of the 19 raters, all of the patterns of response actually given were identified, and each rater was then treated as a meta-item, with the score patterns the rater used treated as response categories of this meta-item; to each of those categories we assigned a score that was equal to the sum of the scores for the individual components of the patterns that defined the categories. An ordered partition model (Wilson & Adams, 1993) was then fitted to this data. The resulting ability estimates (which made no assumption of independence) were then available for comparison with the ability estimates obtained from Analysis 2a (assuming independence) and the two sets of ability estimates were plotted against each other (Figure 6).

Approximate equivalence was established, although the metrics were not identical. That is, it appeared that the failure to model the dependency had little impact on ability estimates, particularly in the middle range of abilities. The slightly curvilinear relationship suggested that allowance should be made for lack of equivalence of the metrics, so the ability estimates were standardized and plotted again, as in Figure 7. Again, a very slightly curvilinear relationship was found, suggesting that ability estimates were little affected by the failure to model the evi-

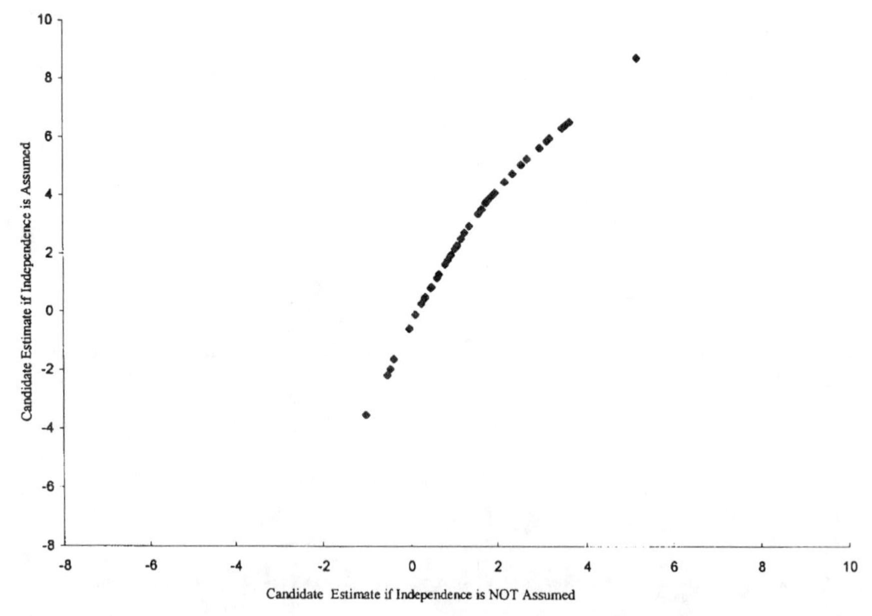

FIGURE 6. Comparison of candidate estimates with and without independence assumptions.

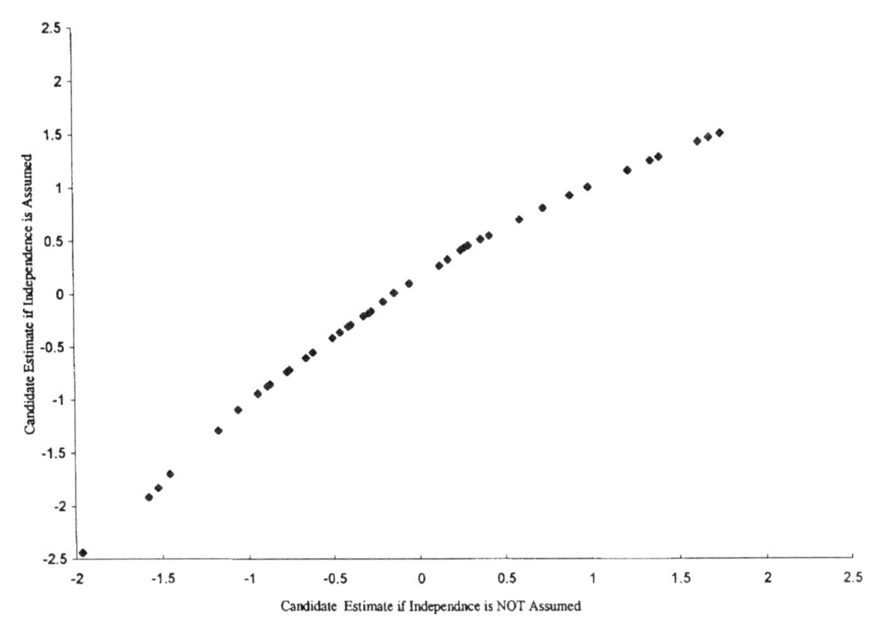

FIGURE 7. Comparison of candidate estimates with and without independence assumptions, both scales standardized.

dent dependency in the data, except for those candidates at the extremes of the ability spectrum.

It was hypothesized that the error of measurement might be increased by modeling the dependency, on the assumption that by acknowledging that any new judgement a rater is making of a candidate is not making an entirely new independent contribution to the picture of his or her ability, the precision of the resulting measures will be reduced. We thus plotted the standard errors under the two analyses. The resulting plot (Figure 8) is to be read in the following way: If an observation is above the identity line, the error assuming independence (that is, where rater dependency is not modeled—Analysis 2a) is greater; if below the line, the error not assuming independence (that is, where rater dependency is modeled—Analysis 2b) is greater. As can be seen, errors tend to cluster below the line, as predicted. As a way of summarizing the effect, we calculated the ratio of the error variance in the independence condition to the error variance in the nonindependence condition and found that it was about .75; that is, if we assume independence, the error variance is about 75% of what it should be if the independence assumption is not made.

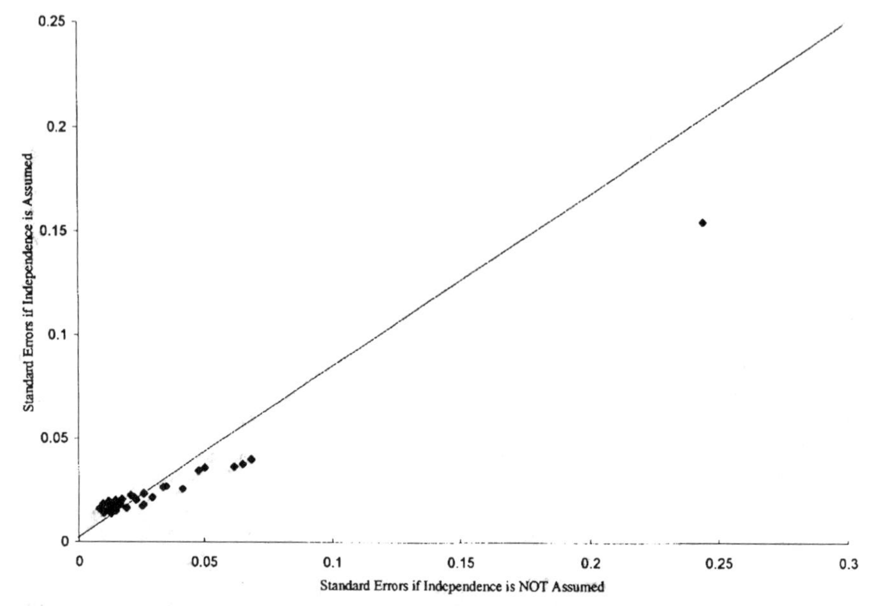

FIGURE 8. Comparison of standard errors for estimates made with and without independence assumptions, both scales standardized.

STUDY 2: RATINGS OF WRITING BY SCHOOLCHILDREN IN ENGLISH AS A MOTHER TONGUE

In a second study, the ratings of 8,296 Year 6 students' response to a single writing task were analysed (Year 6 is the final year of the primary school in Australia; the average age of students in Year 6 is approximately 11 years). The data were gathered as part of a study reported in Congdon and McQueen (1997). Each of the 8,296 scripts was assessed by two randomly chosen raters from a set of 16 raters, the second rating for each script being performed blind. The random allocation of scripts to the raters ensured that links between all raters were obtained. When assessing the scripts each rater was required to provide two ratings:

- One for *Overall Performance (OP)*, a judgment of the task fulfilment, particularly in terms of appropriateness for purpose and audience, conceptual complexity, and organization of the piece
- The other, for *Textual Features (TF)*, a judgment of control over and effective use of syntactic features such as cohesion, subordination, and verb forms and other features of presentation such as punctuation and spelling

The rating of both features was undertaken against a six-point scale with labels G, H, I, J, K, and L used to indicate successively superior levels of performance.

The data were analyzed using a multifaceted model to provide estimates of three sets of parameters: rater, criteria, and step. Figure 9 shows the results of the analysis for raters.

The table shows considerable misfit among the raters: the weighted mean square values, which we would expect to be close to 1, reveal that most of the raters have a fit that is larger than one by a substantial amount.

Figure 10, which presents the results of the analysis for the criteria, shows that the difficulty estimates for the two criteria are very similar. The mean square fit statistics show that there is a lack of independence between the raters' judgments for the two criteria.

In order to investigate this further, the data were analyzed again for a single criterion, *OP*. Figure 11, the rater parameter estimates, shows much better fit for raters.

How are we to interpret these analyses? In the first analysis, when both criteria were included, a strong dependency was observed between the two criteria, *OP* and *TF;* in contrast, the raters appeared not to be consistent. Thus, there is more agreement between the scores for *OP* and *TF* than there is between raters; we note a rater-associated halo effect. If a rater chooses a particular mark on one criterion,

Rater Effects Model One — Fri Oct 18 11:25:34

TABLES OF RESPONSE MODEL PARAMETER ESTIMATES

TERM 1: rater

VARIABLES			UNWGHTED FIT		WGHTED FIT	
rater	ESTIMATE	ERROR	MNSQ	T	MNSQ	T
1 14	0.915	0.03	1.10	3.0	1.13	3.9
2 17	0.116	0.03	1.16	4.8	1.19	5.4
3 18	-0.085	0.03	1.41	10.7	1.46	11.6
4 19	-1.208	0.03	1.21	6.2	1.23	6.7
5 24	0.585	0.03	1.34	9.6	1.36	9.8
6 38	-0.103	0.03	1.08	2.5	1.10	2.8
7 67	0.489	0.03	1.11	3.5	1.14	4.2
8 70	0.102	0.03	1.04	1.3	1.08	2.2
9 73	0.002	0.03	1.13	4.0	1.15	4.5
10 74	-0.203	0.03	1.22	6.9	1.21	6.4
11 78	-0.563	0.03	1.13	4.0	1.17	4.9
12 79	-0.188	0.03	1.05	1.6	1.07	2.0
13 8	0.117	0.03	1.08	2.3	1.09	2.8
14 85	-0.000	0.03	1.26	7.8	1.30	8.7
15 89	0.404	0.03	1.09	2.5	1.11	3.1
16 93	-0.381*					

Most of the raters have a fit that is larger than one by a substantial amount.

An asterisk next to a parameter estimate indicates that it is constrained

FIGURE 9. The parameter estimates for rater harshness—two criteria.

TERM 2: criteria

VARIABLES			UNWGHTED FIT		WGHTED FIT	
criteria	ESTIMATE	ERROR	MNSQ	T	MNSQ	T
1 OP	0.083	0.01	0.87	-8.8	0.85	-9.7
2 TF	-0.083*					

An asterisk next to a parameter estimate indicates that it is constrained

FIGURE 10. Parameter estimates for the criteria.

Rater Effects Model Two — Fri Oct 18 16:54:1

TABLES OF RESPONSE MODEL PARAMETER ESTIMATES

TERM 1: rater

VARIABLES			UNWGHTED FIT		WGHTED FIT	
rater	ESTIMATE	ERROR	MNSQ	T	MNSQ	T
1 14	0.743	0.04	0.89	-3.7	0.92	-2.6
2 17	0.065	0.04	0.93	-2.1	0.96	-1.1
3 18	-0.038	0.04	1.15	4.3	1.18	5.0
4 19	-1.268	0.04	0.99	-0.2	1.00	-0.1
5 24	0.703	0.04	1.08	2.3	1.09	2.6
6 38	0.201	0.04	0.88	-3.8	0.92	-2.4
7 67	0.444	0.04	0.94	-2.0	0.97	-0.8
8 70	-0.093	0.04	0.95	-1.5	0.98	-0.7
9 73	-0.242	0.04	0.87	-4.5	0.87	-4.0
10 74	-0.130	0.04	1.07	2.1	1.04	1.4
11 78	-0.326	0.04	0.89	-3.7	0.93	-2.2
12 79	0.118	0.04	0.87	-4.3	0.91	-2.8
13 8	0.279	0.04	0.88	-4.0	0.90	-3.1
14 85	-0.202	0.04	0.95	-1.5	0.97	-1.0
15 89	0.022	0.04	0.91	-2.8	0.94	-1.7
16 93	-0.277*					

An asterisk next to a parameter estimate indicates that it is constrained

The fit statistics for this model are much better than the corresponding fit statistics for the previous model

FIGURE 11. Rater harshness parameter estimates—single criterion ("overall performance")

let us say *OP*, then we can predict the mark given for *TF* by that rater. In contrast, we cannot predict the mark given by a second rater for the same script. Note that in the analysis done by ConQuest, the entire fit in the data is modeled, unlike in FACETS, where fit is modeled facet by facet. The FACETS analysis would disguise the halo effect that is revealed in the ConQuest analysis.

SUMMARY AND CONCLUSIONS

Both studies reported here reveal the existence of rater-associated dependencies. The nature of dependencies is disguised in analyses fitting models as implemented in FACETS, but is revealed using recently developed software (ConQuest) fitting newer more generalized models. In terms of practical consequences for measurement, however, the results of Study 1 suggest that the measurement implications for the failure to model dependency in the data from performance tests of writing are likely to be modest. This apparently reassuring finding needs to be investigated further for us to be confident it is generalizable; the aim of this chapter is essentially to draw attention to the issue, and to the existence of new tools with which to investigate it. It is obviously necessary to investigate the issues raised in this chapter with larger and more simply structured data sets, and with nonjudgement data; for example, for objectively scored responses to passage dependent items in reading or listening tests. The measurement approach reported in this chapter also allows us to settle long-standing disputes about dimensionality, as it allows analyses to be compared assuming and not assuming unidimensionality in the data; these will be the subject of a subsequent paper.

APPENDIX: DETAILS OF MODELS USED IN THE ANALYSES

ANALYSIS 1A

19 raters; 3 tasks; 4 aspects; 6 categories per aspect.

Model

$$\log P_{ntjik} / P_{ntji(k-1)} = B_n - D_t - C_j - A_i - F_k$$

ANALYSIS 1B

19 raters; 3 tasks; 4^6 possible categories, of which 30 were observed.

Model

$$Pr\ (x = k) = exp\ (b_k q - x_k)\ /\ D$$

where k indexes the different possible response patterns and b_k is the score associated with pattern k.

ANALYSIS 2A

Taking one aspect (for example, task fulfilment): 19 raters; 3 tasks; 6 categories

Models

Step 1

$$B_n - D_t - C_j - I_{tj} - F_k$$

where I_{tj} is a bias term, which the analysis subsequently reveals to be unnecessary.

Step 2

$$B_n - D_t - C_j - F_k$$

ANALYSIS 2B

19 raters; 1 meta-item; 6^3 categories.

Model

As in Analysis 1b.

NOTES

1. Administration of the test is currently the joint responsibility of IDP Australia and Griffith University.

2. Not all 216 notional score pattern possibilities were graphed; the following patterns were plotted: (1) where there was an observation; (2) the model predicted > 0.2 of a candidate for a particular score string.

REFERENCES

Adams, R. J., & Khoo, S. -T. (1993). *Quest* [Computer software]. Camberwell, Victoria: Australian Council for Educational Research.

Adams, R. J., & Wilson, M. R. (1996). A random coefficients multinomial logit: A generalized approach to fitting Rasch models. In G. Engelhard & M. Wilson (Eds.), *Objective measurement: Theory into practice* (Vol. 3). Norwood, NJ: Ablex.

Adams, R. J., Wilson, M. R., & Wang, W. (1997). The multidimensional random coefficients multinomial logit model. *Applied Psychological Measurement, 21,* 1–23.

Adams, R. J., Wilson, M. R., & Wu, M. L. (1997). Multilevel item response modelling: An approach to errors in variables regression. *Journal of Educational and Behavioral Statistics, 22,* 47–76.

Andrich, D. (1978). A rating formulation for ordered response categories. *Psychometrika, 43*, 561–573.

Congdon, P., & McQueen, J. (1997, March) *The stability of rater severity estimates in large scale performance assessment programmes.* Paper presented at the annual meeting of the American Educational Research Association, Chicago.

Fischer, G. H. (1973) The linear logistic model as an instrument in educational research. *Acta Psychologica, 37*, 359–374.

Glas, C. A. W. (1989). *Contributions to estimating and testing Rasch models.* Doctoral dissertation, University of Twente.

Glas, C. A. W., & Verhelst, N. D. (1989). Extensions of the partial credit model. *Psychometrika, 54*, 635–659.

Linacre, J. M. (1989). *Many-faceted Rasch measurement*. Chicago: MESA Press.

Lunt, H., Morton, J., & Wigglesworth, G. (1994, July). *Rater behaviour in performance testing: Evaluating the effect of bias feedback.* Paper presented at the 19th Annual Congress of the Applied Linguistics Association of Australia, Melbourne.

Lynch, B. K., & McNamara, T. F. (1998). Using G-theory and multi-faceted Rasch measurement in the development of performance assessments of the ESL speaking skills of immigrants. *Language Testing, 15*, 158–180.

Masters, G. N. (1982). A Rasch model for partial credit scoring. *Psychometrika, 47*, 149–174.

McNamara, T. F., & Lynch, B.K. (1997). *A generalizability theory study of ratings and test design in the writing and speaking modules of the* ***access:*** *test*. In G. Brindley & G. Wigglesworth (Eds.), ***access:*** *Issues in English language test design and delivery* (pp. 197–213). Sydney, NSW, Australia: National Centre for English Language Teaching and Research, Macquarie University.

Rasch, G. (1960). *Probabilistic models for some intelligent and attainment tests.* Copenhagen: Danmarks Paedogogiske Institut.

Wilson, M. R., & Adams, R. J. (1993). Marginal maximum likelihood estimation for the ordered partition model. *Journal of Educational Statistics*, *18*, 69–90.

Wilson, M. R., & Adams, R. J. (1995). Rasch models for item bundles. *Psychometrika* *60*(2), 181–198.

Wu, M. L., Adams, R. J., & Wilson, M. R. (1997). *ConQuest* [Computer software]. Camberwell, Victoria, Australia: Australian Council for Educational Research.

14

RASCH MEASUREMENT THEORY, THE METHOD OF PAIRED COMPARISONS, AND GRAPH THEORY

Mary Garner
George Engelhard, Jr.
Emory University

INTRODUCTION

Georg Rasch (1966, 1977, 1980) repeatedly pointed out that a key characteristic of objective measurement is that the relative difficulty of any two items should not depend on the characteristics of a particular population. In other words, a comparison between any two items should be independent of a particular population; likewise, a comparison between any two persons should be independent of a particular set of items. Rasch placed great importance on such comparisons; indeed, he developed a theory regarding the generality and validity of scientific statements based on the idea that "comparisons form an essential part of our recognition of our surroundings ... both in everyday life and in scientific studies" (Rasch, 1977, pp. 68–69). He stated that any good measurement model, like any good scientific model, is based on objective comparisons.

A paper presented at the International Objective Measurement Workshop, University of Chicago, March 1997.

> It is my opinion that only through systematic comparisons—experimental or observational—is it possible to formulate empirical laws of sufficient generability to be—speaking frankly—of real value, whether for furthering theoretical knowledge or for practical purposes.
>
> I see systematic comparisons as a central tool in our investigation of the outer world. (p. 74).

The method of paired comparisons and graph theory are both based on comparisons between pairs of objects. Therefore, it seems very appropriate to bring these two areas together with Rasch measurement theory.

The method of paired comparisons is a widely used technique for describing preference behavior, based on the principles described by Thurstone (1927a, 1927b, 1927c) in his law of comparative judgment. The method reduces preference behavior to its most basic and most easily grasped element: a person's choice between two objects. The result is a linear scale along which the objects are ordered. Building from Rasch's work (1960), Choppin (1985) described several methods of estimating the item difficulties of the Rasch model by comparing performance on pairs of items. Andrich (1978) linked Thurstone's law of comparative judgment to Rasch measurement theory, and Engelhard (1984) described the parallels between Thurstone's and Rasch's approaches to measurement. Through the method of paired comparisons, Rasch measurement theory can also be linked with other scaling methods and with graph theory.

Graph theory is a branch of mathematics that provides a language, a set of procedures, and a way of visualizing a system that is built on the relationships between pairs of objects. It has been useful for this reason in assessing the outcome of paired comparisons experiments, and it holds promise as an analytical framework for examining aspects of Rasch measurement theory.

PURPOSE

The purposes of this chapter are to extract from the literature the parallels between the method of paired comparisons and Rasch measurement theory, to describe their intersection in pairwise (PW) algorithms for estimating the parameters of the Rasch model, and to bring forward the graph theoretical concepts that can be used in analyzing links between items that would enhance the use of the pairwise algorithms as well as other methods of parameter estimation. Specifically, the questions addressed are the following: (1) What is the relationship between the method of paired comparisons and Rasch measurement theory? (2) What is the relationship between the method of paired comparisons and graph theory? (3) What can graph theory contribute to our understanding of Rasch measurement theory?

The chapter is divided into four sections. In the first section, the connections between Rasch measurement theory and the method of paired comparisons are presented, along with the pairwise algorithms for estimating parameters of the Rasch model as described by Choppin (1985). The second section includes an introduction to the language of graph theory, a description of what role graph theory has played in applications of the method of paired comparisons, and the potential of graph theory as an analytical tool in using the pairwise algorithm. In the third section, the techniques presented in the previous sections are applied to a small data set. The last section consists of summary, conclusions, and suggestions for additional research.

RASCH MEASUREMENT THEORY AND THE METHOD OF PAIRED COMPARISONS

This section provides an introduction to the Rasch model and to the method of paired comparisons. These two areas are then brought together in pairwise algorithms for estimating parameters of the Rasch model as presented by Choppin (1985).

Rasch Measurement Theory

Rasch measurement theory is based on a mathematical model that describes the probability of a student achieving a certain score on a test as a function of the difference between the student's ability and the difficulty of the items on the test. Specifically, the probability that a person v will score correctly on particular item i ($a_{vi} = 1$) is expressed in terms of the person's ability b_v and the difficulty of the item d_i as follows:

$$\Pr(a_{vi} = 1) = \frac{e^{(b_v - d_i)}}{1 + e^{(b_v - d_i)}}. \tag{1}$$

This model is remarkable for at least two reasons. First of all, it is a stochastic rather than deterministic model; in other words, a student of a certain ability is not predicted to obtain a certain score but may obtain a range of scores with varying probabilities. A second characteristic of the model is what Rasch termed *specific objectivity*; that is, the mathematical structure of the model allows one to eliminate person abilities and be left with a model describing the relationship among item difficulties regardless of the persons involved; conversely, item difficulties can be eliminated to leave a model describing the relationship among person abilities regardless of the items used. Rasch (1966) described specific objectivity as follows:

> ...the comparison of any two subjects can be carried out in such a way that no other parameters are involved than those of the two subjects—neither the parameter of any other subject nor any of the stimulus parameters. Similarly, any two stimuli can be compared

> independently of all other parameters than those of the two stimuli, the parameters of all other stimuli as well as the parameters of the subjects having been replaced with observable numbers. It is suggested that comparisons carried out under such circumstances be designated as "specifically objective." (pp. 104–105)

It is interesting that Rasch chose to define specific objectivity in terms of paired comparisons.

The most frequently used methods for estimating item difficulties and person abilities are maximum likelihood methods (Baker, 1992). These methods involve setting up equations that describe the likelihood of the observed scores in terms of the unknown item difficulties or person abilities, or both. Values for the item difficulties and person abilities are then sought that maximize the likelihood of the observed scores.

The Method of Paired Comparisons

The method of paired comparisons was first suggested by Fechner in 1860. In 1927, Thurstone (1927a, 1927b, 1927c, 1959) popularized the method by providing a rigorous formulation of the method through his law of comparative judgment. Since that time, it has been applied in a variety of fields including dentistry, economics, epidemiology, optics, preference and choice behavior, sensory testing, ecology, acoustics, food science, psychology, medicine, and sociology (David, 1988). In all cases, the method of paired comparisons is used to construct a scale for the measurement of the relative magnitude of some perceived stimulus or non-physical trait, and to assign scale values to the observed phenomena. In 1927, for example, Thurstone (1927b) constructed a scale for the measurement of the perceived seriousness of criminal offenses; the scale value for rape was the highest of all offenses at 3.275, while the value for vagrancy was 0.

In a paired comparisons experiment, a subject is presented with pairs of objects and is asked to indicate a preference for one of the objects according to some characteristic. A *balanced* paired comparison experiment is one in which every judge compares every possible pair of objects. In an *unbalanced* experiment, there are unequal numbers of comparisons between pairs.

Based on the preferences between pairs of objects, a scale is constructed. Noether (1960) presented a simple and very general approach for describing how to obtain scale values from the paired comparison experiment. Noether considered the problem of estimating the true values V_i ($i = 1, 2, \ldots t$) of some set of objects, ordered along a linear scale, when judged pairwise on some characteristic. One restriction must be placed on the set to permit unique estimation of the V_i and the usual restriction is that the V_i sum to zero. The probability of preferring i to j, P_{ij}, is given by:

$$P_{ij} = H(V_i - V_j) \; (i,j = 1,2, \ldots t;\; i \neq j), \qquad (2)$$

where H is the cumulative distribution function (cdf) of the differences, according to the model chosen. David (1988) derived the formula for the V_i regardless of the form of the cdf, and assuming that the sum of the values is zero, as:

$$V_i = \frac{1}{t} \sum_j (V_i - V_j) \,. \tag{3}$$

To see that this is true, note the following:

$$\frac{1}{t} \sum_j (V_i - V_j) = \frac{1}{t} \sum_j (V_i - V_1 + V_i - V_2 + V_i - V_3 + \ldots + V_i - V_t)$$

$$= \frac{1}{t} (tV_i) - \frac{1}{t} \sum_j V_j$$

$$= V_i \text{ (because the sum of the } V_j \text{ is zero).}$$

To estimate the V_i one can estimate the P_{ij} with p_{ij}, the observed proportion of preferences for object i over object j, find approximations $d_{ij} = d_i - d_j$ to $V_i - V_j$ by:

$$d_{ij} = H^{-1}(p_{ij}), \tag{4}$$

and then find the estimates d_i by using the relationship already described in Equation 3 as:

$$d_i = \frac{1}{t} \sum_j d_{ij} \,. \tag{5}$$

David also showed that the d_i obtained in this way are the (unweighted) least squares estimates of the V_i regardless of the cdf used. More specifically, the solution minimizes the expression:

$$\sum_j (d_i - V_i + V_j)^2 \,.$$

What is H^{-1}? According to Case V of Thurstone's law of comparative judgment, the cdf is the normal curve and the d_{ij} is the unit normal deviate. According to the Bradley-Terry-Luce (BTL) model, the cdf takes the form (Bradley, 1953; David, 1988):

$$p_{ij} = H(d_{ij}) = \frac{1}{2}\left[1 + \tanh\left(\frac{1}{2} d_{ij}\right)\right], \tag{6}$$

which is shown in Appendix A to lead to the following relationship between d_{ij} and the probabilities p_{ij} and p_{ji}.

$$d_{ij} = H^{-1}(p_{ij}) = \ln(p_{ij} / p_{ji}). \tag{7}$$

Noether's approach thus provides a least squares estimate of the scale values. This was the approach used by Thurstone (1927b) and by Mosteller (1951) with the normal cdf; consequently, David (1988) called this the Thurstone-Mosteller approach. By using Equation 2, however, the scale values can also be obtained by maximizing the probability of the observed differences. Since the maximum likelihood approach was usually used in conjunction with the BTL model, David (1988) called this approach the BTL approach. However, it is clear from Noether's treatment that either estimation technique is appropriate for either cdf.

To illustrate the least squares method, suppose a paired comparisons experiment is conducted involving four objects. The results of this paired comparisons experiment are summarized in a preference matrix such as matrix B shown in Figure 1. The four rows and four columns correspond to the four objects in the experiment. The entry in the *i*th row and *j*th column corresponds to the number of times *i* was preferred to *j*. So, for example, Object 3 was preferred to Object 2 eight times, whereas Object 2 was prefered to Object 3 only once.

In the first step of the least squares technique, each off-diagonal entry in the matrix can be converted to a proportion p_{ij} as shown in matrix P of Figure 1. Since this is not a balanced experiment, each off-diagonal entry is divided by the total number of comparisons for that pair of objects. Each entry p_{ij} is then divided by p_{ji} as shown in matrix D of Figure 1. Applying the BTL model, we can then apply Equation 5 to the d_{ij} estimates described by Equation 7. The scale values of the objects are obtained by computing the natural logarithm of each entry in matrix D, to produce matrix ln D, and then computing the row means of ln D as shown in Figure 1.

A preference matrix such as B in Figure 1 is also called a tournament matrix. This tournament matrix might reflect the outcome of varying numbers of games (no ties allowed) between every pair of players. Ranking of players in such a tournament is traditionally accomplished by simply summing the rows of the original matrix B. Kendall (1955), among others (Cowden, 1974; David, 1987), described a simple way of accommodating ties and compensating for missing data. Rather than summing the rows of the original tournament matrix, each player can also be given the score of every player that he has beaten. For example, the row sums of the above matrix are:

$$\begin{array}{ccccccccc} 0 & + & 12 & + & 5 & + & 2 & = & 19 \\ 2 & + & 0 & + & 1 & + & 2 & = & 5 \\ 1 & + & 8 & + & 0 & + & 6 & = & 15 \\ 8 & + & 7 & + & 4 & + & 0 & = & 19 \end{array} \tag{8}$$

If we assign to the winner of each game all the wins of his opponent, the scores would change as follows:

$$\begin{array}{ccccccccc} 0 & + & 12(5) & + & 5(15) & + & 2(19) & = & 173 \\ 2(19) & + & 0 & + & 1(15) & + & 2(19) & = & 91 \\ 1(19) & + & 8(5) & + & 0 & + & 6(19) & = & 173 \\ 8(19) & + & 7(5) & + & 4(15) & + & 0 & = & 247 \end{array} \tag{9}$$

Noether's Paired Comparisons Approach	Choppin's Technique for Estimating Item Difficulties

B matrix:
(The result of a paired comparisons experiment with four objects.)

	1	2	3	4
1	0	12	5	2
2	2	0	1	2
3	1	8	0	6
4	8	7	4	0

P matrix:

	1	2	3	4
1	0	12/14	5/6	2/10
2	2/14	0	1/9	2/9
3	1/6	8/9	0	6/10
4	8/10	7/9	4/10	0

D matrix:

	1	2	3	4
1	0	12/2	5/1	2/8
2	2/12	0	1/8	2/7
3	1/5	8/1	0	6/4
4	8/2	7/2	4/6	0

ln D:

	1	2	3	4
1	0	1.79	1.61	−1.39
2	−1.79	0	−2.08	−1.25
3	−1.61	2.08	0	.41
4	1.39	1.25	−.41	0

Row means of ln D:

1	.50
2	−1.28
3	.22
4	.56

B matrix:
(Each entry in the *i*th row and *j*th column represents the number of times item *i* was correct and *j* was incorrect, given that the total score on the pair of items is 1.)

	1	2	3	4
1	0	12	5	2
2	2	0	1	2
3	1	8	0	6
4	8	7	4	0

P matrix:

	1	2	3	4
1	0	12/14	5/6	2/10
2	2/14	0	1/9	2/9
3	1/6	8/9	0	6/10
4	8/10	7/9	4/10	0

D* matrix:

	1	2	3	4
1	0	2/12	1/5	8/2
2	12/2	0	8/1	7/2
3	5/1	1/8	0	4/6
4	2/8	2/7	6/4	0

ln D*:

	1	2	3	4
1	0	−1.79	−1.61	1.39
2	1.79	0	2.08	1.25
3	1.61	−2.08	0	−.41
4	−1.39	−1.25	.41	0

Row means of ln D*:

1	−.50
2	1.28
3	−.22
4	−.56

FIGURE 1. A comparison of Noether's least squares method and Choppin's method.

Thus, Player 4 and Player 1 are no longer tied. Kendall (1955) showed that such reallocation of wins was equivalent to summing the rows of the square of the original preference matrix. He also demonstrated that such a reallocation could take place a second or third time, corresponding to the third or fourth powers of the matrix. For example, if the rows of the third power of the original matrix are summed, then the tie that appears in Equation 9 between Players 1 and 3 is resolved, the row sums being 2451, 1013, 2383, and 2713 respectively. Kendall (1955) observed that if this process continues, as larger and larger powers of the matrix are taken, the vector of scores settles down to the eigenvector associated with the largest eigenvalue of the preference matrix.

Cowden (1974) and Andrews and David (1990) later recommended that Kendall's method should be modified to accommodate unbalanced paired comparison experiments (that is, those experiments in which each pair played a different number of games) and experiments in which comparisons are missing, by using the proportions of games won rather than the count of games won. With this adjustment to Kendall's method, it is possible to see the relationship between Kendall's row-sum approach and Noether's scheme. The key is in the choice of the cdf in Equation 4. The cdf in Kendall's method is simply the identity function, so that $d_{ij} = p_{ij}$ and therefore $d_i = \Sigma p_{ij}$.

Connections Between Rasch Measurement Theory and the Method of Paired Comparisons

The method of paired comparisons and the Rasch measurement model have the same goal: to construct a scale for the measurement of some latent trait, a scale that is independent of the particular items used or the particular group being measured (Rasch, 1980). Rasch (1966, 1980) suggested a pairwise algorithm for obtaining parameters of the Rasch model. A pairwise procedure would take advantage of the specific objectivity that is unique to the Rasch model; indeed, as already noted, Rasch (1966) described specific objectivity in terms of paired comparisons.

Choppin (1985) developed Rasch's suggestion into two techniques for using paired comparisons to estimate item difficulties: a maximum likelihood approach and a least squares approach. In the maximum likelihood approach, the model parameters are chosen so that the probability of the observed test scores is maximized, whereas in the least squares approach, the model parameters are chosen so that the sum of the squared differences between the observed values and the estimated parameters is minimized. The maximum likelihood approach has received much attention in the Rasch literature (Andrich, 1988; Fischer & Tanzer, 1994; Linacre, 1989; van der Linden & Eggen, 1986; Wright & Masters, 1982; Zwinderman, 1995), perhaps because of the original emphasis on maximum likelihood estimation of parameters of the BTL model. On the other hand, the least squares approach is appealingly simple, has been explored extensively outside the Rasch literature, and can be linked to graph theoretical analysis of tournaments

(Cowden, 1974, Kendall, 1955), and to Saaty's (1996) method involving eigenvectors of a certain kind of preference matrix.

Least Squares Pairwise Algorithm for Estimating Parameters of the Rasch Model

Assuming that performance on each item is independent of the performance on any other items, a standard assumption in Rasch measurement, Choppin showed that the person ability parameter can be eliminated entirely from Equation 1. This can be done by using Equation 1 for item i and another for item j, to derive the conditional probability of a person giving a correct response to item i, given that the sum of the scores on item i and item j is 1 ($a_{vi} + a_{vj} = 1$). The result is that:

$$\Pr(a_{vi} = 1 | a_{vi} + a_{vj} = 1) = \frac{e^{d_j}}{e^{d_i} + e^{d_j}}. \quad (10)$$

This probability can be empirically estimated by observing the number of people who respond correctly to item i and incorrectly to item j, b_{ij}, among those that respond correctly either to item i or item j. Then we can write:

$$\Pr(a_{vi} = 1 | a_{vi} + a_{vj} = 1) = \frac{b_{ij}}{b_{ij} + b_{ji}}. \quad (11)$$

Thus,

$$\frac{b_{ij}}{b_{ij} + b_{ji}} = \frac{e^{d_j}}{e^{d_i} + e^{d_j}}. \quad (12)$$

This relationship can be rewritten as:

$$\frac{b_{ij} / b_{ji}}{b_{ij} / b_{ji} + 1} = \frac{e^{(d_j - d_i)}}{e^{(d_j - d_i)} + 1}, \quad (13)$$

and thus:

$$b_{ij} / b_{ji} \text{ estimates } e^{(d_j - d_i)},$$

or:

$$b_{ij} / b_{ji} = e^{d_j} / e^{d_i},$$

which is equivalent to:

$$\ln (b_{ij} / b_{ji}) = d_j - d_i \quad (14)$$

So the difference in item difficulties can be estimated by ln (b_{ij} / b_{ji}), which involves observed values. To obtain a unique solution to Equation 14, we must add one more constraint; a common constraint is that the item difficulties sum to zero.

Thus, Rasch's goal of achieving measurement that does not depend on the abilities of the people measured is demonstrated mathematically. Furthermore, this method of pairwise comparisons for obtaining item difficulties arises naturally from a consideration of the properties of the model.

In order to solve the system of equations described in Equation 14, Choppin (1985) recommended setting up a matrix B, with entries b_{ij} representing the number of people who got item i right and item j wrong, as shown in Figure 1. The result is an asymmetrical matrix of entries, with zeros on the diagonal. The matrix B is then converted to a matrix D* with entries d_{ij} equal to b_{ji} / b_{ij}. D* is then converted to ln D* with entries ln (b_{ji} / b_{ij}). These entries in ln D* represent the log-odds of getting item i correct, given that either item i or item j is correct but not both. Choppin (1985) then showed that the item difficulties can be calculated from the matrix ln D* using the following formula:

$$d_i = \frac{1}{t} \sum_j \ln(b_{ji} / b_{ij}) \,, \tag{15}$$

where t is the number of items. Equation 15 amounts to obtaining the means of the rows of the natural logarithm of the matrix D*. Once the item difficulties have been calculated, the original model Equation 1 can be used to set up another set of equations to solve for the ability parameters.

The preceding approach is exactly the same as the approach described by Noether for obtaining scale values from paired comparisons experiments using the BTL model, as is illustrated in Figure 1. Equation 14 is the same as Equation 7, and Equation 15 is the same as Equation 5, except for a factor of –1, which can be attributed to the fact that the scales are reversed; that is, choosing an item more often means it is easier, and therefore lower, on the difficulty scale, whereas the usual case in a paired comparison experiment is that an item that gets chosen more often would have a higher value on the scale. Thus, Choppin's method is equivalent to a least squares estimate of item difficulties using the BTL model for an unbalanced paired comparisons experiment.

The only difficulty in using this approach for estimating Rasch item difficulties arises when any of the B matrix entries are zero, which must be expected when the same person does not take two items or when both items are always right or both are always wrong. Noether suggested that 0 be replaced by $1/(2N)$ where N is the number of items. Choppin, on the other hand, showed algebraically that the entries of B^2 rather than B may be used in Equation 15. This technique is equivalent to Kendall's (1955) approach of reallocating wins in a tournament. Choppin (1985) showed algebraically that this technique essentially replaces the results of the direct comparisons between i and j with the sum of the indirect comparisons of i and j through an intermediate, k. If the items are adequately linked, all off-diagonal entries of the squared matrix will be nonzero. Rasch provided support for this approach in the following "rule of transitivity":

> The rule of transitivity seems to generalize one of the most fundamental properties of measurement. If, for instance, we wish to measure the distance between two points A and C on a straight line we may do it directly or we may interpose a third point B, measure the distance AB, and on top of that measure the distance BC to obtain the total AC. (Rasch, 1961, p. 332)

Extensions of Choppin's Least Squares Algorithm and Connection to the Analytic Hierarchy Method

Choppin's use of the square of the B matrix is equivalent to Kendall's (1955) technique of reallocating wins in a tournament. As Kendall pointed out, this reallocation could be repeated by using higher powers of the matrix. As higher powers of the matrix are used, the solution converges to the eigenvector associated with the largest eigenvalue; in fact, this approach is equivalent to using the power method for obtaining the dominant eigenvalue and associated eigenvector (Saaty, 1996). Cowden (1974) points out that this convergence will result if, in every possible partition of the players into two nonempty sets, some player in each set has won at least once from some player in the other set. This requirement is exactly the same as the requirement for the convergence of the maximum likelihood algorithm for paired comparisons, usually called the Zermelo-Ford condition (David, 1988).

There is thus a connection between the item difficulties of the Rasch model and the eigenvectors of the paired comparisons matrix B. This approach is further justified by consideration of Saaty's (1996) analytic hierarchy process, which also makes use of eigenvectors. In the analytic hierarchy process, subjects are asked to indicate not just a preference for two objects, but they are asked to estimate the strength of the preference in terms of pairwise ratios. The resulting comparisons matrix, called a reciprocal matrix, looks like the one shown below:

$$\begin{bmatrix} w_1/w_1 & w_1/w_2 & w_1/w_3 & \cdots & w_1/w_N \\ w_2/w_1 & w_2/w_2 & w_2/w_3 & \cdots & w_2/w_N \\ w_3/w_1 & w_3/w_2 & w_3/w_3 & \cdots & w_3/w_N \\ \vdots & \vdots & \vdots & & \vdots \\ w_N/w_1 & w_N/w_2 & w_N/w_3 & \cdots & w_N/w_N \end{bmatrix}$$

where w_i is the scale value of the ith object. The following equation must be true:

$$\begin{bmatrix} w_1 / w_1 & w_1 / w_2 & w_1 / w_3 & \cdots & w_1 / w_N \\ w_2 / w_1 & w_2 / w_2 & w_2 / w_3 & \cdots & w_2 / w_N \\ w_3 / w_1 & w_3 / w_2 & w_3 / w_3 & \cdots & w_3 / w_N \\ \cdot & \cdot & \cdot & & \cdot \\ \cdot & \cdot & \cdot & & \cdot \\ \cdot & \cdot & \cdot & & \cdot \\ w_N / w_1 & w_N / w_2 & w_N / w_3 & \cdots & w_N / w_N \end{bmatrix} \begin{bmatrix} w_1 \\ w_2 \\ w_3 \\ \cdot \\ \cdot \\ \cdot \\ w_N \end{bmatrix} = N \begin{bmatrix} w_1 \\ w_2 \\ w_3 \\ \cdot \\ \cdot \\ \cdot \\ w_N \end{bmatrix}$$

By definition of eigenvectors and eigenvalues, the solution vector of w_i's is the eigenvector associated with the eigenvalue N.

To connect the preceding system with the pairwise algorithm, recall that each entry in the D* matrix, b_{ji} / b_{ij}, estimates e^{d_j} / e^{d_i}. Thus, D* is a reciprocal matrix as described by Saaty and the item difficulties we seek are the natural logarithm of the eigenvector association with the largest eigenvalue of the D* matrix, as shown below.

$$\begin{bmatrix} e^{d_1} / e^{d_1} & e^{d_1} / e^{d_2} & e^{d_1} / e^{d_3} & \cdots & e^{d_1} / e^{d_N} \\ e^{d_2} / e^{d_1} & e^{d_2} / e^{d_2} & e^{d_2} / e^{d_3} & \cdots & e^{d_2} / e^{d_N} \\ e^{d_2} / e^{d_1} & e^{d_2} / e^{d_2} & e^{d_2} / e^{d_3} & \cdots & e^{d_2} / e^{d_N} \\ \cdot & \cdot & \cdot & & \cdot \\ \cdot & \cdot & \cdot & & \cdot \\ \cdot & \cdot & \cdot & & \cdot \\ e^{d_{1N}} / e^{d_1} & e^{d_N} / e^{d_2} & e^{d_N} / e^{d_3} & \cdots & e^{d_N} / e^{d_N} \end{bmatrix} \begin{bmatrix} e^{d_1} \\ e^{d_2} \\ e^{d_3} \\ \cdot \\ \cdot \\ \cdot \\ e^{d_N} \end{bmatrix} = N \begin{bmatrix} e^{d_1} \\ e^{d_2} \\ e^{d_3} \\ \cdot \\ \cdot \\ \cdot \\ e^{d_N} \end{bmatrix}$$

This application of the analytic hierarchy process to the scaling of choice preferences can only be accomplished through the BTL or Rasch models, because only those models transform a difference in scale values to a ratio.

THE METHOD OF PAIRED COMPARISONS AND GRAPH THEORY

Graph Theory

A *digraph* consists of a set of vertices or nodes, and a set of edges or arcs connecting those vertices and oriented in a specific direction (Berge, 1985). A digraph with 10 vertices and 13 edges is shown in Figure 2. The vertices may represent cities on an airline route, or phones in a telephone network, or tasks in a production line, or items in an item bank. Given a set of vertices and arcs, graph theory provides answers to questions such as: Are every pair of vertices connected

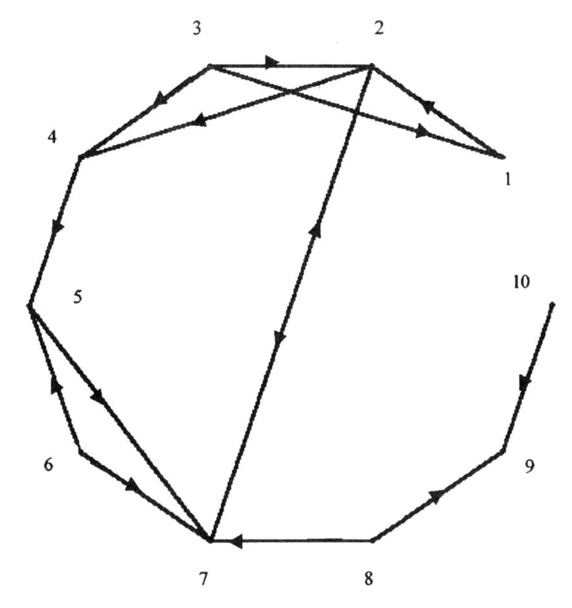

FIGURE 2. A digraph.

through some sequence of arcs? What is the shortest route between two vertices? Could the digraph be disconnected by eliminating just one edge? Graph theorists have built a set of computer algorithms that may be used to answer such questions. The user has to only supply substantive meaning to the vertices and the connections between them.

> Graphs and multigraphs often appear under other names: sociograms (psychology), simplexes (topology), electrical networks, organizational charts, communication networks, family trees, etc. It is often surprising to learn that these diverse disciplines use the same theorems. The primary purpose of graph theory was to provide a mathematical tool that can be used in all these disciplines. (Berge, 1985, p. 3)

One way of representing a digraph is through an adjacency matrix. An adjacency matrix is a square matrix A with n rows and n columns, where n corresponds to the number of vertices in the digraph. For each entry $a_{ij} = 1$, there is an arc from vertex i to vertex j. If $a_{ij} = 0$ or if $i = j$, there is no arc. The adjacency matrix associated with the digraph in Figure 2 is as follows:

$$
\begin{matrix}
0&1&0&0&0&0&0&0&0&0\\
0&0&0&1&0&0&1&0&0&0\\
1&1&0&1&0&0&0&0&0&0\\
0&0&0&0&1&0&0&0&0&0\\
0&0&0&0&0&0&1&0&0&0\\
0&0&0&0&1&0&1&0&0&0\\
0&1&0&0&0&0&0&0&0&0\\
0&0&0&0&0&0&1&0&1&0\\
0&0&0&0&0&0&0&0&0&0\\
0&0&0&0&0&0&0&0&1&0
\end{matrix}
$$

A well-established property of adjacency matrices is that the entries a_{ij}^{m} of the powers of the original matrix A^m provide the number of distinct walks of length m between the vertices i and j. A *walk* is an alternating sequence of vertices and arcs in which each are in the sequence joins the vertex that comes before the arc to the vertex that comes after the arc in the sequence (Bollobas, 1979). The walk may repeat arcs or vertices, or both. A *path* between vertices is a walk in which arcs and vertices are visited only once. A *cycle* or *circuit* in a digraph is a path that begins and ends with the same vertex.

To see that the powers of the adjacency matrix, A^m, provide the number of distinct walks of length m between the vertices i and j, consider the entries a_{ij}^{2} of A^2 for example. Each entry a_{ij}^{2} of A^2 is formed by summing the products $a_{ik} \times a_{kj}$ over all k; but this product is 0 if either term is 0 (that is, if there is no arc from vertex i to vertex k or no arc from k to j), and the product is 1 if both terms are 1 (that is, if there is an arc from i to k and one from k to j). Thus, a_{ij}^{2} represents the number of times there is an arc from i to k and one from k to j, where k is any other vertex besides i or j. In other words, a_{ij}^{2} is the number of walks of length 2 from vertex i to vertex j.

Each arc in a digraph may have a number associated with it. This number may represent a weight, cost, or distance associated with crossing that arc, or some allowed flow through the arc. The entries in the adjacency matrix could then contain the numbers associated with each arc.

The following characteristics of graphs are important in answering questions regarding the connectivity of a set of vertices:

1. A digraph is *strongly connected* if you can find a path between any two vertices by following the direction of the arcs. If you can find a path only by disregarding the direction of the arcs, then the graph is *weakly connected.* The digraph in Figure 2 is weakly but not strongly connected. A digraph is disconnected if it is not weakly or strongly connected; in other words, there is clearly a physical space between sets of vertices with no arcs between the sets.
2. A *component* of a digraph is a maximal connected subgraph; that is, the subgraph is maximal in the sense that it is as large as possible without decreasing its con-

nectivity. A weakly connected digraph or a disconnected digraph may have strongly connected components.

3. A strongly connected digraph is *k-connected* if a minimum of k vertices must be deleted to decrease the connectivity of the digraph.
4. A strongly connected digraph is *k-edge-connected* if a minimum of k edges is required to decrease the connectivity of the digraph.
5. A *cut vertex* (or articulation vertex) is a vertex whose deletion increases the number of components in a graph; that is, the graph may change from strongly connected to weakly connected or disconnected upon deletion of the vertex. A *bridge* (or isthmus) is an arc whose deletion increases the number of components in a graph.
6. The *indegree* of a vertex is the number of arcs that terminate in a vertex, and the *outdegree* is the number of arcs that begin at the vertex. Note that the column sums of the adjacency matrix are the indegrees of the vertices and the row sums are the outdegrees.

Connections Between the Method of Paired Comparisons and Graph Theory

Kendall (Kendall, 1955; Kendall & Smith, 1940) used digraphs to visualize the results of a paired comparisons experiment. Only one arc existed between any two edges, and that arc pointed to the loser in the comparison. The paired comparisons matrix is the adjacency matrix for this graph, with edges weighted according to how many times the vertex at the initial end of the arc won against the vertex at the terminal end. The vertices can be ordered according to indegree or outdegreee, and this ordering goes from least able player (highest indegree, lowest outdegree, most losses) to most able player (lowest indegree, highest outdegree, fewest losses). The reallocation of wins that takes place by squaring the adjacency matrix can be visualized as utilizing all the walks of length 2 between every pair of vertices in the digraph.

The Use of Graph Theory in the Pairwise Algorithm

A connection between Rasch measurement theory and graph theory has been made on two occasions, through a discussion of the pairwise algorithm for estimating parameters of the Rasch model. Fischer and Tanzer (1994) and van der Linden and Eggen (1986) used digraphs to provide an interpretation of the Zermelo-Ford condition for uniqueness of the maximum likelihood solution. The digraph is defined by the original paired comparison matrix B, with a directed edge from item i to item j if there is a nonzero entry in the matrix for b_{ij}. The B matrix can thus be considered an adjacency matrix for a digraph. If the digraph is strongly connected, the maximum likelihood estimates are unique. The digraph must also be strongly connected for the matrix powers to converge to the eigenvector associated with the largest eigenvalue, as indicated by Cowden (1974) and Saaty (1996).

If the digraph is not strongly connected, there is at least one item (or set of items) that has incident arcs in only one direction; in other words, there is at least one item that is always the correct one out of every pair or always the incorrect one out of every pair. Such a situation has always been recognized as unacceptable in Rasch measurement. Items on which all persons have succeeded or on which all persons have failed should be eliminated from consideration since their position on the item difficulty scale is indeterminable except to say that these items are beyond the item difficulty range calibrated by the persons. It would seem that any item that always matches performance on some other paired item adds nothing to the scale.

If a digraph associated with a paired comparison matrix is not strongly connected, the strongly connected components of the graph may be easily identified. There is, however, more that graph theory can provide, especially in the case of data sets with missing data. The connectivity of the digraph can be determined and used to indicate how well connected the system is. For example, if a digraph is two-connected or biconnected, then it would imply that the graph remains strongly connected even when any one item is removed from the system; this is also equivalent to the condition that there are two unique paths comparing any pair of items. For a graph that is one-connected, identification of the cut vertices, the vertices that could break the graph into a weakly connected system with strongly connected components, would allow examination of the quality of those items that are crucial for the connectivity of the whole system. A parallel analysis could be built from the determination of edge connectivity and identification of bridges. Furthermore, two items might be compared via different paths to assess the consistency of the system, or what Rasch might have described as adherence to the rule of transitivity.

DATA ANALYSIS

Data from a study by Monsaas and Engelhard (1996) are used to illustrate the techniques described in this chapter.

Instrument

An 11-item subtest of the Home Observation for Measurement of Environment (HOME) instrument was used. The subtest is designed to describe the type of learning stimulation available in a child's home. Each item is scored dichotomously. Two thirds of the items were scored by a teacher who was trained in the use of the test and who visited the child's family and observed the environment. About one third of the items were scored on the basis of parental reports.

Participants

The data shown in Table 1 reflect the results of the HOME subtest for 40 preschool children who had been defined as being at risk for school failure, as described in Monsaas and Engelhard (1996).

TABLE 1.
HOME Data

	Items										
Persons	1	2	3	4	5	6	7	8	9	10	11
1	1	1	1	1	1	1	1	1	1	1	1
2	1	0	0	1	0	1	1	1	1	0	1
3	1	1	0	1	1	1	1	1	0	0	1
4	1	0	1	1	1	1	1	1	0	1	1
5	1	0	1	1	1	0	1	0	0	1	1
6	1	1	0	1	1	1	1	1	1	1	1
7	1	1	0	1	0	0	0	0	0	0	0
8	1	1	1	1	1	1	1	0	1	1	1
9	1	1	1	1	1	1	1	1	1	1	1
10	1	0	0	1	1	1	0	1	1	0	1
11	1	1	0	1	0	1	1	1	0	0	1
12	1	0	0	1	1	1	1	1	0	0	1
13	1	0	1	1	0	1	0	1	0	0	1
14	1	0	1	1	0	1	0	1	0	0	1
15	1	1	1	1	1	1	1	1	0	0	1
16	1	0	0	1	0	1	0	0	1	0	1
17	1	0	0	1	1	1	1	0	0	0	1
18	1	1	0	1	0	1	1	0	1	0	1
19	0	0	0	0	0	1	0	0	0	0	1
20	1	0	0	1	0	1	0	1	0	0	1
21	1	0	0	1	1	0	0	0	0	0	1
22	0	0	1	1	0	1	1	1	1	0	1
23	1	0	1	1	1	0	1	1	0	1	1
24	0	0	1	1	1	0	0	1	1	0	1
25	1	1	0	1	1	1	1	1	0	1	1
26	0	0	0	0	0	0	1	0	1	0	1
27	1	1	1	1	1	1	1	1	0	1	1
28	1	1	1	1	1	1	1	1	1	1	1
29	1	1	1	1	1	1	1	1	0	0	1
30	1	1	0	1	0	1	1	0	1	0	1
31	1	1	0	1	1	1	1	1	1	0	1
32	1	1	0	1	1	1	1	0	0	0	1
33	0	0	0	1	0	0	0	0	0	0	1
34	1	1	0	1	1	1	1	1	1	1	1
35	1	0	0	1	0	1	0	1	0	1	1
36	1	0	0	0	0	1	0	0	0	0	0
37	1	1	0	1	1	1	0	0	0	0	1
38	1	0	0	1	1	1	1	0	0	1	1
39	1	1	1	1	1	1	1	1	1	0	1
40	1	1	0	1	0	1	1	1	1	0	1

Procedures

SAS routines shown in Appendix B were designed to estimate item difficulties according to the following methods: (1) Choppin's pairwise least squares algorithm using the B matrix of paired comparisons, with zero entries replaced with

TABLE 2.
HOME Data Set with Missing Values

	Items										
Persons	1	2	3	4	5	6	7	8	9	10	11
1	1	1	1	1	1						
2	1	0	0	1	0						
3	1	1	0	1	1						
4	1	0	1	1	1						
5	1	0	1	1	1						
6	1	1	0	1	1						
7	1	1	0	1	0						
8	1	1	1	1	1						
9	1	1	1	1	1						
10	1	0	0	1	1						
11							1	1	0	0	1
12							1	1	0	0	1
13							0	1	0	0	1
14							0	1	0	0	1
15							1	1	0	0	1
16							0	0	1	0	1
17							1	0	0	0	1
18							1	0	1	0	1
19							0	0	0	0	1
20							0	1	0	0	1
21	1	0	0							0	1
22	0	0	1							0	1
23	1	0	1							1	1
24	0	0	1							0	1
25	1	1	0							1	1
26	0	0	0							0	1
27	1	1	1							1	1
28	1	1	1							1	1
29	1	1	1							0	1
30	1	1	0							0	1
31	1	1	0	1	1	1	1	1		0	1
32	1	1	0	1	1	1	1	0		0	1
33	0		0	1	0	0	0	0	0	0	1
34	1	1	0	1		1	1	1	1	1	1
35	1	0	0		0	1	0	1	0	1	1
36	1	0	0	0	0	1		0	0	0	0
37	1	1		1	1	1	0	0	0	0	1
38	1	0	0	1	1	1	1	0	0	1	
39	1		1	1	1	1	1	1	1	0	1
40	1	1	0	1	0	1	1		1	0	1

1/(2*N*) (PW least squares – B) described by Equation 15; (2) Choppin's least squares algorithm using the *n*th power of the B matrix (PW least squares – B^n). The FACETS computer program (Linacre & Wright, 1992) was used to obtain estimates of the item difficulties and standard errors using joint maximum likelihood (JML) estimation.

Using the B matrix obtained from the HOME data as an adjacency matrix, the connectivity of the system of items was explored using Mathematica (Wolfram, 1993). Specifically, the following were obtained: (1) strongly connected components, (2) biconnected components, (3) cut vertices, (4) bridges, (5) vertex connectivity, and (6) edge connectivity. This analysis was also performed on an incomplete version of the HOME data set, in order to illustrate the results of the very simple pairwise algorithm on incomplete data and to illustrate the application of graph theory to analyzing the connectivity of the system of items. Table 2 shows an incomplete data set.

Results

Table 3 shows the B matrix for the HOME data shown in Table 1. Table 4 shows the item difficulty estimates obtained through JML estimation, pairwise least squares using the matrix B, and pairwise least squares using successively higher powers of B. The estimates using the least squares algorithm on the B matrix are often more than one standard error from the JML estimates. It appears that the method of handling the missing data in matrix B is inadequate. As Kendall observed, the item difficulties appear to settle down with successive powers of the B matrix. The only dramatic difference in values occurred in using B^2 rather than B; perhaps the dramatic change can be attributed to the fact that B^2 was the first matrix with no zero entries. Consistent with Saaty's analytic hierarchy method, the

TABLE 3.
Comparison Matrix B for HOME Data

	Items										
Items	1	2	3	4	5	6	7	8	9	10	11
1	0	15	22	1	12	4	10	12	21	22	2
2	0	0	12	0	5	1	2	6	9	12	1
3	2	7	0	0	3	3	3	2	8	7	0
4	3	17	22	0	13	6	11	12	21	24	1
5	1	9	12	0	0	4	4	7	14	12	0
6	2	14	21	2	13	0	9	10	18	22	1
7	2	9	15	1	7	3	0	8	13	15	0
8	2	11	12	0	8	2	6	0	13	15	0
9	3	6	10	1	7	2	3	5	0	11	0
10	0	5	5	0	1	2	1	3	7	0	0
11	5	19	23	2	14	6	11	13	21	25	0

solution converges to the natural logarithm of the eigenvector associated with the maximum eigenvalue of the D* matrix derived from B^2.

Figure 3 shows the digraph associated with the B matrix. This digraph was strongly connected and was characterized by four-vertex-connectivity and five-

TABLE 4.
Item Difficulty Estimates Based on Choppin's Pairwise Least Squares Algorithm Compared to Joint Maximum Likelihood Estimates for HOME Data

		PW Least Squares Algorithm Estimates					
Item	JML (*SE*)	B	B^2	B^3	B^4	B^5	B^{10}
10	2.34(.43)	3.05	2.20	2.25	2.23	2.24	2.23
3	1.99(.41)	2.23	1.70	1.76	1.73	1.74	1.74
9	1.66(.40)	1.59	1.47	1.42	1.43	1.43	1.43
2	1.35(.39)	1.89	1.49	1.28	1.30	1.29	1.30
5	.58(.40)	1.20	.60	.63	.62	.62	.62
8	.42(.40)	1.13	.52	.53	.53	.53	.53
7	.09(.42)	.34	.08	.08	.08	.08	.08
6	–1.15(.51)	–.86	–.65	–.69	–.68	–.68	–.68
1	–1.73(.57)	–2.13	–1.25	–1.29	–1.28	–1.28	–1.28
4	–2.51(.68)	–3.91	–2.74	–2.74	–2.73	–2.73	–2.73
11	–3.05(.80)	–4.51	–3.41	–3.23	–3.23	–3.23	–3.23
Mean	0	0	0	0	0	0	0
SD	1.86	2.53	1.84	1.80	1.80	1.80	1.80

Note. The item difficulties constitute the natural logarithm of the eigenvector associated with the maximum eigenvalue of the D* matrix for the HOME data.

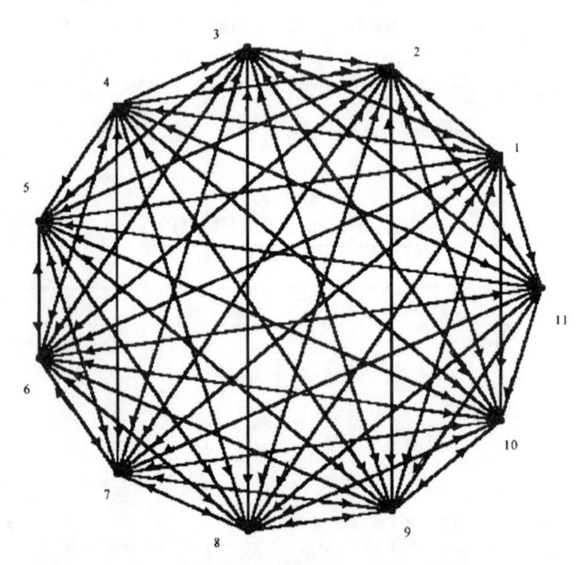

FIGURE 3. The digraph associated with the adjacency matrix shown in Table 3 and the HOME data shown in Table 1.

edge-connectivity, indicating that it would take the deletion of at least four items or five comparisons to decrease the connectivity of the digraph. In other words, comparisons between items can be made through at least four independent paths through the digraph. For example, there is a direct comparison between Items 1 and 9, but these items can also be compared through Item 2, through Item 11, or through Items 3 and 8.

The paired comparisons matrix of the incomplete HOME data set corresponds to the digraph shown in Figure 4 and the item difficulties obtained using the least squares algorithm are shown in Table 5. The system was so poorly connected that the fourth power of the B matrix was the first matrix to contain no zero entries. It appears that not until this fourth power did the estimates of the item difficulties settle down. The system illustrated by the digraph in Figure 4 is strongly connected, but only one-connected. There are two cut vertices, Items 1 and 3; in other words, deletion of either of these items would change the strongly connected graph to a weakly connected graph and prevent proper parameter estimation. Items 1 and 3 would have to be examined to determine whether the connectivity of the system should rest with either of these items. If Item 1 is deleted, for example, the system breaks into two strongly connected components, one including items 2, 3, 5, 7, 8, 9, and 10, and the other including items 4, 6, and 11. Figure 5 shows how the graph shown in Figure 4

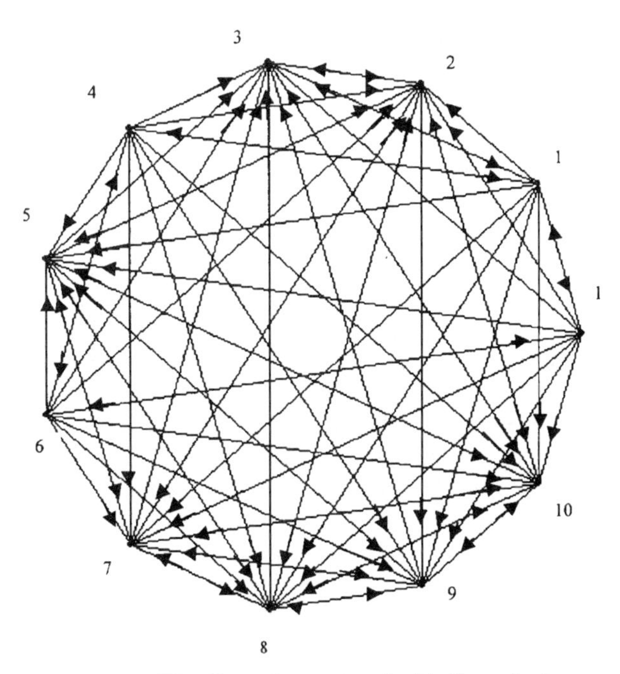

FIGURE 4. The digraph associated with the paired comparison matrix for the data shown in Table 2.

TABLE 5.
Item Difficulty Estimates for the HOME Data with Missing Values, Based on Choppin's Pairwise Least Squares Algorithm Using Powers of the B Matrix

		PW Least Squares Algorithm Estimates							
Item	JML (*SE*)	B	B^2	B^3	B^4	B^5	B^6	B^{10}	B^{11}
10	3.33(.57)	3.04	3.65	3.22	2.89	2.90	2.90	2.90	2.90
3	2.89(.53)	4.10	3.01	3.69	3.21	2.64	2.65	2.64	2.64
9	2.23(.62)	2.36	3.01	2.68	2.31	2.29	2.29	2.28	2.28
2	1.68(.51)	1.69	2.86	2.28	2.01	1.99	1.98	1.98	1.98
5	.74(.64)	1.15	1.75	1.38	.98	1.00	.99	.99	.99
8	.82(.56)	1.51	2.12	1.78	1.31	1.27	1.26	1.26	1.26
7	.74(.54)	1.08	1.81	1.33	.93	.90	.89	.89	.89
6	–3.32(1.34)	–3.64	–5.67	–7.85	–5.78	–5.25	–5.17	–5.18	–5.18
1	–1.47(.77)	–3.31	–.59	–.34	–.71	–.71	–.71	–.71	–.71
4	–3.44(1.28)	–3.73	–5.65	–3.81	–3.28	–3.19	–3.21	–3.20	–3.20
11	–4.20(1.12)	–4.25	–6.29	–4.36	–3.89	–3.85	–3.86	–3.86	–3.86
Mean	0	0	0	0	0	0	0	0	0
SD	2.68	3.09	3.93	3.72	3.02	2.85	2.84	2.83	2.83

Note. The items are ordered according to the difficulty of the items determined using the complete HOME data set.

breaks into components when Item 1 is deleted, illustrating the opportunity to examine connections among subsets of the items. Clearly, the component involving Items 4, 6, and 11 is minimally connected. All comparisons between Items 4, 6, and 11 and other items, must be mediated by Item 1 because of the connectivity.

SUMMARY AND CONCLUSIONS

This study was motivated by three questions. The first question was: What is the relationship between the method of paired comparisons and Rasch measurement theory? The method of paired comparisons and Rasch measurement theory have the same goal: to construct a linear scale along which a set of objects or items can be located. Rasch measurement theory has the additional goal of placing persons on that scale after the calibration of objects or items. Through Choppin's work it was shown that item difficulties in the Rasch model could be estimated by methods that are equivalent to least squares or maximum likelihood estimation of item difficulties using the BTL model for an unbalanced paired comparisons experiment. Applying the least squares algorithm to powers of the paired comparison matrix appeared to be more effective than arbitrarily filling in values for missing data in the comparison matrix, as shown in Table 4. This power method was tied to Saaty's analytic hierarchy method in which the scale values are components of the eigenvector associated with the maximum eigenvalue of the appropriate

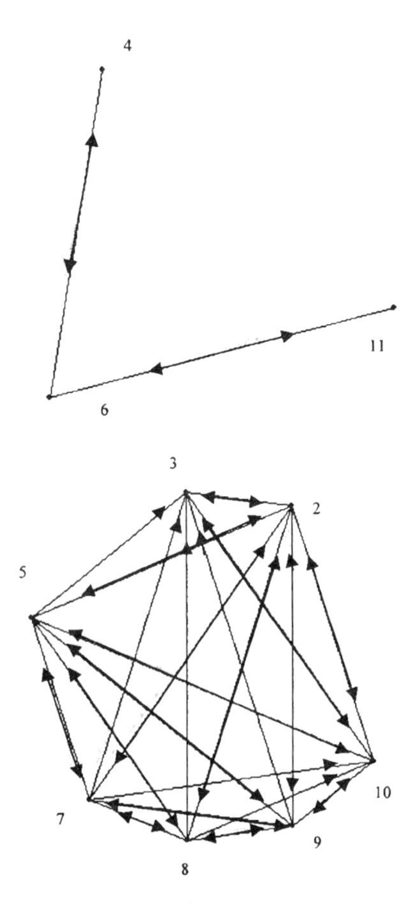

FIGURE 5. The strongly connected components of the digraph shown in Figure 4 with Item 1 deleted.

matrix. The item difficulties obtained are similar to JML estimates. The connectivity required in the system of paired comparisons is parallel to the situation in Rasch measurement theory in which items that are always correct or always incorrect cannot be properly placed on the scale with the other items in the system.

The second question was: What is the relationship between the method of paired comparisons and graph theory? It was shown that the paired comparisons matrix is an adjacency matrix for a digraph with edges weighted according to how many times the vertex at the initial end of the arc won against the vertex at the terminal end. Using graph theory, the connectivity of a system built from pairs of objects or items may be made explicit.

The third question was: What can graph theory contribute to our understanding of Rasch measurement theory? This chapter just began to answer this question. It was shown that graph theory is useful in analyzing the connectivity of the system produced by the paired comparisons algorithm. Graph theory provides a well-established language and framework for discussing any systems based on pairwise comparisons. The influence of different degrees of connectivity must be explored. Network flow algorithms may provide new graph theoretical means of analyzing and estimating parameters of the Rasch model. Other applications of graph theory might be in determining goodness of fit measures for the item difficulties produced by the pairwise algorithm.

The pairwise methods of estimating item difficulties are important in that they provide a way of utilizing the specific objectivity of the Rasch model. By separating estimation of item difficulties from person abilities, it becomes possible to establish a measuring instrument that can be used consistently across different populations. Choppin (1968) was particularly interested in pairwise estimation for this reason. It is an ideal procedure for setting up item banks.

> The idea (of an item bank) is that a large collection of test items, the characteristics of which are known, be made available at some central place so that individuals who wish to construct achievement tests, but who lack the resources to carry out detailed standardization and validation procedures, can select items from the bank to form a test of known characteristics. (p. 870)

The usefulness of a pairwise algorithm was expressed as follows:

> The advantages of these procedures over classical item analysis techniques are several. First, because the model allows the separation of person and item parameters, we can make the estimation for any pair of items, without much regard for which set of individuals provides the data. People who score one on the item pair contribute to the estimation. People who score two or zero contribute nothing but do not spoil it. (p. 872)

Choppin also pointed out how simple the least squares algorithm is. In Appendix B is shown the few lines of code that are necessary to generate item difficulties, even in the presence of missing data. Choppin lamented that the technique was so easy that it allows one to produce item difficulties even when the data are not sufficiently interconnected. However, by extending his technique to powers of the comparison matrix, exploiting the link to Saaty's technique using eigenvectors, and applying graph theoretical analysis of the paired comparison matrix, the simple technique might be successfully applied and thoroughly analyzed.

This chapter sets the foundation necessary for a comprehensive treatment of a very simple procedure for calibrating achievement items according to the Rasch model. The statistical properties of the method must be explored further and the method must be extended to situations in which items are not dichotomously scored, but graded on a scale, and to situations involving raters. In addition, the applications of graph theory must be explored further and can certainly be extended to other graph theoretical constructs.

APPENDIX A

Conversion of the cdf for the BTL model to inverse form. Shown here is the conversion of the cumulative distribution function:

$$p_{ij} = H(d_{ij}) = \frac{1}{2}\left[1 + \tan h(\frac{1}{2} d_{ij})\right], \tag{6}$$

(where $d_{ij} = d_i - d_j$) to the form:

$$d_{ij} = H^{-1}(p_{ij}) = \ln(p_{ij}/p_{ji}) . \tag{7}$$

By definition of tan h:

$$p_{ij} = \frac{1}{2}\left[1 + \tan h(\frac{1}{2}(d_i - d_j)\right] = \frac{1}{2}\left[1 + \frac{\sin h\ \frac{1}{2}(d_i - d_j)}{\cos h\ \frac{1}{2}(d_i - d_j)}\right],$$

which is equal by definition of sin h and cos h to:

$$= \frac{1}{2}\left[1 + \frac{e^{\frac{1}{2}(d_i - d_j)} - e^{\frac{-1}{2}(d_i - d_j)}}{e^{\frac{1}{2}(d_i - d_j)} + e^{\frac{-1}{2}(d_i - d_j)}}\right]$$

$$= \frac{1}{2}\left[1 + \frac{1 - e^{-(d_i - d_j)}}{1 + e^{-(d_i - d_j)}}\right]$$

$$= \frac{1}{2}\left[1 + \frac{2}{1 + e^{-(d_i - d_j)}}\right].$$

So,

$$p_{ij} = \frac{1}{1 + e^{-(d_i - d_j)}} = \frac{e^{(d_i - d_j)}}{1 + e^{(d_i - d_j)}} .$$

To obtain H^{-1} note that:

$$p_{ij} = \frac{e^{(d_i - d_j)}}{1 + e^{(d_i - d_j)}} \Rightarrow e^{(d_i - d_j)} = \frac{p_{ij}}{1 - p_{ij}} .$$

Thus,

$$d_i - d_j = \ln\left(\frac{p_{ij}}{1 - p_{ij}}\right),$$

which implies that:

$$d_i - d_j = \ln\left(\frac{p_{ij}}{p_{ji}}\right)$$

if we assume that $p_{ji} = 1 - p_{ij}$, which would be true as long as ties are not allowed.

APPENDIX B

SAS Routines

Routine #1: Input: X matrix shown in Table 1 or Table 2.
Output: B matrix shown in Figure 1.

```
NITEM = NCOL(X);              * NITEM IS THE NUMBER OF ITEMS;
B = J(NITEM,NITEM,0.0);       * INITIALIZE THE COMPARISON MATRIX;

* CREATE THE B MATRIX OF PAIRED COMPARISONS;
   DO K=1 TO N;
   DO I=1 TO NITEM;
   DO J=1 TO NITEM;
   IF X[K,I] ^= 9 & X[K,J] ^= 9 THEN DO;          *9 indicates missing value;
   IF X[K,I] > X[K,J] THEN B[I,J] = B[I,J] + 1.0;
   END; END; END; END;
```

Routine #2:	Input: B matrix or power of B matrix with no zero entries. Output: Item difficulties according to least squares routine.

```
D = B` / B;                    * See Figure 2 for description of D matrix.
LOGIT = LOG(D);
G = LOGIT[,:];
```

REFERENCES

Andrews, D. M., & David, H. A. (1990). Nonparametric analysis of unbalanced paired-comparison or ranked data. *Journal of the American Statistical Association, 85*(412), 1140–1146.

Andrich, D. (1978). Relationships between the Thurstone and Rasch approaches to item scaling. *Applied Psychological Measurement, 2*(3), 449–460.

Andrich, D. (1988). *Rasch models for measurement.* Newbury Park, CA: Sage.

Baker, F. B. (1992). *Item response theory. Parameter estimation techniques.* New York: Marcel Dekker.

Berge, C. (1985). *Graphs.* New York: North Holland.

Bollobas, B. (1979). *Graph theory.* New York: Springer-Verlag.

Bradley, R. A. (1953). Some statistical methods in taste testing and quality evaluation. *Biometrics, 9,* 22–38.

Choppin, B. (1968). Item banking using sample-free calibration. *Nature, 219*(5156), 870–872.

Choppin, B. (1985). A fully conditional estimation procedure for Rasch model parameters. *Evaluation in Education, 9,* 29–42.

Cowden, D. J. (1974). A method of evaluating contestants. *American Statistician, 29*(2), 82–84.

David, H. A. (1987). Ranking from unbalanced paired-comparison data. *Biometrika, 74*(2), 432–436.

David, H. A. (1988). *The method of paired comparisons.* New York: Oxford.

Engelhard, G. (1984). Thorndike, Thurstone and Rasch: A comparison of their methods of scaling psychological tests. *Applied Psychological Measurement, 8,* 21–38.

Fischer, G. H., & Tanzer, N. (1994). Some LBTL and LLTM relationships. In G. H. Fischer & D. Laming (Eds.), *Contributions to mathematical psychology, psychometrics, and methodology* (pp. 277–303). New York: Springer Verlag.

Kendall, M. G. (1955). Further contributions to the theory of paired comparisons. *Biometrics, 11,* 43–62.

Kendall, M. G., & Smith, B. (1940). On the method of paired comparisons. *Biometrika, 31,* 324–345.

Linacre, J. M. (1989). *Many-facet Rasch measurement.* Chicago: MESA Press.

Linacre, J. M., & Wright, B. D. (1992). *FACETS* (Version 2.5) [Computer software]. Chicago: MESA Press.

Monsaas, J., & Engelhard, G. (1996). Examining changes in the home environment with the Rasch measurement model. In G. Engelhard & M. Wilson (Eds.), *Objective measurement: Theory into practice* (Vol. 3, pp. 127–140). Norwood, NJ: Ablex.

Mosteller, F. (1951). Remarks on the method of paired comparisons: I. The least squares solution assuming equal standard deviations and equal correlations. *Psychometrika, 16*(1), 3–9.

Noether, G. E. (1960). Remarks about a paired comparison model. *Psychometrika, 25*(4), 357–367.

Rasch, G. (1966). An individualistic approach to item analysis. In P. F. Lazarfeld & N. W. Henry (Eds.), *Readings in mathematical social science* (pp. 89–107). Chicago: Science Research Associates.

Rasch, G. (1977). On specific objectivity: An attempt at formalizing the request for generality and validity of scientific statements. *Danish Yearbook of Philosophy, 14*, 58–94.

Rasch, G. (1980). *Probabilistic models for some intelligence and attainment tests.* Chicago: University of Chicago Press. (Original work published 1960)

Saaty, T. L. (1996). *Multicriteria decision making: The analytic hierarchy process.* Pittsburgh, PA: RWS Publications.

Thurstone, L. L. (1927a). A law of comparative judgment. *Psychological Review, 34,* 278–286.

Thurstone, L. L. (1927b). The method of paired comparisons for social values. *Journal of Abnormal and Social Psychology, 21,* 384–400.

Thurstone, L. L. (1927c). Psychophysical analysis. *American Journal of Psychology, 38,* 368–389.

Thurstone, L. L. (1959). *The measurement of values.* Chicago: University of Chicago Press.

van der Linden, W. J., & Eggen, T. J. H. M. (1986). An empirical Bayesian approach to item banking. *Applied Psychological Measurement, 10*(4), 345–354.

Wolfram, S. (1993). *Mathematica* (Version 2.2) [Computer programming language]. Champaigne, IL: Wolfram Research.

Wright, B. D., & Masters, G. N. (1982). *Rating scale analysis.* Chicago: MESA Press.

Zwinderman, A. H. (1995). Pairwise parameter estimation in Rasch models. *Applied Psychological Measurement, 19*(4), 369–375.

15

A PROCEDURE FOR DETECTING PATTERN CLUSTERING IN MEASUREMENT DESIGNS

George A. Marcoulides
Zvi Drezner
California State University at Fullerton

INTRODUCTION

The two most common approaches for analyzing measurement designs are generalizability theory (Cronbach, Gleser, Nanda, & Rajaratnam, 1972) and multifaceted Rasch measurement (Linacre, 1988, 1989). Recently, several researchers have provided a comparison of generalizability theory and the multifaceted (FACETS) Rasch model (see, for example, Bachman et al., 1993; Marcoulides, 1997, 1998; Stahl, 1994; Stahl & Lunz, 1993) and illustrated how both approaches can be used to provide information with respect to measurement designs that involve judges using defined rating scales. Although both methods were found to provide similar information, G theory was criticized for its focus on group behavior and the interaction of groups. According to Stahl & Lunz (1993), very little information is provided about individual examinees included in a design or about the individual judges providing the performance ratings. Instead, "G theory focuses on the performance of groups" (Stahl & Lunz, 1993; p. 4).

In response to the preceding criticism, Marcoulides and Drezner (1997) introduced an extension to G theory that can be used to provide specific information on the reliability of individual ability estimates and diagnostic information at the individual and group levels. As illustrated by Marcoulides and Drezner (1997), this extension to G theory can be used to analyze performance assessments to provide three important pieces of information[1]: (1) a diagnostic scatter diagram, which can be examined for unusual patterns for each examinee and each judge in the measurement design; (2) an index for examinees, which can be used to examine the ability levels of examinees; and (3) an index for judges, which can be used to examine judge severity. Marcoulides (1997) emphasized that the aforementioned extension to G theory can be considered a special type of item response theory, or IRT, model (similar to FACETS) capable of estimating latent traits such as examinee ability estimates, rater severity, and item difficulties (for a complete discussion, see Marcoulides, 1998).

An interesting extension to the information provided by the Marcoulides and Drezner (MD) model is to segment the observed measurements into constituent groups (or clusters) so that the members of any one group are similar to each other according to some selected criterion. In the statistical literature, the most commonly used term for techniques that seek to separate data into constituent groups is cluster analysis (although other names such as Q-analysis, typology, grouping, clumping, classification, numerical taxonomy, and unsupervised pattern recognition are becoming popular, see Marcoulides & Hershberger, 1997). And, although many algorithms have been proposed in the literature for conducting cluster analysis, there is no generally accepted "best" method (Manly, 1994). In fact, it has been shown on several occasions that many of the available algorithms tend to produce conflicting results on a given set of data (Everitt, 1974; Manly, 1994).

The purpose of this chapter is to introduce a new procedure that can be used to detect pattern clustering in data obtained from measurement designs, and that avoids many of the pitfalls of other algorithms. Although the clustering procedure was originally developed to examine results generated by the MD model, it can also be applied to results from multifaceted Rasch measurement. The analysis problem is basically one of detecting and identifying clusters (or geometric relationships among subsets of the observed data) which may be present in the data. Since the most commonly accepted test of any algorithm is to take a set of data with a known structure and see whether the algorithm is able to reproduce the a priori structure, the proposed procedure will be subjected to several such tests.

AN OVERVIEW OF THE MD MODEL

To develop a perspective from which to view the proposed procedure for detecting pattern clustering, we briefly review the MD model and its approach to measurement designs.

The MD model is a variance decomposition model that is based on the assumption that n points in a measurement design (that is, examinees, judges, items, or any other set of observations that define a facet of interest) are located in an m-dimensional space. Weights (w_{ij}) between each pair of two points are given for $i,j = 1,\ldots n$. The weights express the importance of the proximity between points in space (for example, the similarity in examinee ability level estimates, judge severity estimates, or item difficulty). In the MD model, the weights (with $w_{ii} = 0$ for $i = 1,\ldots,n$ and $w_{ij} = w_{ji}$) are determined by using:

$$w_{ij} = \frac{1}{D^{p}ij},$$

where D is the m-dimensional distance between points i and j, and the power p is a parameter that varies according to the measurement design (for further details, see Marcoulides & Drezner, 1992, 1993, 1997) The model is then formulated as the minimization of the objective function:

$$f(\mathbf{X}) = \frac{\sum_{i,j=1}^{n} w_{ij}d^{2}ij}{\sum_{i,j=1}^{n} d^{2}ij},$$

where **X** is a vector of values for the points (defined according to the latent trait of interest—either examinee ability, judge severity, or item difficulty) that ensures that points with an associated large weight are close to each other, d^{2}_{ij} is the squared Euclidean distance between points i and j. The actual values of **X** (that is, the observed values on the latent trait of interest) are determined by calculating the eigenvectors of the second smallest eigenvalues of a matrix S (whose elements are defined as $s_{ii} = \sum_{j} w_{ij}$ and $s_{ij} = -w_{ij}$). The eigenvectors associated with the second and third smallest eigenvalues of S also provide coordinates of a diagnostic scatter plot for examining the various observations and conditions within a facet in any measurement design.

Figure 1 presents a diagnostic scatter plot generated by the MD method using an example measurement design (the data for the example are presented in Table 1). The example design comes from a study of the generalizability of the U.S. Department of Labor's ratings of the educational requirements of occupations (for a complete discussion of the analysis, see Marcoulides & Drezner, 1997; Webb & Shavelson, 1981; Webb, Shavelson, Shea, & Morello, 1981). In this example study, three job analysts were given on two occasions written descriptions of 27 jobs published in the *Dictionary of Occupational Titles* and asked to rate the jobs using the language component of the U.S. Department of Labor's General Education Development (GED) scale. The language component of the GED is measured on a six-point scale and covers the ability to "learn job duties from oral

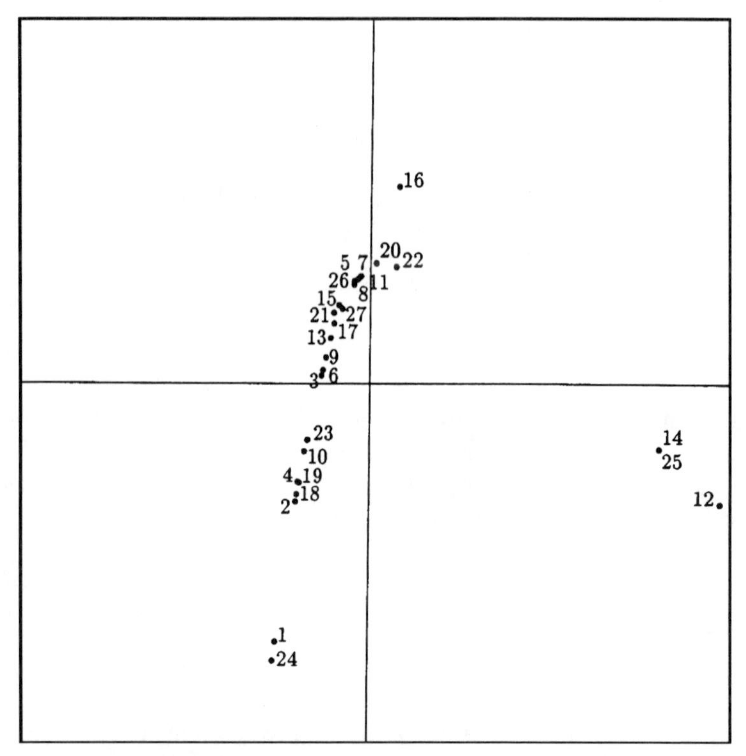

FIGURE 1. MD plot for a study of GED ratings of jobs.

instructions or demonstration..." (1 point) to the ability to "report, write, or edit articles for publications..." (6 points) (U.S. Department of Labor, 1965).

Although the plot merely highlights results from more detailed analyses generated by the MD model, it can be used to visualize the distribution of language abilities needed to perform each job. As can be seen in Figure 1, the plot illustrates that the 27 jobs are spread out in the language requirements, and one can quickly identify that job 24, job 1, job 12, job 14, and job 25 require different language ability levels than the other jobs.

Table 2 presents the language ability indices that are also provided as standard output by the MD model. The MD model (somewhat similar to FACETS[2]) independently calibrates the facets of interest so that all observations are positioned on the same scale: the scales range from +1 to –1. Thus, negative values of the language ability index indicate low ability levels and positive values higher ability levels. As can be seen, the indices corroborate the observations made through examining the plot. The ability estimates for job 12 (0.5910), job 14 (0.4864), and job 25 (0.4864) are the highest and the ability estimates for job 24 (–0.1646) and job 1 (–0.1581) are the lowest.

TABLE 1.
Data from a Study of GED Ratings of Jobs

	Occasion 1			Occasion 2		
Rater:	1	2	3	1	2	3
Job	Language Development Score					
1	1	1	2	1	1	2
2	2	2	2	2	1	2
3	2	3	2	3	3	2
4	2	2	1	3	1	1
5	3	3	3	4	3	4
6	2	3	2	3	3	3
7	3	4	3	4	3	4
8	4	3	3	4	4	4
9	3	3	2	3	3	2
10	2	2	1	3	2	2
11	4	4	3	4	3	4
12	5	6	5	5	6	6
13	4	3	2	3	3	2
14	5	5	5	5	5	5
15	4	3	3	4	2	2
16	5	3	5	4	3	5
17	4	2	3	3	3	2
18	1	2	2	2	2	1
19	2	2	2	2	2	1
20	4	4	4	3	4	4
21	4	2	2	4	2	3
22	4	4	4	4	4	5
23	2	2	2	2	3	2
24	1	1	1	1	1	1
25	5	5	5	5	5	5
26	4	3	4	4	3	3
27	4	3	3	3	3	3

TABLE 2.
Job language ability estimates

Job	Estimate
1	–0.1581
2	–0.1253
3	–0.0827
4	–0.1222
5	–0.0275
6	–0.0800
7	–0.0213
8	–0.0156
9	–0.0750
10	–0.1112

TABLE 2. (continued)

Job	Estimate
11	–0.0167
12	0.5905
13	–0.0668
14	0.4864
15	–0.0532
16	0.0481
17	–0.0613
18	–0.1237
19	–0.1199
20	0.0103
21	–0.0618
22	0.0437
23	–0.1061
24	–0.1624
25	0.4864
26	–0.0271
27	–0.0475

As indicated previously, an interesting extension to the information provided by the MD method is to segment the jobs into constituent groups (or clusters) so that the members of any one group are similar to each other according to some selected criterion (in this case language ability). In the next section, we present a procedure that can be used to detect and identify clusters that may be present in measurement data.

PROPOSED METHOD OF CLUSTERING

Suppose that we have observed the vector **X** generated by the MD method (that is, the eigenvector $\mathbf{v_1} = \{v_1, \ldots, v_n\}$ associated with the second smallest eigenvalue of the matrix S described previously). Find $v_{min} = min\{v_i\}$ and $v_{\max} = max\{v_i\}$. Next, we standardize the elements of the vector $\mathbf{v_1}$ using:

$$s_i = \frac{v_i - v_{min}}{v_{max} - v_{min}}.$$

The vector of s_i is now standardized between 0 and 1. Because the relative standing of the terms in vector of s_i are the same as those of the vector $\mathbf{v_i}$, we sort the vector $\mathbf{s_i}$ in order to obtain $s_{(i)}$ such that $s_{(i)} \leq s_{(i+1)}$. Based on this sort, it follows that $s_{(1)} = 0$ and $s_{(n)} = 1$. Next we calculate:

$$\Delta_i = s_{(i+1)} - s_{(i)}$$

for $i = 1, \ldots, n-1$. The values of Δ_i represent the gaps between two consecutive values in the sorted vector of s_i. It is important to note that if the standardization of the vector $\mathbf{v_i}$ was not done, the relative gaps between all values would have been the same. Finally, the vector Δ is sorted in decreasing order:

$$\Delta_{(1)} \geq \Delta_{(2)} \geq \Delta_{(3)} \geq \Delta_{(4)} \geq \ldots\ldots$$

The largest term $\Delta_{(1)}$ divides the n points into two clusters with the widest possible cluster. When the first k Δ's are selected, $k+1$ clusters are defined maximizing the smallest gap between any two clusters. Thus, the number of clusters (identified in terms of the number of gaps between clusters) can be determined by examining the percent contribution of Δ_i. One may select, for example, the first Δ's that add up to 90% of the gaps. By the standardization of the vector $\mathbf{v_i}$ we know that the total of all gaps is exactly 1, and therefore, 90% is achieved when the cumulative value of the selected Δ's reaches 0.9.

EXAMPLE AND COMPUTATIONAL RESULTS

We have tried the proposed clustering algorithm on numerous real and artificial data sets to test and evaluate the utility of the procedure in detecting and identifying clusters in observed measurement designs. For demonstration purposes, we examine the results generated by our proposed procedure using two artificially generated data sets with known data structure. In the first example the data form four clusters, whereas in the second example the data do not form any clusters.

Example 1

We selected a problem in which 40 individuals are represented using a five-dimensional profile and form four clusters. The coordinates of these points are given in Table 3. The four clusters are:

Cluster 1: 1,3,5,13,17,20,26,40 (8 members)
Cluster 2: 2,10,11,15,19,24,25,31,38 (9 members)
Cluster 3: 7,8,12,14,16,23,30,35,37,39 (10 members)
Cluster 4: 4,6,9,18,21,22,27,28,29,32,33,34,36 (13 members)

Figure 2 provides the results generated by the MD model using the example data discussed above. The four clusters are clearly visible. The first cluster at the top (Cluster 2) is more spread out than the other three clusters. The individual estimates that are provided as standard output by the MD method are also provided in Table 4. As can be seen, the indices corroborate the observations made through examining the two-dimensional plot.

TABLE 3.
The Coordinates for the Example 1 Problem

Individual	x_1	x_2	x_3	x_4	x_5
1	28	19	81	20	81
2	29	38	59	40	61
3	28	20	81	21	80
4	29	79	22	80	20
5	29	20	81	20	81
6	29	79	22.	81	20
7	30	60	38	60	37
8	30	58	41	57	39
9	30	79	19	80	21
10	31	40	62	41	57
11	30	39	61	39	57
12	28	59	40	61	38
13	29	20	81	19	79
14	30	62	42	61	37
15	28	38	60	39	60
16	29	59	40	59	41
17	28	18	81	18	81
18	30	80	22	81	18
19	32	41	58	40	57
20	28	19	81	21	77
21	31	80	20	80	18
22	28	81	18	80	20
23	28	61	39	57	41
24	32	39	60	41	57
25	29	39	60	37	61
26	29	20	82	19	79
27	29	82	19	77	17
28	31	78	22	79	21
29	31	81	19	81	18
30	58	38	60	39	30
31	28	41	62	40	57
32	28	81	22	77	19
33	31	78	22	81	20
34	31	79	18	79	21
35	29	62	39	60	41
36	28	80	19	77	17
37	28	59	40	59	41
38	31	38	60	41	59
39	28	60	42	57	38
40	28	19	78	19	79

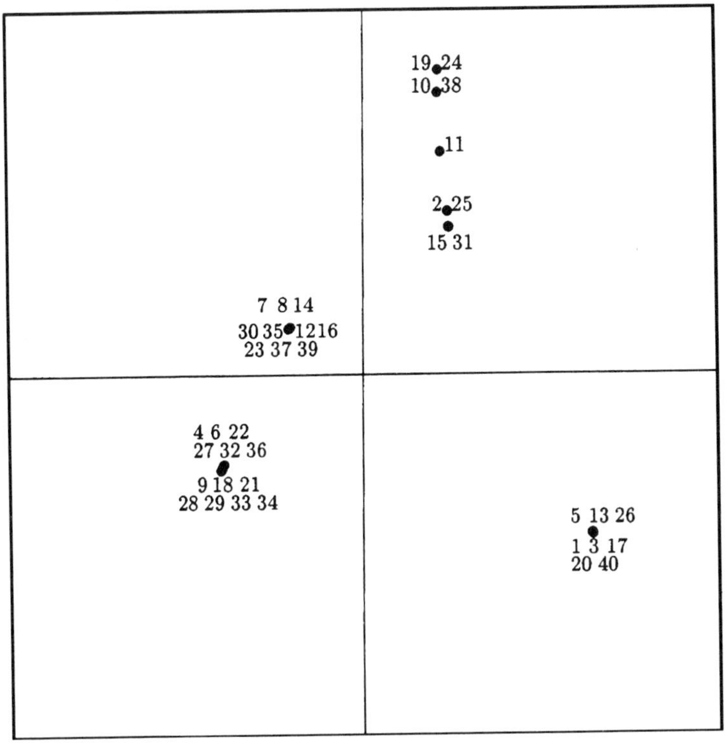

FIGURE 2. MD plot for the Example 1 problem.

TABLE 4.
Individual Estimates for the Example 1 Problem

Job	Estimate
1	0.2580
2	0.0966
3	0.2580
4	–0.1568
5	0.2576
6	–0.1568
7	–0.0838
8	–0.0835
9	–0.1584
10	0.0851
11	0.0885
12	–0.0815
13	0.2576
14	–0.0838
15	0.0983

TABLE 4. (continued)

Job	Estimate
16	–0.0823
17	0.2580
18	–0.1584
19	0.0854
20	0.2579
21	–0.1598
22	–0.1568
23	–0.0814
24	0.0854
25	0.0968
26	0.2576
27	–0.1569
28	–0.1597
29	–0.1598
30	–0.0837
31	0.0978
32	–0.1567
33	–0.1597
34	–0.1597
35	–0.0825
36	–0.1568
37	–0.0814
38	0.0851
39	–0.0814
40	0.2580

Table 5 presents an Excel spreadsheet constructed for the calculation of the clusters based on our proposed procedure. As can be seen by examining the last column of Table 5 (the cumulative percent contribution of Δ_i), the first two gaps cover approximately 40% and 38% of the range. The third gap adds another 17% for a total of 95% of the range. All other gaps are much smaller, which clearly indicates that the data are separable into four clusters. Evidence that the data are separable into four clusters is also provided by examining Figure 3 (which displays the graphical representation of the vector $\mathbf{s}_\mathbf{i}$).

Example 2

We selected a problem in which 40 individuals are represented in a circle. Figure 4 provides a plot of the results generated by the MD model. As can be seen in Figure 4, there are no clusters in this measurement data. The individual indices generated by the MD model for this example data are provided in Table 6, whereas Table 7 presents the results from the constructed Excel spreadsheet. As can be seen by

TABLE 5.
Excel Spreadsheet for the Example 1 Problem

Obs	v	S	Sorted S	Δ	Sorted Δ	Cum.Δ
1	0.2580	1	0	0	0.398516	0.398516
2	0.0966	0.613691	0	0.000239	0.381283	0.779799
3	0.2580	1	0.000239	0	0.174485	0.954284
4	–0.1568	0.00718	0.000239	0	0.019387	0.973672
5	0.2576	0.999043	0.000239	0.003112	0.00742	0.981091
6	–0.1568	0.00718	0.003351	0	0.00359	0.984682
7	–0.0838	0.181905	0.003351	0.00359	0.003112	0.987793
8	–0.0835	0.182623	0.006941	0.000239	0.002393	0.990187
9	–0.1584	0.003351	0.00718	0	0.002393	0.99258
10	0.0851	0.586166	0.00718	0	0.001915	0.994495
11	0.0885	0.594303	0.00718	0	0.001197	0.995692
12	–0.0815	0.18741	0.00718	0.000239	0.000718	0.99641
13	0.2576	0.999043	0.00742	0.174485	0.000718	0.997128
14	–0.0838	0.181905	0.181905	0	0.000479	0.997607
15	0.0983	0.61776	0.181905	0.000239	0.000479	0.998085
16	–0.0823	0.185495	0.182145	0.000479	0.000479	0.998564
17	0.2580	1	0.182623	0.002393	0.000239	0.998803
18	–0.1584	0.003351	0.185017	0.000479	0.000239	0.999043
19	0.0854	0.586884	0.185495	0.001915	0.000239	0.999282
20	0.2579	0.999761	0.18741	0.000239	0.000239	0.999521
21	–0.1598	0	0.18765	0	0.000239	0.999761
22	–0.1568	0.00718	0.18765	0	0.000239	1
23	–0.0814	0.18765	0.18765	0.398516	0	1
24	0.0854	0.586884	0.586166	0	0	1
25	0.0968	0.614169	0.586166	0.000718	0	1
26	0.2576	0.999043	0.586884	0	0	1
27	–0.1569	0.006941	0.586884	0.00742	0	1
28	–0.1597	0.000239	0.594303	0.019387	0	1
29	–0.1598	0	0.613691	0.000479	0	1
30	–0.0837	0.182145	0.614169	0.002393	0	1
31	0.0978	0.616563	0.616563	0.001197	0	1
32	–0.1567	0.00742	0.61776	0.381283	0	1
33	–0.1597	0.000239	0.999043	0	0	1
34	–0.1597	0.000239	0.999043	0	0	1
35	–0.0825	0.185017	0.999043	0.000718	0	1
36	–0.1568	0.00718	0.999761	0.000239	0	1
37	–0.0814	0.18765	1	0	0	1
38	0.0851	0.586166	1	0	0	1
39	–0.0814	0.18765	1	0	0	1
40	0.2580	1	1			

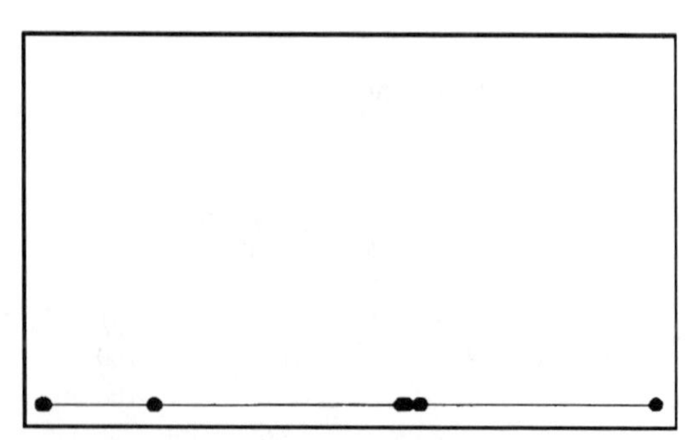

FIGURE 3. A graphical representation of s_i for Example 1.

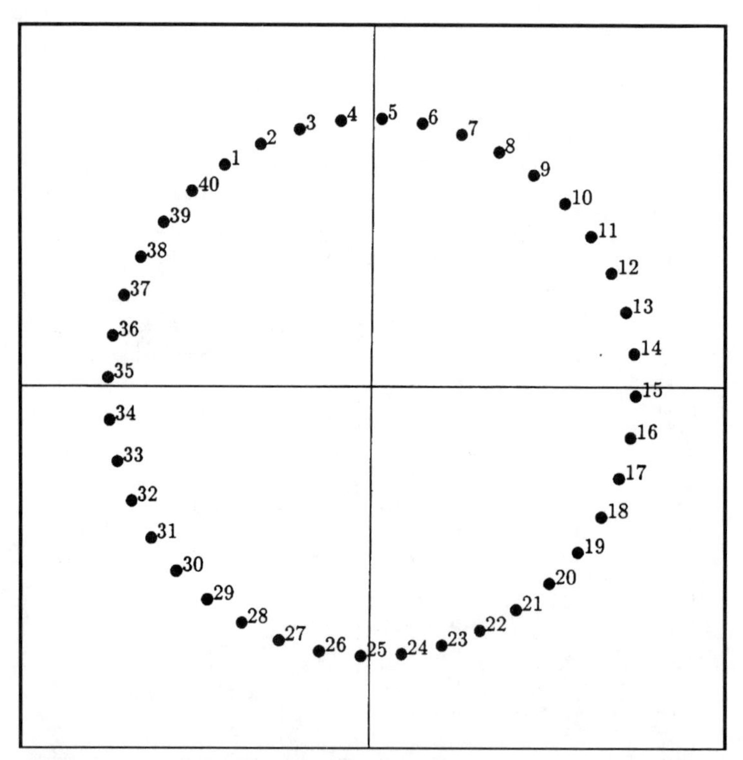

FIGURE 4. MD plot for the Example 2 problem.

TABLE 6.
Individual Estimates for the Example 2 Problem

Job	Estimate
1	–0.1255
2	–0.0950
3	–0.0622
4	–0.0278
5	0.0073
6	0.0421
7	0.0760
8	0.1079
9	0.1372
10	0.1632
11	0.1851
12	0.2024
13	0.2148
14	0.2219
15	0.2235
16	0.2196
17	0.2103
18	0.1958
19	0.1765
20	0.1529
21	0.1255
22	0.0950
23	0.0622
24	0.0278
25	–0.0070
26	–0.0421
27	–0.0760
28	–0.1079
29	–0.1372
30	–0.1632
31	–0.1851
32	–0.2024
33	–0.2148
34	–0.2219
35	–0.2235
36	–0.2196
37	–0.2103
38	–0.1958
39	–0.1765
40	–0.1529

TABLE 7.
Excel Spreadsheet for the Example 2 Problem

Obs	v	S	Sorted S	Δ	Sorted Δ	Cum.Δ
1	–0.1255	0.2192	0.0000	0.0036	0.0459	0.0459
2	–0.095	0.2875	0.0036	0.0051	0.0459	0.0917
3	–0.0622	0.3609	0.0087	0.0107	0.0450	0.1367
4	–0.0278	0.4378	0.0195	0.0101	0.0450	0.1817
5	0.0073	0.5163	0.0295	0.0177	0.0425	0.2242
6	0.0421	0.5942	0.0472	0.0148	0.0425	0.2667
7	0.076	0.6700	0.0620	0.0239	0.0394	0.3060
8	0.1079	0.7414	0.0859	0.0192	0.0394	0.3454
9	0.1372	0.8069	0.1051	0.0298	0.0351	0.3805
10	0.1632	0.8651	0.1349	0.0230	0.0351	0.4157
11	0.1851	0.9141	0.1579	0.0351	0.0327	0.4483
12	0.2024	0.9528	0.1931	0.0262	0.0320	0.4803
13	0.2148	0.9805	0.2192	0.0394	0.0320	0.5123
14	0.2219	0.9964	0.2586	0.0289	0.0309	0.5432
15	0.2235	1.0000	0.2875	0.0425	0.0309	0.5740
16	0.2196	0.9913	0.3300	0.0309	0.0298	0.6038
17	0.2103	0.9705	0.3609	0.0450	0.0298	0.6336
18	0.1958	0.9380	0.4058	0.0320	0.0289	0.6624
19	0.1765	0.8949	0.4378	0.0459	0.0289	0.6913
20	0.1529	0.8421	0.4837	0.0327	0.0262	0.7174
21	0.1255	0.7808	0.5163	0.0459	0.0262	0.7436
22	0.095	0.7125	0.5622	0.0320	0.0239	0.7676
23	0.0622	0.6392	0.5942	0.0450	0.0239	0.7915
24	0.0278	0.5622	0.6391	0.0309	0.0230	0.8145
25	–0.0073	0.4837	0.6700	0.0425	0.0230	0.8376
26	–0.0421	0.4058	0.7125	0.0289	0.0192	0.8568
27	–0.076	0.3300	0.7414	0.0394	0.0192	0.8761
28	–0.1079	0.2586	0.7808	0.0262	0.0177	0.8937
29	–0.1372	0.1931	0.8069	0.0351	0.0177	0.9114
30	–0.1632	0.1349	0.8421	0.0230	0.0148	0.9262
31	–0.1851	0.0859	0.8651	0.0298	0.0148	0.9409
32	–0.2024	0.0472	0.8949	0.0192	0.0107	0.9517
33	–0.2148	0.0195	0.9141	0.0239	0.0107	0.9624
34	–0.2219	0.0036	0.9380	0.0148	0.0101	0.9725
35	–0.2235	0.0000	0.9528	0.0177	0.0101	0.9826
36	–0.2196	0.0087	0.9705	0.0101	0.0051	0.9877
37	–0.2103	0.0295	0.9805	0.0107	0.0051	0.9928
38	–0.1958	0.0620	0.9913	0.0051	0.0036	0.9964
39	–0.1765	0.1051	0.9964	0.0036	0.0036	1.0000
40	–0.1529	0.1579	1.0000			

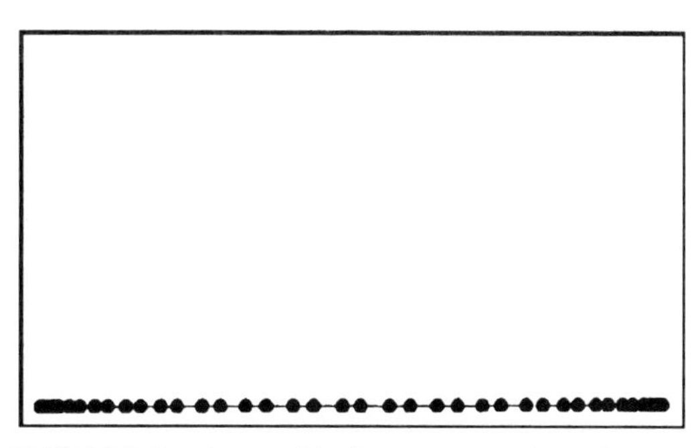

FIGURE 5. A graphical representation of s_i for Example 2.

examining the last column (that is, the cumulative percent contribution of Δ_i), no significant gaps are revealed. For example, even after considering the first 10 gaps, only about 41% of the range is covered. These results clearly indicate that the data are not separable into any clusters. In addition, this result is again evident by examining Figure 5, which displays the graphical representation of the vector $\mathbf{s_i}$.

CONCLUSIONS

This chapter introduced a new procedure that can be used to detect pattern clustering in data obtained from measurement designs. As illustrated, the procedure is able to reproduce a priori defined structures in example data sets. Although only two example cases were presented, the procedure can be utilized with all types of measurement designs that might be encountered in practical applications.

NOTES

1. Although Marcoulides and Drezner (1997) illustrated their method using performance assessment designs, the model can be used with any type of design and with any number of conditions of facets and observations.

2. It is important to note that the indices in the MD model are somewhat similar to the log-odds units or "logits" in the FACETS model (for a complete discussion, see Marcoulides, 1998).

REFERENCES

Bachman, L., Boodoo, G., Linacre, J. M., Lunz, M. E., Marcoulides, G. A., & Myford, C. (1993, April). *Generalizability theory and many-faceted Rasch modeling.* Invited presentation at the joint annual meeting of the American Educational Research Association and the National Council on Measurement in Education, Atlanta, GA.

Cronbach, L. J., Gleser, G. C., Nanda, H., & Rajaratnam, N. (1972). *The dependability of behavioral measurements: Theory of generalizability scores and profiles.* New York: Wiley.

Drezner, Z. (1987). A heuristic procedure for the layout of a large number of facilities. *Management Science, 33*(7), 907–915.

Everitt, B. (1974). *Cluster analysis.* London: Heinemann.

Linacre, J. M. (1988). *FACETS.* Chicago: MESA Press.

Linacre, J. M. (1989). *Many-faceted Rasch measurement.* Chicago: MESA Press.

Manly, B. F. J. (1994). *Multivariate statistical methods: A primer.* New York: Chapman & Hall.

Marcoulides, G. A. (1997). *Generalizability theory: Models and applications.* Invited session at the annual meeting of the American Educational Research Association, Chicago.

Marcoulides, G. A. (1998). Generalizability theory: Picking up where the Rasch IRT model leaves off? In S. Embretson & S. L. Hershberger (Eds.), *The new rules of measurement: What every psychologist and educator should know.* Mahwah, NJ: Erlbaum.

Marcoulides, G. A., & Drezner, Z. (1992). A procedure for transforming data in multidimensional space. *Proceedings of the Western Decision Sciences Institute, 1*, 834–835.

Marcoulides, G. A., & Drezner, Z. (1993). A procedure for transforming data in multidimensional space to a two-dimensional representation. *Educational and Psychological Measurement, 53*(4), 933–940.

Marcoulides, G. A., & Drezner, Z. (1997). A method for analyzing performance assessments. In M. Wilson, G. Engelhard, Jr., & K. Draney (Eds.), *Objective measurement: Theory into practice* (Vol. 4, pp. 261–277). Norwood, NJ: Ablex.

Marcoulides, G. A., & Hershberger, S. L. (1997). *Multivariate statistical methods: A first course.* Mahwah, NJ: Erlbaum.

Stahl, J. A. (1994). What does generalizability theory offer that many-facet Rasch measurement cannot duplicate? *Rasch Measurement Transactions, 8*(1), 342–343.

Stahl, J. A., & Lunz, M. E. (1993). *A comparison of generalizability theory and multifaceted Rasch measurement.* Paper presented at the annual meeting of the American Educational Research Association, Atlanta, GA.

U.S. Department of Labor. (1965). *Dictionary of occupational titles* (3rd ed., Vol. 2). Washington, DC: U.S. Government Printing Office.

Webb, N. M., & Shavelson, R. J. (1981). Multivariate generalizability of general education development ratings. *Journal of Educational Measurement, 18,* 13–22.

Webb, N. M., Shavelson, R. J., Shea, J., & Morello, E. (1981). Generalizability of general educational development ratings of jobs in the U.S. *Journal of Applied Psychology, 66*, 186–191.

16

EXAMINING REPLICATION EFFECTS IN RASCH FIT STATISTICS

Richard M. Smith
University of Florida

Randall E. Schumacker
University of North Texas

M. Joan Bush
Irving Independent School District

INTRODUCTION

Numerous studies have reported on the properties of the statistics used to test the fit of items and persons to the Rasch measurement models. The literature includes studies related to the effect of (1) test length, (2) sample size, (3) item difficulty distribution, (4) person ability distribution, and (5) the number of steps in each item on the fit statistics. In most cases, computer-simulated data with 2 to 100 replications were used. No research, however, has determined if these results reported would have been affected by the number of replications used in the study.

RASCH FIT STATISTICS

Rasch fit statistics are used to provide a frame of reference for judging the appropriateness of the responses to a given item or by a given person on an objectively measured variable. Rasch (1960/1980) suggested several methods for assessing item and person fit to his model. Unfortunately, no computer programs were available at the time that were capable of computing these statistics. As a result, his suggested fit indices have not been widely used. Even so, his influence is reflected in the subsequent work related to fit indices.

The first fit statistics were based upon the overall Pearsonian chi-square approach (Wright & Panchapakesan, 1969) or the likelihood-ratio chi-square approach (Anderson, 1973). Later, the Pearsonian chi-squares were converted to weighted or unweighted mean squares. Weighted mean squares were referred to as "infit," since the information function is used to assign more weight to items or persons near the estimated logit measure. In contrast, unweighted mean squares were referred to as "outfit," since the tendency for these statistics is give more weight to unexpected responses far from the expected item or person logit measure. More recently, a cube-root transformation has been used to convert the mean squares to an approximate unit normals with an expected mean of zero and an expected standard deviation of approximately one (Smith, 1982, 1991a).

In the recent Rasch measurement computer programs (Smith 1991b; Wright & Linacre, 1991), the total (infit and outfit) and less often the between fit statistics are reported for items and persons. These statistics are sensitive to different types of measurement disturbances. The total fit statistic is sensitive to measurement disturbances such as guessing, start-up fluctuations, sloppiness, and unexpected correct and incorrect responses. In addition, the total fit statistic is sensitive to changes in the slope of the person or item characteristic curves. Thus, the total fit statistic is sensitive to unsystematic measurement disturbances. In contrast, the between fit statistic is sensitive to systematic measurement disturbances such as bias or differential item functioning (Smith, 1991b).

Smith (1982) found that the means and standard deviations of weighted and unweighted between fit statistics were almost identical, with a correlation of .99. Both the BICAL (Wright, Mead, & Bell, 1979) and IPARM (Smith, 1991b) Rasch programs use the unweighted version of the between fit statistic, however, the IPARM program is less restrictive in the number of ability groups and number of persons per sample. Also, with IPARM, item invariance differences can be tested by demographic subgroups.

A number of simulation studies have been conducted related to Rasch fit statistics. For example, Mead (1975) examined the use of the unweighted total and between fit statistics. Smith has examined (1) the robustness of fit statistics (Smith, 1985), (2) person fit (Smith, 1986), (3) the distributional properties of Rasch standardized residuals (Smith, 1988a), (4) power comparisons of Rasch total and between item fit statistics (Smith, 1994), (5) the distributional properties of Rasch item fit statistics (Smith, 1991a), and (6) separate calibration versus between fit statistics in detecting item bias (Smith, 1996). These findings, however, were typically based on only 2 to 20 replications, depending of the generality of the situation. It is therefore important to determine if a larger number of replications might have affected the results reported. Therefore, the purpose of this study is to investigate the differences in the total weighted and unweighted fit statistics and the between fit statistic as a function of varying numbers of replications.

METHODOLOGY

In this study, simulated data sets were used which varied in test length and sample size. Two test lengths (20 and 50 items) and three sample sizes (150, 500, and 1,000 persons) were completely crossed for a total of six replication sets. Each set was replicated 100 times.

The data sets were constructed using SIMTEST 2.1 (Luppescu, 1992). For each replication, person abilities were normally distributed and item difficulties were uniformly distributed. BIGSTEPS (Wright & Linacre, 1992), a Rasch calibration program, was used to calibrate each of the 600 data sets. Next, the Rasch-based item statistics were calculated with IPARM (Smith, 1991b), a Rasch item and person analysis program. This program calculated the (1) item weighted total fit statistic, (2) item unweighted total fit statistic, (3) item unweighted between fit statistic, (4) person weighted total fit statistic, and (5) person unweighted total fit statistic.

Summary statistics, which included the means, standard deviations, and Type I error rate, were calculated for each fit statistic (five) for each replication set (six) after 10, 25, 50, and 100 replications. The comparison of the stability of the means, standard deviations, and Type I error rates over varying numbers of replications should give an indication of the stability of these statistics over replications and help to answer the question of how many replications are really needed in this type of study.

RESULTS

The data presented in the tables are organized to aid interpretation. In Tables 1 to 5, the first four rows of numbers under the *mean* column heading for 20 items indicates the average t-value. For example, in Table 1 the values for the 150 person simulations means are –.01, –.06, .01, and –.02 with an expected value of 0.0. The next four values in the same column are the mean standard deviations of the t-values just reported. For example, the .95, .94, .94, and .94 values for the 150 person simulations in Table 1 have an expected values of 1.0. The IPARM program that calculated these values uses n rather than $n - 1$ in calculating the mean squares used in the t calculation; hence, these values have not been corrected for degrees

TABLE 1.
Means and Standard Deviations of Item Unweighted Between Fit Statistics

	Number of Items			
	20		50	
	Mean	*SD*	Mean	*SD*
150 Persons				
Mean				
10 reps	−.01	.15	−.04	.15
25 reps	−.06	.24	−.01	.13
50 reps	.01	.22	.02	.14
100 reps	−.02	.21	.02	.14
SD				
10 reps	.95	.11	.97	.08
25 reps	.94	.12	.95	.07
50 reps	.94	.14	.97	.09
100 reps	.94	.13	.97	.09
500 Persons				
Mean				
10 reps	.06	.27	.07	.11
25 reps	.05	.25	.07	.12
50 reps	.05	.23	.04	.14
100 reps	.05	.22	.03	.14
SD				
10 reps	.90	.13	1.00	.13
25 reps	.93	.13	.97	.11
50 reps	.90	.14	.97	.11
100 reps	.92	.14	.97	.10
1000 Persons				
Mean				
10 reps	.19	.14	.02	.14
25 reps	.22	.18	.01	.14
50 reps	.18	.18	.03	.14
100 reps	.17	.18	.03	.13
SD				
10 reps	.84	.09	.98	.09
25 reps	.87	.13	1.00	.09
50 reps	.88	.12	.98	.09
100 reps	.88	.12	.97	.09

TABLE 2.
Means and Standard Deviations of Item Unweighted Total Fit Statistics

	Number of Items			
	20		50	
	Mean	*SD*	Mean	*SD*
150 Persons				
Mean				
10 reps	−.17	.06	−.05	.09
25 reps	−.16	.08	−.05	.07
50 reps	−.18	.08	−.05	.07
100 reps	−.18	.09	−.05	.06
SD				
10 reps	.75	.14	.85	.06
25 reps	.78	.16	.85	.10
50 reps	.83	.17	.87	.10
100 reps	.84	.16	.87	.10
500 Persons				
Mean				
10 reps	−.29	.12	−.13	.04
25 reps	−.27	.11	−.13	.05
50 reps	−.31	.11	−.14	.05
100 reps	−.32	.10	−.14	.05
SD				
10 reps	.83	.15	.89	.08
25 reps	.86	.18	.91	.11
50 reps	.87	.16	.89	.11
100 reps	.86	.17	.88	.10
1000 Persons				
Mean				
10 reps	−.49	.14	−.20	.06
25 reps	−.48	.13	−.19	.05
50 reps	−.48	.11	−.19	.05
100 reps	−.48	.10	−.19	.06
SD				
10 reps	.79	.17	.89	.11
25 reps	.84	.17	.88	.10
50 reps	.83	.14	.88	.09
100 reps	.81	.14	.88	.10

TABLE 3.
Means and Standard Deviations of Item Weighted Total Fit Statistics

	Number of Items			
	20		50	
	Mean	*SD*	Mean	*SD*
150 Persons				
Mean				
10 reps	–.23	.04	–.09	.03
25 reps	–.24	.04	–.09	.03
50 reps	–.23	.04	–.09	.03
100 reps	–.23	.04	–.09	.03
SD				
10 reps	.79	.10	.77	.07
25 reps	.79	.12	.77	.08
50 reps	.82	.14	.80	.08
100 reps	.83	.12	.80	.09
500 Persons				
Mean				
10 reps	–.45	.04	–.18	.02
25 reps	–.44	.04	–.18	.02
50 reps	–.44	.04	–.18	.02
100 reps	–.44	.05	–.17	.02
SD				
10 reps	.82	.18	.77	.08
25 reps	.80	.16	.79	.10
50 reps	.83	.15	.78	.10
100 reps	.82	.15	.78	.09
1000 Persons				
Mean				
10 reps	–.59	.03	–.23	.02
25 reps	–.59	.04	–.24	.02
50 reps	–.59	.04	–.24	.02
100 reps	–.60	.04	–.24	.02
SD				
10 reps	.76	.14	.84	.13
25 reps	.80	.14	.81	.11
50 reps	.81	.13	.80	.10
100 reps	.81	.13	.79	.09

TABLE 4.
Means and Standard Deviations of Person Unweighted Total Fit Statistics

	Number of Items			
	20		50	
	Mean	*SD*	Mean	*SD*
150 Persons				
Mean				
10 reps	–.03	.02	–.04	.02
25 reps	–.04	.02	–.03	.02
50 reps	–.04	.02	–.04	.02
100 reps	–.04	.02	–.04	.02
SD				
10 reps	.89	.04	.94	.07
25 reps	.89	.05	.92	.05
50 reps	.89	.05	.91	.05
100 reps	.89	.05	.92	.05
500 Persons				
Mean				
10 reps	–.04	.01	–.03	.01
25 reps	–.04	.01	–.03	.01
50 reps	–.04	.01	–.04	.01
100 reps	–.03	.01	–.04	.01
SD				
10 reps	.89	.02	.93	.03
25 reps	.89	.03	.92	.03
50 reps	.89	.03	.92	.03
100 reps	.89	.03	.92	.03
1000 Persons				
Mean				
10 reps	–.03	.01	–.04	.01
25 reps	–.04	.01	–.03	.01
50 reps	–.04	.01	–.03	.01
100 reps	–.04	.01	–.03	.01
SD				
10 reps	.88	.02	.88	.02
25 reps	.89	.02	.89	.02
50 reps	.89	.02	.89	.02
100 reps	.89	.02	.89	.02

TABLE 5.
Means and Standard Deviations of Person Weighted Total Fit Statistics

	Number of Items			
	20		50	
	Mean	*SD*	Mean	*SD*
150 Persons				
Mean				
10 reps	−.07	.02	−.05	.02
25 reps	−.07	.02	−.05	.02
50 reps	−.07	.02	−.05	.02
100 reps	−.07	.02	−.05	.02
SD				
10 reps	.89	.05	.91	.05
25 reps	.89	.05	.90	.04
50 reps	.90	.06	.89	.05
100 reps	.90	.05	.90	.05
500 Persons				
Mean				
10 reps	−.07	.01	−.05	.01
25 reps	−.07	.01	−.05	.01
50 reps	−.07	.01	−.05	.01
100 reps	−.07	.01	−.04	.01
SD				
10 reps	.89	.03	.92	.04
25 reps	.90	.03	.91	.03
50 reps	.89	.03	.90	.03
100 reps	.89	.03	.90	.03
1000 Persons				
Mean				
10 reps	−.06	.01	−.04	.00
25 reps	−.06	.01	−.04	.01
50 reps	−.06	.01	−.04	.01
100 reps	−.06	.01	−.04	.01
SD				
10 reps	.89	.02	.88	.02
25 reps	.90	.02	.89	.02
50 reps	.89	.02	.89	.02
100 reps	.89	.02	.89	.02

of freedom. (For a more complete discussion of this point, see Smith, 1991b.) The second column of values, under the *SD* heading, contains the standard deviations of the mean values reported in the previous column. The standard deviation of the mean *t*-values for the 150 person simulations for 20 items were .15, .24, .22, and .21. The next four rows in the second column of values contain the standard deviations of the mean *t* standard deviations: .11, .12, .14, and .13. These eight rows are repeated for 500 persons and 1,000 persons. The last two columns contain the same information for the 50 item simulations.

Tables 1, 2, and 3 contain results for the three item fit statistics. Tables 4 and 5 contain results for the two person fit statistics. Tables 6, 7, and 8 contain the Type I error rate analysis results for the three item fit statistics. Tables 9 and 10 contain the Type I error rate analysis results for the person total fit statistics. Values in Tables 6 through 10 indicate the number of items falling in the extreme tails of the distribution and should be compared to the expected percentage of false positive associated with the critical value chosen. This analysis uses a critical value of 2.0, which is based on a two-tailed hypothesis at the .05 level resulting in 2.5% expectation in each tail of the distribution.

TABLE 6.
Frequency of Extreme Values for Item Unweighted Between Fit Statistics

	Number of Items							
	20				50			
	t>+2	%	t>–2	%	t>+2	%	t>–2	%
150 Persons								
10 reps	1	0.5	1	0.5	7	1.4	11	2.2
25 reps.	4	0.8	7	1.4	21	1.7	18	1.4
50 reps	18	1.8	12	1.2	57	2.3	38	1.5
100 reps	30	1.5	30	1.5	115	2.3	75	1.5
500 Persons								
10 reps	4	2.0	2	1.0	15	3.0	5	1.0
25 reps	11	2.2	6	1.2	39	3.1	13	1.0
50 reps	20	2.0	7	0.7	78	3.1	37	1.5
100 reps	42	2.1	16	0.8	133	2.7	77	1.5
1,000 Persons								
10 reps	2	1.0	1	0.5	9	1.8	12	2.4
25 reps	13	2.6	2	0.4	29	2.3	26	2.1
50 reps	16	1.6	3	0.3	58	2.3	44	1.8
100 reps	35	1.8	11	0.6	120	2.4	73	1.5

TABLE 7.
Frequency of Extreme Values for Item Unweighted Total Fit Statistics

	Number of Items							
	20				50			
	t >+2	%	*t* >–2	%	*t* >+2	%	*t* >–2	%
150 Persons								
10 reps	0	0.0	0	0.0	10	2.0	1	0.2
25 reps	5	1.0	0	0.0	26	2.1	3	0.2
50 reps	15	1.5	1	0.1	47	1.9	12	0.5
100 reps	35	1.8	7	0.4	92	1.8	23	0.5
500 Persons								
10 reps	3	1.5	1	0.5	6	1.2	2	0.2
25 reps	7	1.4	7	1.4	19	1.5	10	0.8
50 reps	11	1.1	22	2.2	36	1.4	25	1.0
100 reps	21	1.1	35	1.8	70	1.4	42	0.8
1,000 Persons								
10 reps	1	0.5	2	1.0	4	0.4	9	1.8
25 reps	2	0.4	13	2.6	11	0.9	21	1.7
50 reps	4	0.4	27	2.7	23	0.9	37	1.5
100 reps	7	0.4	40	2.0	47	0.9	68	1.4

TABLE 8.
Frequency of Extreme Values for Item Weighted Total Fit Statistics

	Number of Items							
	20				50			
	t >+2	%	*t* >–2	%	*t* >+2	%	*t* >–2	%
150 Persons								
10 reps	0	0.0	2	1.0	3	0.6	2	0.4
25 reps	1	0.2	5	1.0	6	0.5	6	0.5
50 reps	6	0.6	10	1.0	19	0.8	21	0.8
100 reps	12	0.6	25	1.3	42	0.8	40	0.8
500 Persons								
10 reps	0	0.0	6	3.0	1	0.2	2	0.4
25 reps	1	0.2	14	2.8	7	0.6	11	0.9
50 reps	3	0.3	34	3.4	15	0.7	25	1.0
100 reps	6	0.3	65	3.3	28	0.6	45	0.9
1,000 Persons								
10 reps	0	0.0	6	3.0	3	0.6	9	1.8
25 reps	0	0.0	18	3.6	7	0.6	16	1.3
50 reps	1	0.1	38	3.8	13	0.5	37	1.5
100 reps	3	0.2	63	3.2	24	0.5	65	1.3

TABLE 9.
Frequency of Extreme Values for Person Unweighted Total Fit Statistics

	Number of Items							
	20				50			
	t >+2	%	t >–2	%	t >+2	%	t >–2	%
150 Persons								
10 reps	23	1.5	12	0.8	30	2.0	18	1.2
25 reps	62	1.7	26	0.7	61	1.6	37	1.0
50 reps	141	1.9	45	0.6	118	1.6	79	1.1
100 reps	264	1.8	96	0.6	242	1.6	163	1.1
500 Persons								
10 reps	89	1.8	38	0.8	75	1.5	80	1.6
25 reps	230	1.8	91	0.7	200	1.6	164	1.3
50 reps	430	1.7	189	0.8	403	1.6	335	1.3
100 reps	854	1.7	381	0.8	819	1.6	639	1.3
1,000 Persons								
10 reps	158	1.6	80	0.8	143	1.4	109	1.1
25 reps	418	1.7	223	0.9	372	1.5	284	1.1
50 reps	800	1.6	423	0.8	761	1.5	579	1.2
100 reps	1593	1.6	815	0.8	1535	1.5	1160	1.1

TABLE 10.
Frequency of Extreme Values for Person Weighted Total Fit Statistics

	Number of Items							
	20				50			
	t >+2	%	t >–2	%	t >+2	%	t >–2	%
150 Persons								
10 reps	11	0.7	25	1.7	18	1.2	32	2.1
25 reps	33	0.9	61	1.6	40	1.1	69	1.8
50 reps	79	1.1	123	1.6	88	1.2	129	1.7
100 reps	170	1.1	251	1.7	180	1.2	254	1.7
500 Persons								
10 reps	62	1.2	99	2.0	79	1.6	114	2.3
25 reps	150	1.2	243	1.9	164	1.3	241	1.9
50 reps	255	1.0	474	1.9	332	1.3	466	1.9
100 reps	516	1.0	942	1.9	625	1.3	892	1.8
1,000 Persons								
10 reps	97	1.0	188	1.9	106	1.1	179	1.8
25 reps	281	1.1	480	1.9	291	1.2	430	1.7
50 reps	535	1.1	951	1.9	856	1.7	573	1.1
100 reps	1097	1.1	1862	1.9	1684	1.7	1150	1.2

Expected Values

The item unweighted between fit statistics in Table 1 are sensitive to systematic differences owing to bias or differential item functioning. The results in the table indicate that increasing the number of items, from 20 to 50, reduced the standard deviations of the mean *t*-values. This indicates that the means were more stable over replications for the longer test. The change across the four numbers of replications (10, 25, 50, and 100) showed little variability for the two summary statistics across the six cells (3 number of persons × 2 test lengths). The largest sample size (1,000 persons)resulted in the highest mean *t*-values for the shorter test (.19, .22, .18, and .17) and the lowest mean standard deviations (.84, .87, .88, and .88). However, the values were much closer to the expected values for the 50 item test.

The item unweighted total fit statistic (outfit) in Table 2 is affected by random disturbances such as guessing and sloppiness in responding. These results in the table are not as close to the expected values of 0.0 for means and 1.0 for standard deviations as the results for the unweighted item between fit statistic. Dependencies between persons and items as well as the use of n rather than $n - 1$ in calculating the mean square values affects these results. The longer test means were closer to the expected values than the means for the shorter test. In contrast to the item between fit statistic, the variation across means for the different numbers of replications for the six cells was less.

The item weighted total fit statistic (infit) in Table 3 is less affected by unexpected responses by persons who are far from the item's difficulty. The results reported in this table are similar to those reported in Table 2 except that the values for the mean and the standard deviation are a little further from the expected values. The predominance of negative means in the tables suggest that the results are conservative and should indicate a lower Type I error rate than normally expected. Again, there is little variability in the mean values across the number of replications.

Smith (1991a, 1991b) suggests several corrections that can be used with the unweighted and weighted item fit statistic to improve the correspondence of the results for simulated data to the expected values. When these adjustments are made, the values more closely approximate expected values. Two corrections are necessary: One for the restriction in the average mean square and another for the restriction in the standard deviation of the mean squares. Once the mean squares are corrected, the distribution of the *t*-transformation is much closer to the expected values and the Type I error rate.

The results for the person unweighted and weighted total fit statistics are reported in Tables 4 and 5. The values in Table 4 for the person unweighted total fit statistic indicate a good fit between the mean and the expected value of 0.0. However, the standard deviation is approximately 0.1 lower than the expected value of 1.0 across the replication sets. The variability in mean and standard deviation values across the number of replications is less than in the previous three tables.

Table 5, which contains the results for the person weighted total fit statistic, also indicates a good fit between the mean and the expected value of 0.0. Again, the standard deviation is approximately 0.1 lower than the expected value of 1.0 across the replication sets. The variability in values across the number of replications is similar to that in Table 4 and less than in the previous three tables.

The outfit expected values were in Table 4 were closer to the expected value of 0.0 than the infit values reported in Table 5. For example, compare –.03, –.04, –.04, and –.04 (Table 4) versus –.07, –.07, –.07, and –.07 (Table 5) for 20 items. Notice that values across the number of replications were almost identical, suggesting that the number of replications does not affect the person fit statistics.

Type I Error Rates

Tables 6, 7, and 8 indicate the number of items falling in the extreme tails of a two-tailed distribution for item unweighted between, item unweighted total, and item weighted total fit statistics. The number of items expected in each tail can be easily computed. For example, given a two-tailed test at the .05 level, one would expect .025% of the items to fall in each tail ($t = \pm 2.00$). Therefore, .025 times 20 items = .5, which taken over 10 replications yields five items in each tail. For the other numbers of replications with 20 items, the number of items expected would be 12.5; 25; and 50, respectively. These findings can be interpreted by comparing the actual number of items reported versus the expected number of items or by using the percent of items falling into each tail, as shown in Tables 6 through 10. For example, given 20 items and 100 replications, there are 2,000 replications. Since only 30 items were identified with $t > | 2 |$, the tails of the distribution contain only 1.5% false positive values (30/2,000 = .015).

The differences in these findings across the number of replications in each replication set were negligible. For example, given 20 items and 150 persons, the percentages ranged from .05% (10 replications) to 1.8% (50 replications). For the three item fit statistics (Tables 6, 7, and 8), the largest range of percentages over the four levels of replications was 1.8% in Table 7 for 20 items and 150 persons.

In comparing the number of items indicated in these tables, the Type I error rate is closer to the expected values for the item between fit statistics in Table 6. The largest discrepancy between expected and observed Type I errors is with the item weighted total fit statistic (Table 8), where the frequency of Type I errors was seldom above 1% until the 1,000 person, 50 item replications. Thus it would appear that a ±2 critical value for the item weighted total fit statistic would yield a Type I error rate of approximately 0.015.

The two person fit statistics in Tables 9 and 10 show a similar pattern. There is little variability in the percentage of Type I error across the number of replications. The largest range is 0.6% in Table 10 for 1,000 persons and 50 items. The person fit statistic Type I errors were more consistent across number of replications than were the item fit statistics. The average Type I error rates are less than expected with the person unweighted total fit being in the 3% range and the person weighted total fit being in the 2% to 3% range. But the observed Type I error rate is closer to the expected values than for the item fit statistics.

CONCLUSIONS

These simulations indicate that the number of replications included in the study introduces less variability in the summary statistics and the Type I error rates for the five fit statistics studied than did variations in the number of items, number of persons, and type of statistic. In most cases, the variability over the number of replications—10, 25, 50, and 100—was so small as to suggest that there is little, if any, benefit in extending the number of simulations in these types of studies beyond the 10 to 25 replications range.

REFERENCES

Andersen, E. B. (1973). Goodness of fit test for the Rasch model. *Psychometrika, 38,* 123–140.

Luppescu, S. (1992). *SIMTEST 2.1. A computer program for simulating test data.* Chicago: MESA Press.

Mead, R. J. (1975). Analysis of fit to the Rasch model. Doctoral dissertation, University of Chicago.

Rasch, G. (1960). *Probabilistic models for some intelligence and attainment tests.* Copenhagen: Danish Institute for Educational Research. (Expanded edition, Chicago: University of Chicago Press, 1980)

Smith, R. M. (1982). *Detecting measurement disturbances with the Rasch model.* Doctoral dissertation, University of Chicago.

Smith, R. M. (1985). A comparison of Rasch person analysis and robust estimators. *Educational and Psychological Measurement, 45*, 433–444.

Smith, R. M. (1986). Person fit in the Rasch model. *Educational and Psychological Measurement, 46*, 359–372.

Smith, R. M. (1988). The distributional properties of Rasch standardized residuals. *Educational and Psychological Measurement, 48*, 657–667.

Smith, R. M. (1991a). The distributional properties of Rasch item fit statistics. *Educational and Psychological Measurement, 51*, 541–565.

Smith, R. M. (1991b). *IPARM: Item and person analysis with the Rasch model.* Chicago: MESA Press.

Smith, R. M. (1994). A comparison of the power of Rasch total and between item fit statistics to detect measurement disturbances. *Educational and Psychological Measurement*, 54, 886–896.

Smith, R. M. (1996). A comparison of the Rasch separate calibration and between fit methods of detecting item bias. *Educational and Psychological Measurement, 56,* 403–418.

Wright, B. D., & Linacre, J. M. (1991). *BIGSTEPS: Rasch analysis for all two-facet models*. Chicago: MESA Press.

Wright, B. D., Mead, R. J., & Bell, S. R. (1979). *BICAL: Calibrating items with a Rasch model.* Research Memorandum No. 23B, MESA Psychometric Laboratory, Department of Education, University of Chicago.

Wright, B. D., & Panchapakesan, N. (1969). A procedure for sample-free item analysis. *Educational and Psychological Measurement. 29*, 23–48.

Author Index

Subject Index